An Activity-Based Learning Approach

Book 1: Isolated Intersections

First Edition

by Michael Kyte
Tom Urbanik

published by

Pacific Crest
Plainfield, IL

Traffic Signal Systems Operations and Design

An Activity-Based Learning Approach

Book 1: Isolated Intersections

First Edition

by Michael Kyte
Tom Urbanik

Layout and Production by Denna Hintze, Cover Design by Denna Hintze

Published by

Pacific Crest
13250 S. Route 59, Unit 104
Plainfield, IL 60585
815-676-3470
www.pcrest.com

ISBN: 978-1-60263-420-6

Photo Credits

The following photographs are courtesy of:

Kevin Lee (Figures 2 and 3)

Econolite Control Products (Figures 16 and 50)

Joe Pallen (Figures 12-15, 17-22, 37, 49, 176, and 177)

FHWA NGSIM project (Figures 24 and 46)

Velela, "Supermarket checkout", en. Wikipedia (Figure 37)

Table of Contents

Acknowledgements

This book is intended to help university civil engineering students and practicing traffic engineers to better understand how isolated actuated traffic control systems work and to provide both groups with an opportunity to complete a signal timing design for an isolated signalized intersection. Creating a book like this one you are now reading, with a unique focus on activity-based learning, is a complex process and has only been possible through the efforts of a number of valued colleagues.

The Federal Highway Administration funded the development of the MOST (Mobile Signal Timing Training) simulation environment and the accompanying 37 experiments (many of which are included in this book). Some of the people who contributed to this project include Michael Dixon, Ahmed Abdel-Rahim, Paul Olson, Enas Amin, Darcy Bullock, Eddie Curtis, Rick Denny, Milan Sekulic, Kiel Ova, Gary Duncan, Hua Wang, Azizur Rahman, Anuj Sharma, Matt Wiesenfeld, Mike Inerowicz, Chris Day, Howard Cooley, and Jim Pline.

The Federal Highway Administration funded the TransEd project (through its Transportation Education Development Pilot Program) that has developed four activity-based textbooks including this book. Clark Martin from FHWA was the program manager. Most of the 63 activities in this book were developed as part of this project. A number of people played a major role in this project and in the development of these activities. Steve Beyerlein provided invaluable feedback and insightful ideas on the pedagogy on which this book is based. J.J. Peterson, Kevin Lewis, and Matt Ricks worked closely with students from the University of Idaho to test and evaluate many of these activities. Peter Koonce provided ideas and feedback on the readings, used some of the material in his classes, and helped to prepare some of the "in my practice" material. Peter Furth provided significant ideas on the choice point covered in Chapter 9. Anuj Sharma, Mike Hunter, Ed Smaglik, and Dave Hurwitz tested some of this material in their classes and provided useful feedback. Richard Wall helped to focus ideas on controller system technology. Marti Ford and Maria Tribelhorn reviewed a final draft of the manuscript and provided valuable suggestions. Cindy Urbanik provided continuing support as the ideas for this book were developed over a number of years. Denna Hintze was the final editor, did the design and layout of the text, and provided a number of helpful ideas on pedagogy. James Colyar, John Halkias, and Vassili Alexiadis provided information on the NGSIM data sets. Joe Pallen and Kevin Lee provided many of the photographs used in the text. Finally, a number of students at the University of Idaho and other universities have reviewed and suffered through early iterations of this material and provided invaluable feedback. We sincerely thank all of those who have helped in this effort.

This book is dedicated to the memory of a valued colleague and friend.

"The way to be outstanding in your field is to be out standing in the field."

Bill Kloos, City of Portland, Manager, Signals and Street Lighting Division

1951-2009

Preface

Overview

> "When I was in Mrs. Lavender's kindergarten class, in a suburban public school in Los Angeles, we were visited one day by a local police officer. The officer brought a traffic signal, mounted on a short pole. The signal had displays that could be seen by two lines of students, as if we were pedestrians waiting to cross an intersection. We were instructed to form two lines and to follow directions: walk when it was green and stop when it was red." – the author

We learn the lessons of traffic control early in life, and with good reason. There are nearly 300,000 traffic signals today in the United States. Each traffic signal performs the task of regulating whose turn it is to go and who must wait. Some signal systems, known as fixed time systems, provide the same amount of time to serve each group of users and in the same sequence. Other systems respond to the volume of vehicles and pedestrians present at the intersection and provide varying amounts of time to serve these users. Some signals operate independently and respond to the traffic demands at that intersection alone, while others operate together in a system so that traffic can be moved with as few stops or as little delay as possible. A traffic signal system at its core has two major tasks: move as many users through the intersection as possible and do it with as little conflict between these users as possible. The first task relates to efficiency and capacity while the second relates to safety. Both tasks are performed by first clearly defining which group of users has the right of way at a given time and second by determining how long the group has the right of way.

The traffic signal system is probably the most important kind of transportation facility in operation today, considering the perspectives of both safety and efficiency. Two-thirds of all miles driven each year in the U.S. are on roadways controlled by traffic signals. In some urban areas, signals at busy intersections control the movement of more than 100,000 users each day (Koonce, et al., 2008). The signal system also has a great impact on energy usage and the environment. The more times a vehicle stops, the higher the level of pollutants that it emits. And, twenty percent of the fuel used by automobiles traveling along urban arterials is consumed while waiting at a red light at a signalized intersection.

According to a 2007 report from the National Highway Traffic Safety Administration, 20 percent of all motor vehicle fatalities in the United States each year occur at an intersection. Between 1997 and 2004, this figure represented 76,162 lost lives. In addition, tens of thousands of drivers, cyclists, and pedestrians are injured each year in traffic accidents at intersections (Subramanian & Lombardo, 2007).

Any one of these traffic signals, or a system of several signals, can cause a motorist to wait unnecessarily. The Federal Highway Administration estimates that half of these traffic signals need some sort of timing or operational improvement (Koonce, et al., 2008). A recent national report card gave the nation's traffic signal systems poor grades (National Transportation Operations Coalition, 2007). While there are a number of reasons for this poor assessment, we believe that there are three major contributing factors. First, there is a lack of high quality and comprehensive references defining good practice. While many states and local jurisdictions do have standards that guide their signal timing design practices, often these standards are not based on good science or sound theory that allow the standards to be transferable to new situations or conditions. Second, university textbooks do not cover traffic signal systems in a comprehensive and realistic manner. Too often, the systems are assumed to be fixed time (rarely the case in the field) while the traffic controller itself is not covered at all. Most university laboratories do not have traffic signal controllers and are thus not able to give their students experience in their use. Third, traffic engineers often have little direct experience with traffic controllers since their university experience is often limited to using models that often poorly emulate the operation of a traffic controller. This results in a problematic dichotomy. A signal engineer designs the signal system and timing plan, but the implementation of the timing plan (and many of the important timing details) is left to the technician. The former understands how the system should work while the latter understands how the traffic controller works but often without the same broad perspective that the engineer brings to the problem.

So, how do we overcome these problems and provide systems of learning that will produce transportation engineers who understand how traffic control systems work and have the ability to design the components of these systems? Happily, there are signs that things are changing in the right direction. The Federal Highway Administration has produced a new *Traffic Signal Timing Manual* that brings together a broad array of information that can be used by traffic engineers to design traffic signal systems. FHWA has also produced a new guidebook on intersections, both signalized and unsignalized, that provides basic guidance on intersection design and operation. We hope this current book and the one to follow on coordinated traffic signal systems will also help.

The Approach

Our motivation in writing this book is to provide a learning environment and the necessary materials for a senior or graduate level university course in the design and operation of one important part of the traffic signal control system: the isolated intersection. A future book will cover coordinated traffic control systems. Some of the material in this book may also be appropriate for portions of advanced classes in transportation engineering.

We have made four assumptions that have guided the preparation of the material in this book. These assumptions are listed below and elaborated on in the following pages:

1. You must understand the traffic control system and its component parts, especially the traffic controller.
2. You should work in a learning environment with an activity-based learning approach, in which you learn by doing and observing.
3. You need to learn to see, interpret, and integrate what you see. You also need to connect the theory that you learn with models that integrate the theory and connect both with what actually happens in the real world of signalized intersections and traffic flow.
4. You must understand the traffic control system and its operation in order to design the system and its components. You need to have an understanding of the design process and gain experience in the application of this process.

Understanding the Traffic Signal Control System and the Traffic Controller

The traffic control system includes four interrelated subsystems or components: the user, the detector, the controller, and the display. Each component directly affects another component: for example, the detector responds to the user, while the controller responds to the detector. You will use tools that will allow you to visualize these relationships and more thoroughly understand them. The system and these components are more fully described and illustrated later in this book.

It has been our observation that you (engineering students) have, in recent years, gained considerable experience with simulation models of various systems, including transportation systems. Simulation allows engineers to test and observe the performance of a system under a wide variety of conditions, without disturbing the operation of an actual system. However, at the same time, you are getting less experience with the fundamental devices and equipment that are the basis for the operation of many transportation systems.

This is certainly the case with the traffic controller, the most ubiquitous and fundamental device of today's urban transportation system. We believe that in order for an engineer to design and operate a traffic system, understanding the operation of a traffic controller, and how its various settings affect the flow of traffic at an intersection, is critical. It is the task of the engineer, not the technician, to establish the policy and guidelines for the operation of city streets and rural highways, and the control of these streets and highways must be based on the engineer's knowledge of the controller itself, how it functions, and how its various settings result in varying levels of performance at an intersection.

Creating an Active Learning Environment

If you learn to play the cello, you learn music theory. But most of your time is spent playing the cello: practicing and performing. Also, your cello instructor provides periodic (sometimes constant!) assessment of how you are doing: holding the bow correctly, placing your hand on the finger board, and listening to the quality of the tone that you produce.

Educational research points to a hands-on active learning environment as the best approach to improving student understanding of important concepts. The highly successful Traffic Signal Summer Workshop developed at the University of Idaho has shown how students benefit directly from this approach. This workshop included one week of hands-on experiences with traffic signal controllers and the supporting hardware and software (Kyte, Abdel-Rahim, & Lines, 2003). Two quotes, one from a student and one from an instructor, illustrate the benefits of the workshop experience.

> "The best parts of the week were the hands-on work and introductory lectures to the more advanced technologies of video detection and hardware-in-the-loop simulation. Exposure to this technology was worth the trip alone."

> "I think the valuable part is that students don't just look at pictures or mathematical equations. They get a chance to tinker, make mistakes, and ultimately get various components up and running… much like they will have to in the real world. This means when they are on their first job and things don't work exactly as expected during a [system] turn-on, they will have their wits about them and know how to debug the system and get it running."

Unlike many courses that emphasize an instructor focus (with lectures presented to students), this book emphasizes a student focus in which you will learn by doing an experiment, analyzing data that you collect, and drawing conclusions about what makes good signal timing practice. This approach requires a more active preparation for each class on your part and a readiness to actively participate in the work of each class. This focus on "you" means more work but it rewards you with the potential for deeper learning and understanding of this material.

Integrating What You See and Learn: Theory, Models, and the Real World

Learning to see the traffic system, to interpret and integrate what you see, to connect a model with what you observe in the field, and to understand the system and its operation, are all important attributes for a good traffic engineer. To help you in this process, you will be given a number of visualization tools: videos, simulation, charts, and figures. Study the ones that are provided to you and we will try to help you focus on their key points.

Using this book is not about learning how to use a specific simulation model, though the experiments that you will complete are conducted using the VISSIM microsimulation model. Nor is it about a specific traffic signal controller, even though you will sometimes use Econolite's ASC/3 controller emulator. Finally you will not be presented with guidelines or standards that you should follow even though you will learn to use the *Traffic Signal Timing Manual*.

Rather, you will use the simulation environment provided here to directly see the results of the phasing plans and timing parameters that you select. Using VISSIM's animation and movie files, you will visualize the duration of a green interval, the length of a queue, or the delay experienced by vehicles travelling through a signalized intersection with the phasing and timing plan that you design. You will use this information to make judgments about the quality of intersection performance and whether you need to make further adjustments to the signal timing to improve intersection operations. It is almost as good as standing out at an intersection, with one eye on the traffic and the other on what is happening in the controller cabinet. And, you will use the guidelines from the *Traffic Signal Timing Manual* as a framework to validate your design as well as to explore topics from the perspective of the professional.

Understanding the System in Order to Design It

There is a logical design process that should be followed when preparing a signal timing design. Your learning process should provide an environment in which you can design components of the system and then test the design using a realistic simulation environment.

We attempt to provide this same framework as you complete the design project that is the endgame for this book. You will be given a set of design requirements and asked to produce a design that meets these requirements. You will be provided a means to assess the quality of your design, working with your instructor. The feedback that you get from this assessment is a powerful means to improve the quality of your work as well as to ensure that you have mastered the principles that are presented.

It is also our experience that many transportation engineering classes are based on simplistic and clean problems, ones with a single solution path, and one "right" solution. Simplistic problems tend to give you a biased and inaccurate view of practice and often do not provide the complexity and challenge that most engineering students look for. In practice, however, problems are messy and complex, often with multiple solutions. These problems include challenges that stimulate you, providing you with a greater understanding of what engineers do in practice, with all of the uncertainty that this entails.

We believe this book addresses these two important issues by providing the details of the operation of traffic controllers with a set of applications that provide a more realistic setting for the development and design of traffic signal control parameters.

Opening the Book

This book is divided into 10 chapters and includes 63 activities. It is intended primarily for a one semester course in traffic signal systems design either as part of a senior design class or one at the graduate level. The first four chapters of the book (see Figure 1) provide a base level of knowledge. Chapters 5 through 9 address specific system components, providing first an understanding of how these components function and second how to design them. Finally, Chapter 10 integrates the components together into a final design in which you prepare a report and make a presentation covering your work.

Most chapters have a similar structure. Each chapter begins with a Reading that provides important information on the topics covered in the chapter. A series of activities follow, each providing hands-on experiences with the chapter topic.

- Assessment activities give you the chance to test and apply what you learned in the reading
- Discovery activities provide you with the opportunity to discover new factors or perspectives about the chapter topic by observing animations, collecting or analyzing data, or making calculations
- Field activities allow you to explore traffic flow and control conditions directly in the field and connect your field observations with the theory that you learned in other activities

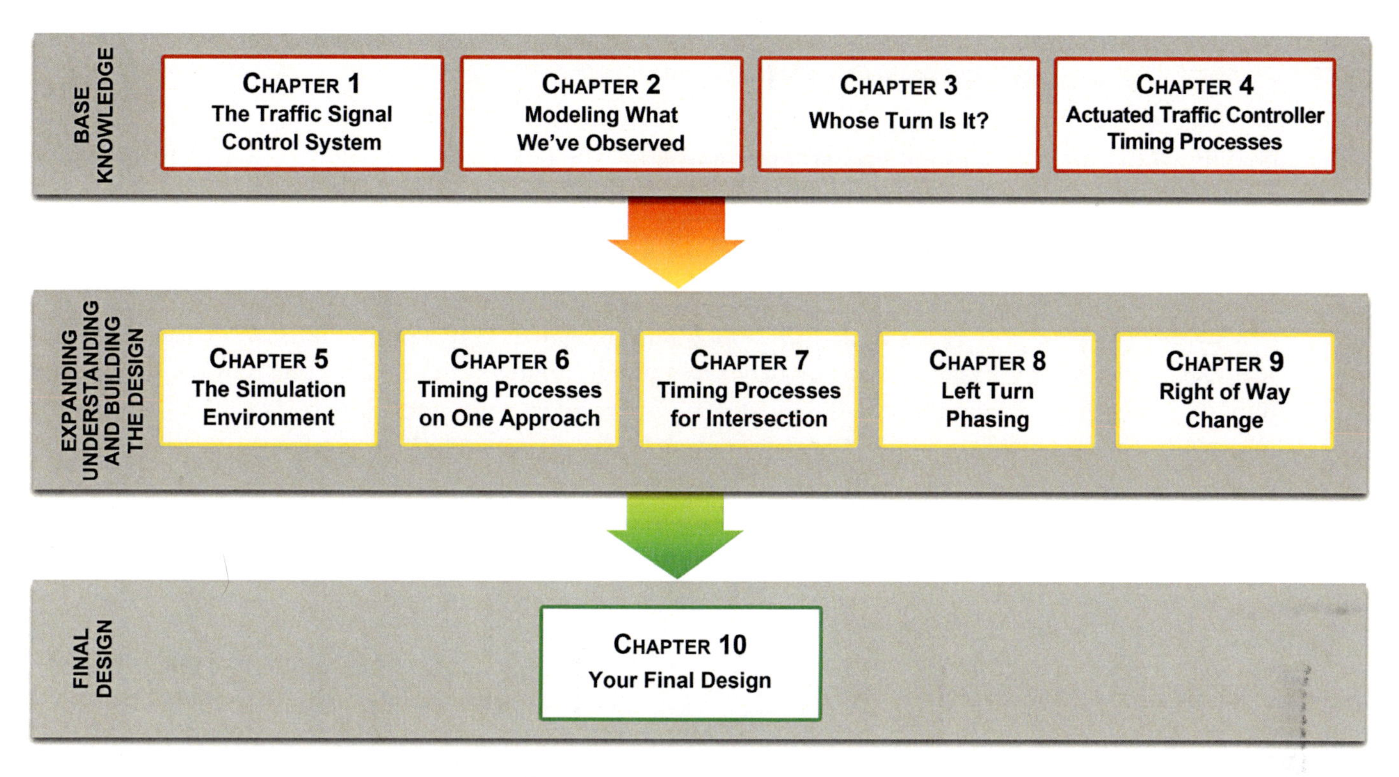

Figure 1. Organization of book

Each chapter concludes with:

- A design activity in which you will determine one component of your design and evaluate its performance
- An In Practice activity in which you will compare your design component with recommended practice from the *Traffic Signal Timing Manual*

Each activity is based on a consistent format:

- The **Purpose** lets you know what the activity is about and why it is worth doing
- The **Learning Objectives** describe what you will know or be able to do when you have completed the activity
- The **Required Resources** identify what you will need as you complete the activity
- The **Deliverables** describe what you are expected to produce
- The **Critical Thinking Questions** provide you with a chance to deeply and thoughtfully consider what you learned
- The **Information** provides additional ideas or notes that will help you complete the activity
- The **Tasks** list the specific steps that you will need to follow

Throughout these activities, you will use several learning tools such as concept maps and preparation of glossaries. Both of these tools will help you, as a learner, to move beyond the level of basic knowledge (think "memorizing" or "filling in a blank") and towards comprehension, application, and analysis of ideas and concepts. This progression of learning culminates in the kind of working expertise that professionals demonstrate, as they inspect, model, problem-solve, test, examine, and design. It is not a coincidence that these are all skills that you will practice throughout the activities in this book (Bobrowski, 2007).

- The **GLOSSARY** portion of each Reading activity asks you to define terms and variables. What is sought is not an authoritative definition you find in this book (or another book or even online); the authors know the definitions, as do your instructor(s), and professionals. Further, there is no doubt that you have the skills required to look up the terms and copy the definitions you do find. Therefore, what is actually sought is your comprehension of the idea or dynamic represented by the term. When you are able to provide a definition or explanation, in your own words, you have moved beyond memorization or filling in a blank to comprehending the idea.
- The **CONCEPT MAP** portion of each In Practice activity takes your comprehension of discrete ideas (represented by your glossary responses) and asks you to relate them to each other. Creating a concept map is an excellent way to not only demonstrate higher order comprehension, but to actually build and increase it. For more information about creating concept maps, including some helpful examples, see The Theory Underlying Concept Maps and How to Construct and Use Them (Novak & Cañas, 2008) at http://cmap.ihmc.us/Publications/ResearchPapers/TheoryUnderlyingConceptMaps.pdf. Other references for concept maps are provided on the companion web site for this book.

Resources

The primary resource that you will use, in addition to this book, is the *Traffic Signal Timing Manual*, produced by the Federal Highway Administration. Many of the readings that you will be assigned are from this manual. But rather than use the manual as the basis for your design, it will be a framework for you to use to review the design that you develop using the experiments that you complete, the observations that you make, and the insights that you gain from the activities in this book. As part of the design activity concluded in Chapter 10, you will be asked to compare your design results with those values recommended in the *Traffic Signal Timing Manual*.

In addition to the *Traffic Signal Timing Manual*, you will use several resources that are available on the companion website for this book. Your instructor will provide you with the URL for the site that includes these resources:

- Data sets and analysis tools in Excel spreadsheets that you will use in Activities #10, #21, #36, and #54
- VISSIM data files for your initial base network in Activity #28
- Movie files for Activities #3, #4, #19, #20, #27, #32, #33, #34, #35, #41, #42, #47, #48, and #49
- *Traffic Analysis Tools*, Volumes I and III
- Example professional traffic signal timing reports
- Tips and tutorials for using VISSIM and Excel

Focus: The Intersection and Vehicles Users

The traffic control system is a complex one that serves a variety of users. In the broadest view, the system consists of a network of arterials, and each arterial includes a set of signalized intersections. The intersections may be interconnected so that vehicles may progress from one intersection to the next with as little delay as possible. The operation and design of signalized arterial systems is the subject of a future book. To make sure that you will learn and understand the fundamental aspects of traffic control, this current book focuses on the operation of a single signalized intersection, one that is isolated, and not connected to a larger network.

A traffic control system also serves a variety of users including those driving passenger cars or trucks, riding a bus or light rail train, walking, or riding a bicycle. Many systems prioritize the manner in which each of these users is served. For example, an intersection that serves a high speed rail line will preempt service to all other users so that the train can proceed safely through the intersection. Or, to meet local policy, priority may be given to pedestrians and bicyclists so that environmental or energy goals are achieved. Again, so that you can learn the fundamental aspects of traffic control, we will focus solely on vehicles (users who drive passenger cars or trucks).

A Concluding Thought

We believe that providing you with this hands-on learning environment will allow you to participate in a very important experiment: how much can we reduce fuel consumption, reduce vehicle emissions, and better manage traffic flow in our cities by improving the timing of the nation's nearly 300,000 traffic signals? It will also transform your perspective from that of a user of the system (pedestrian, bicyclist, or driver of a passenger car) to that of a transportation engineer, where you will now see the traffic control system in all of its complexity. Best wishes on your journey to keep us all moving, or at least not stopping for long.

References

Alexiadis, V., Jeannotte, K., and Chandra, A. (2004). *Traffic analysis toolbox, volume I: Traffic analysis tools primer.* Federal Highway Administration (FHWA-HRT-04-038).

Bobrowski, P. (2007). Bloom's taxonomy—expanding its meaning. In S. Beyerlein, C. Holmes, & D. Apple (Eds.), *Faculty guidebook.* Lisle, Il: Pacific Crest.

Bonneson, J.A. and McCoy, P.T. (2005). *Manual of traffic detector design, (2nd ed.).* Washington, D.C.: Institute of Transportation Engineers.

Byrne, A., de Laski, A., Courage, K., and Wallace, C. (1982). *Handbook of computer models for traffic operations analysis.* Technology Sharing Report. Federal Highway Administration (FHWA-TS-82-213).

Dowling, R., Skabardonis, A., and Alexiadis, V. (2004). *Traffic analysis toolbox, volume III: Guidelines for applying traffic microsimulation software.* Federal Highway Administration (FHWA-HRT-04-040).

Koonce, P., Rodegerdts, L., Lee, K., Quayle, S., Beaird, S., Braud, C., Bonneson, J., Tarnoff, P., and Urbanik, T. (2008). *Traffic signal timing manual.* Federal Highway Administration (FHWA-HOP-08-024).

Kyte, M., Abdel-Rahim, A., and Lines, M. (2003). Traffic signal operations education through hands-on experiences: Lessons learned from a workshop prototype, *The Journal of the Transportation Research Board*, Transportation Research Record 1848, pp.50-56.

Kyte, M., Urbanik, T., and Amin, E. (2007). Foundation for joint determination of passage time and detection zone length using stop bar presence detection, *The Journal of the Transportation Research Board*, Transportation Research Record 2035, pp 114-121.

May, A.D. (1989). *Traffic flow fundamentals.* Englewood Cliffs, NJ: Prentice-Hall.

National Transportation Operations Coalition (2007). *National traffic signal report card technical report, 2007.* Retrieved from http://www.ite.org/reportcard/2007/default.asp

Novak, J.D. & Cañas, A.J. (2008). The theory underlying concept maps and how to construct and use them, Technical Report IHMC CmapTools 2006-01 Rev 01-2008, Florida Institute for Human and Machine Cognition. Available at: http://cmap.ihmc.us/Publications/ResearchPapers/TheoryUnderlyingConceptMaps.pdf

Subramanian, R. and Lombardo, L. (2007). A*nalysis of fatal motor vehicle traffic crashes and fatalities at intersections, 1997 to 2004.* Washington, D.C.: National Highway Traffic Safety Administration.

Transportation Research Board (2010). *Highway capacity manual 2010.* Washington, D.C.: National Academies.

Tufte, E. R. (2006). *Beautiful evidence.* Cheshire, CT: Graphics Press.

Tufte, E. R. (2001). *The visual display of quantitative information (2nd ed.).* Cheshire, CT: Graphics Press.

University of Idaho, Purdue University, University of Tennessee, Pline Engineering, Federal Highway Administration, PTV America, and Econolite Control Products (2009). *MOST: A hands-on approach to traffic signal timing education*, Federal Highway Administration. Retrieved from http://www.webs1.uidaho.edu/most/index.htm

The Traffic Signal Control System: Its Pieces and How They Fit Together

Purpose

In this first chapter, we set the stage for learning about (understanding, then designing) signal timing for the traffic signal control system.

Learning Objectives

When you have completed the activities in this chapter, you will be able to

- Describe the basic components and the operation of the traffic control system
- Determine your current level of competency with traffic signal system concepts
- Describe how drivers respond to signal displays
- Identify and describe various physical components of a signalized arterial
- Assess the realism of a simulation environment by comparing it with a video of actual field operations
- Develop your ability to "see" and "observe" video and animation of traffic flow at a signalized intersection and relate these observations to traffic flow theory and principles
- Identify essential team behaviors that lead to successful completion of activities and the design project
- Explain how team behaviors support different team roles
- Develop group consensus on how the team will work with, treat, and communicate with each other
- Describe the content, scope, and organization of the *Traffic Signal Timing Manual*

Chapter Overview

This chapter begins with a *Reading* activity (Activity #1) that describes the system itself and its components and how they fit together. The basic prerequisite to a course using this book is an introductory course in transportation engineering, offered in most undergraduate programs in civil engineering. Activity #2 is designed to review some of the basic concepts that you learned in this introductory course that are relevant to traffic signal timing, identifying what you know (and don't know) about this subject. Activity #3 explores a system that you already, in many respects, know quite well: a signalized arterial. You will begin to "see" this system in a new way, using the terms and concepts commonly used by the transportation engineer. Activity #4 introduces you to the notion of seeing what is important at a signalized intersection. We often describe the experienced traffic signal engineer as having one eye on the traffic flow and one eye in the cabinet, looking at the controller and its timing processes. You will learn to use a simulation environment that includes a realistic traffic signal controller emulator. While you will use a specific simulation model and traffic controller, your work is about neither. Rather, they are both tools to help you learn about traffic operations and signal timing. It may be a surprise to you to see two activities about working in teams in a book about traffic signal timing. Yet most engineers work in teams and explicitly learning to do so is a critical skill for the transportation engineer. Activity #5 is based on three readings on team building, while Activity #6 takes you through the process of team building and creating a team agreement that will provide a context for your work as a team.

The chapter concludes with an activity (Activity #7) called *In Practice* in which you will be introduced to the *Traffic Signal Timing Manual*. This manual, developed by the Federal Highway Administration, is a compilation of guidance for signal timing. You will have other readings throughout this book from the *Traffic Signal Timing Manual* that will provide a basis for checking your understanding of traffic signal timing and the results from your design activities.

Activity List

Number and Title		Type
1	Exploring the System and Providing a Framework	*Reading*
2	What Do You Know About Traffic Signal Systems?	*Assessment*
3	Exploring the System: Driving Along an Arterial and Noting What You See	*Discovery*
4	Learning to See: The Simulation Environment in Which We Will Work	*Discovery*
5	Working Together – Team Building for Effective Learning and Design	*Discovery*
6	Team Agreement	*Design*
7	Introduction to the *Traffic Signal Timing Manual*	*In Practice*

ACTIVITY 1

Exploring the System and Proving a Framework

PURPOSE

The purpose of this activity is to help you develop a base knowledge of the introductory concepts relating to traffic control systems.

LEARNING OBJECTIVE

- Describe basic components and operations of the traffic control system

DELIVERABLES

- Define the terms and variables in the Glossary
- Prepare a document that includes answers to the Critical Thinking Questions

GLOSSARY

Provide a definition for each of the following terms and variables. Paraphrasing a formal definition (as provided by your text, instructor, or another resource) demonstrates that you understand the meaning of the term or phrase.

actuated control	
detector	
display	
fixed time control	
movement	
queue	
user	

CRITICAL THINKING QUESTIONS

When you have completed the reading, prepare answers to the following questions.

1. In contrast to fixed time control, in what situations is actuated control appropriate?

2. Describe the interrelationships that are shown in Figure 5 and Figure 6 between the components of the traffic control system.

3. How would you measure the performance of the traffic control system and what data would you need to make these measurements?

4. What are the primary physical elements of a signalized intersection?

5. What are the discrete periods of traffic flow during one signal cycle? Briefly describe the manner in which vehicles arrive and depart during each of these periods.

6. List any other questions that you have on the reading material.

INFORMATION

The Traffic Signal Control System and Its Components

The primary purpose of a traffic signal control system is safety, to avoid conflicts by providing a time-separation between the movements of people and vehicles traveling through the intersection. An important, but secondary purpose of the system is to provide priority to certain groups or users to achieve goals or objectives that have been established for the performance of the system.

Often the transportation engineer addresses a traffic problem for one or more arterials, with each arterial consisting of a number of individual intersections that are controlled by traffic signals. Many arterials are designed to move large numbers of users, including auto drivers, transit riders, bicyclists, and pedestrians, through the system with as few stops as possible. These coordinated systems often give priority to certain users (for example, transit riders) based on the performance objectives or desired outcomes that have been set for the system. Figure 2 shows a view of a street in downtown Portland, Oregon, that operates under a coordinated system.

Figure 2. Aerial view of a street in Portland, Oregon

At other times, the focus of the transportation engineer is the operation of an individual intersection. Figure 3 shows a signalized intersection serving three user groups: vehicles and bicyclists traveling through the intersection and pedestrians crossing the street.

Intersection traffic control can either be fixed time or actuated (see Figure 4). In fixed time or pretimed systems, the green interval and cycle length are fixed and do not vary even as traffic demand varies though there may be plans for different periods of the day. Actuated control systems respond to traffic demand by extending the green interval by a specified amount of time each time a new vehicle arrives on an approach. The green interval will last at least a minimum specified time but no longer than a pre-established maximum time. More advanced control strategies are based on adapting timing plans to changing traffic patterns during the day.

Figure 3. Intersection in Portland, Oregon

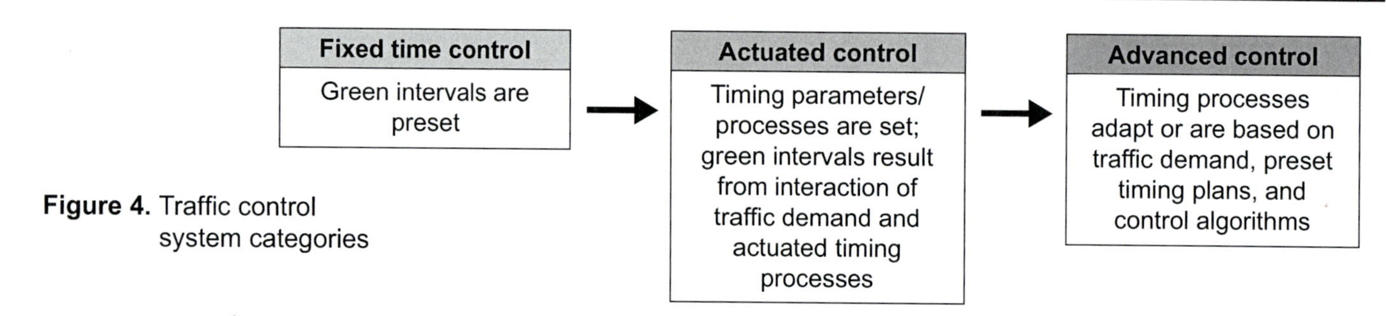

Figure 4. Traffic control system categories

The focus of this book is on the operation and design of a single intersection, operating under actuated control, serving only vehicles (or people driving passenger cars and trucks). The primary benefit of this approach is that it allows you to learn about the basic components of the operation of an actuated control system, without the complexity of considering more advanced control strategies and the needs of other users (such as transit riders or pedestrians). This focus may be criticized because of the importance of considering all modes or users, and the growing interest in providing priority to non-automobile modes at a signalized intersection. However, we believe that the benefits to you in learning the fundamentals of signal timing and operations in this context (single intersection, vehicle user) will allow you to take the next steps of considering the more complex issues of pedestrian timing, transit priority, signal coordination, and others, using the knowledge and skills gained from this book as a solid foundation. So the operating environment that you will address is a single intersection with vehicle drivers as the user class.

One way to view an actuated traffic control system is to consider the components shown in Figure 5. The inputs to the system are the level of vehicle demand and the physical geometry or layout of the intersection. The traffic control system itself is composed of four components: the users of the system, detectors or sensors, the traffic controller, and the display. These components work together in pairs, in a linear fashion, as shown in Figure 5. Each component is connected directly with two of the other subsystems, dependent on or responsive to one and directly influencing the other. These four components are, of course, affected by the geometry of the intersection itself and the volume of the users that demand service at the intersection. Sometimes signal timing can't completely address a performance problem and the only practical solution may be a change in the intersection geometry, as shown in the bottom feedback loop in Figure 5. The change in geometry could be an increase in the number of lanes.

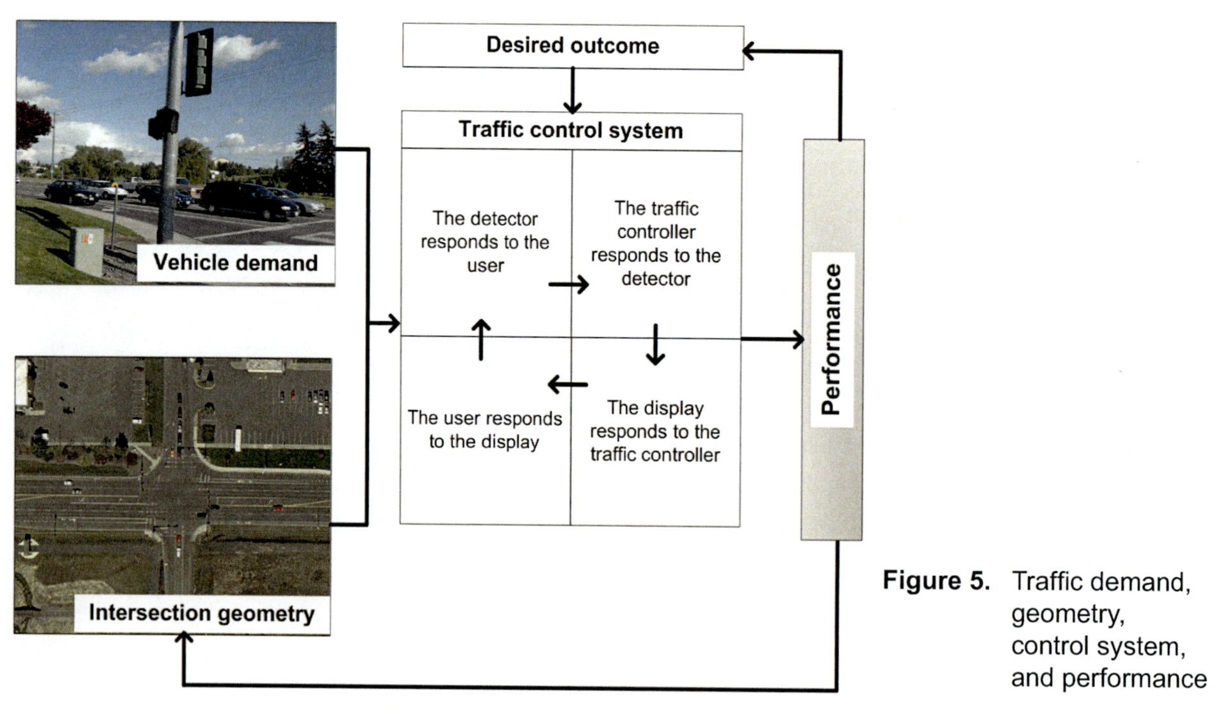

Figure 5. Traffic demand, geometry, control system, and performance

While we as engineers may be most interested in the design of the intersection itself, most people who travel through the intersection are not. They are interested in traveling through the intersection without stopping, and if they have to stop, they don't want to have to stop for long. An important part of the design process is to establish the desired outcome, or what do we want to accomplish with our design. For example, one objective could be to minimize the delay for all users, and this objective could be accomplished by limiting the maximum amount of green time that is given to each user. We can determine how well we meet the desired outcome by establishing performance measures. We measure the performance of the intersection, from the perspective of the users, based on how often people or vehicles must stop when traveling through the intersection, how long they stop, and the length of queues that form when they do stop. We define these measures as number of stops, delay, and queue length. These measures form the basis for evaluating how well the intersection performs.

Figure 6 shows an expanded view of the traffic control system and its components. Each of these components (the users, the detectors or sensors, the traffic controller, and the display) will now be discussed in greater detail.

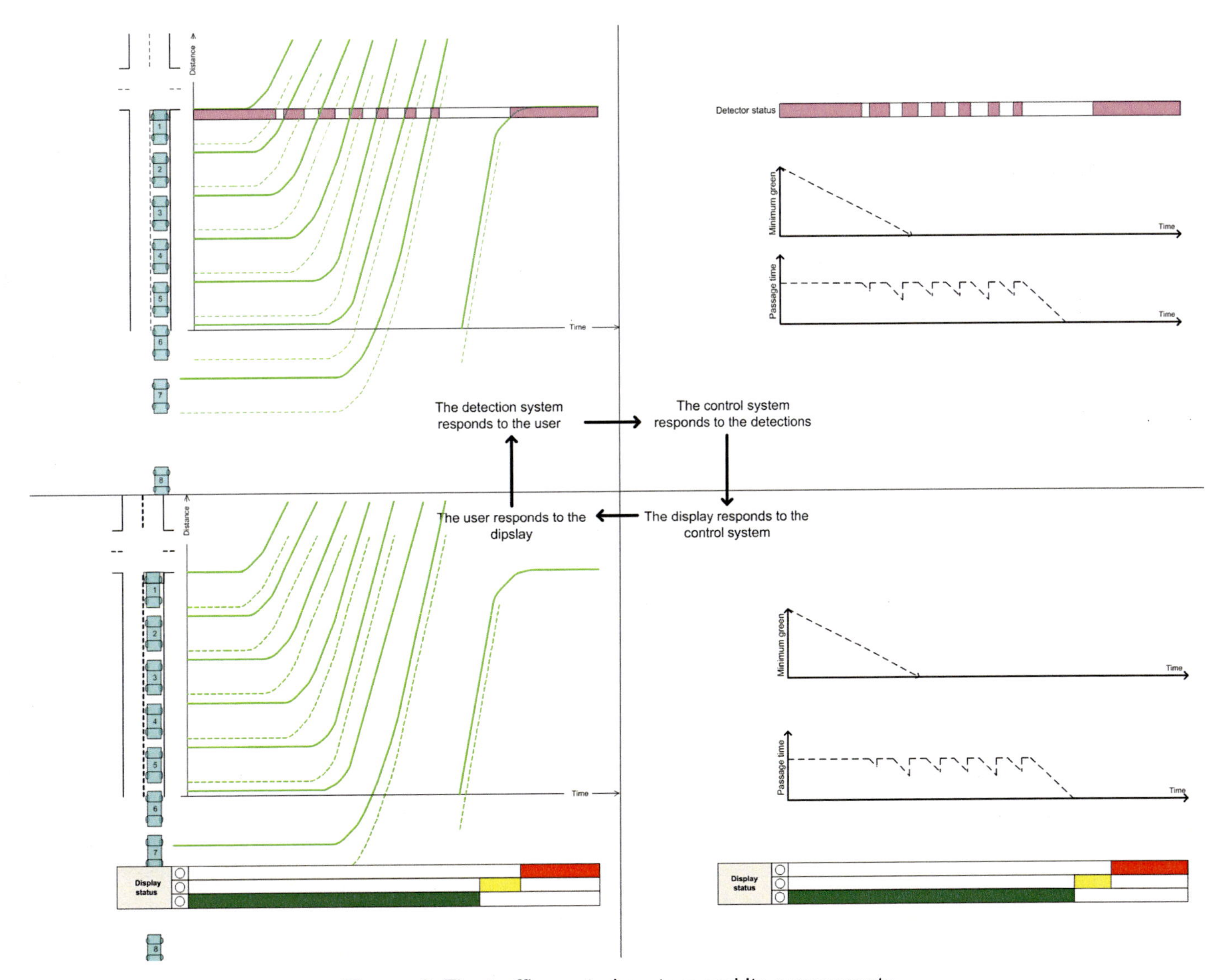

Figure 6. The traffic control system and its components

Users

A vehicle user can include the driver of a passenger car, a truck, or other commercial vehicle that desires service at (desires to travel through) the intersection. Each user category has a set of attributes such as length, width, acceleration and deceleration capabilities, and capacity that affect the operation of the intersection and its required signal timing. Users respond to the display with various possible actions, depending on the display state, the distance that the user is upstream of the intersection stop bar, the speed that the user is traveling, and other factors.

Figure 7 shows a time-space diagram that illustrates how users respond to the display. The figure shows the trajectories for the front and rear of eight vehicles in the format of a time space diagram. The first seven vehicles are part of a queue that has formed during the red interval. The vehicles respond to the change in the display (from red to green), perceiving the change, responding to the change by accelerating and traveling through the intersection. The last vehicle (vehicle 8) responds to the change in the display from green to yellow by decelerating and stopping.

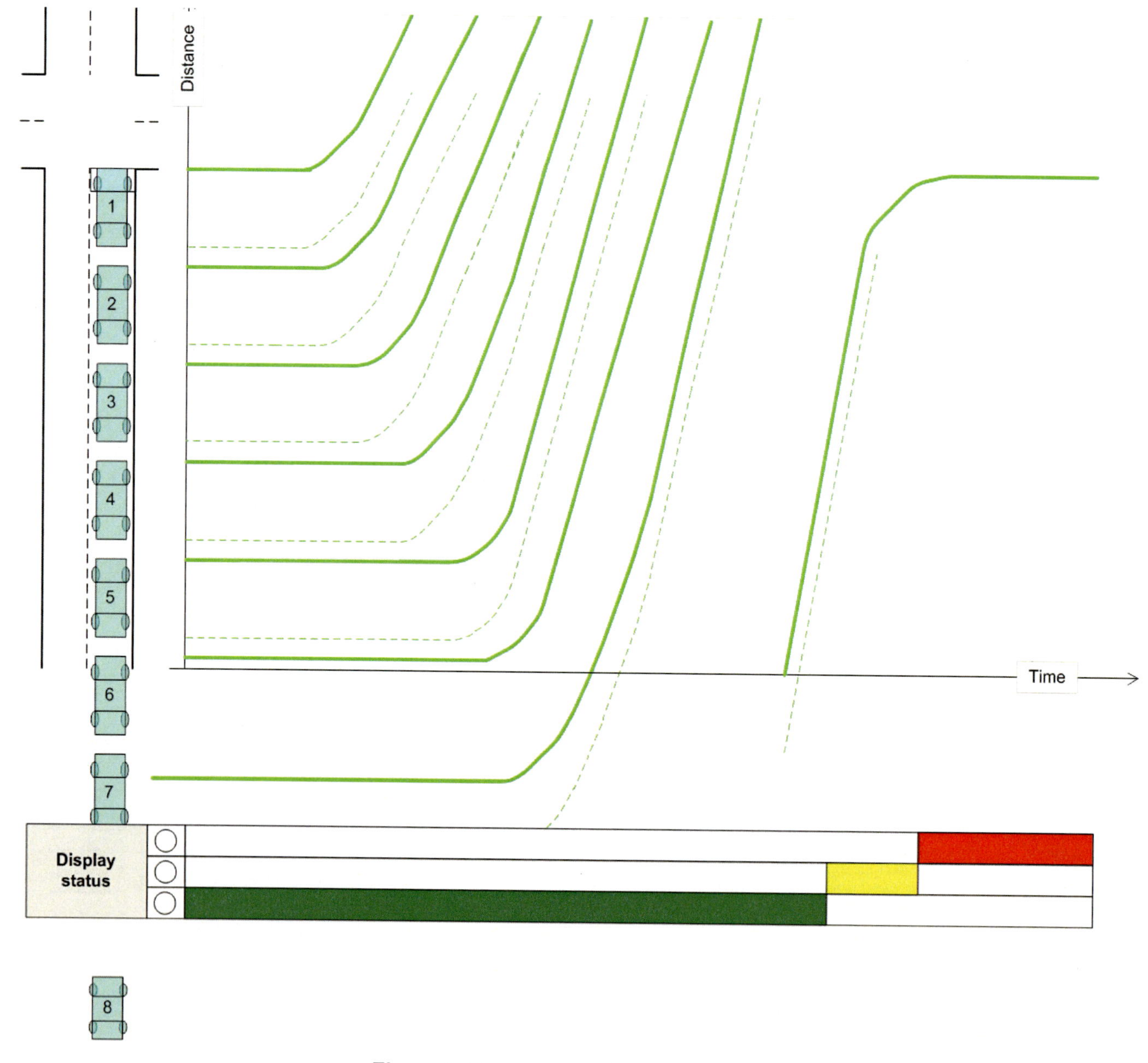

Figure 7. Users responding to the display

Detectors

There are a variety of detectors or sensors used at signalized intersections including inductive loops, video, microwave, and others. Inductive loops include pulse and presence detection types. We will focus on presence loop detectors that are "active" or send a "call" to the controller as long as a vehicle is within the detection zone.

Figure 8 shows the interaction between vehicles and the detector. The vehicle trajectories that we saw in the previous figure (showing both the front and back of each vehicle) are repeated here. The detection zone shown here is located at the intersection stop bar. We can follow the state of the detector over time. The presence of the vehicle in the detection zone is noted by the solid color. When no vehicle is present in the zone, the detector status is off and no color is shown.

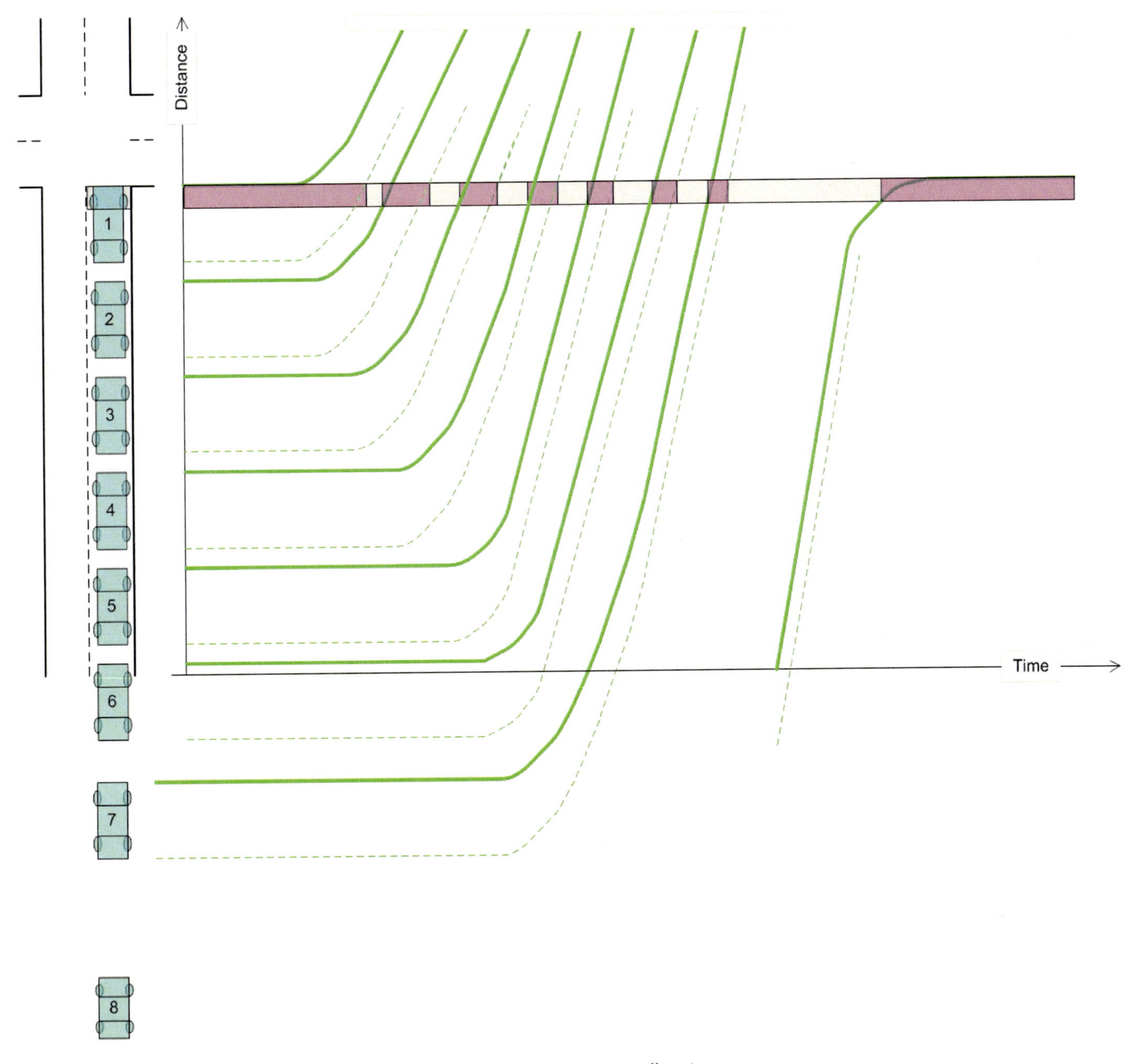

Figure 8. Detectors responding to users

Controller

The traffic controller receives calls or requests for service from the detectors. The controller determines which group of users (movements) are to be served at any given time (the order or sequence of service) and the duration that each group of users is served. The duration of service is determined by the status of a set of timing processes and a decision making framework that responds to this status to determine when service to a particular group of users will end. Examples of these timing processes are the minimum green time, the vehicle extension or passage time, and the maximum green time.

Figure 9 shows the response of two controller timing processes to the detector status. The first timing process, the minimum green timer, becomes active at the beginning of the green interval, and continues timing until the timer reaches zero and expires. In this case, there is no direct link between the detector and the timing process. The second timing process, the passage timer (sometimes called the *vehicle extension timer*), also begins timing at the beginning of the green interval but is continually reset as long as the detector is active (a vehicle is present in the detection zone). When the detection zone is not occupied, the passage timer begins to time down. The timer is reset six times in the example shown in the figure, each time representing when a vehicle enters the zone. The timer finally expires when the time that the zone is unoccupied exceeds the value of the passage time set in the controller. You will learn more about these timing processes in Chapter 4.

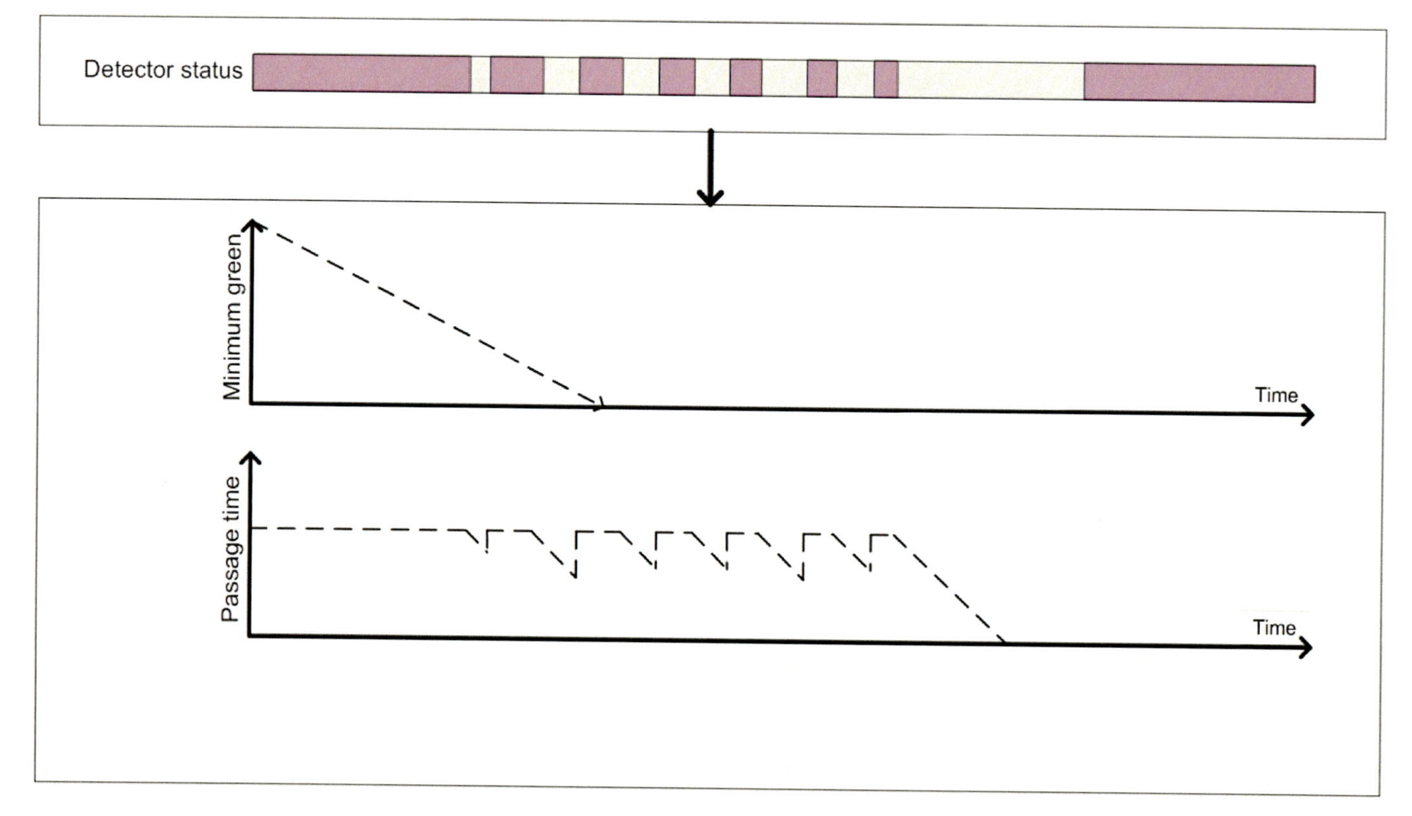

Figure 9. Controller responding to detector

Display

The display is a set of indications that provide information to users on what to do as they approach or are stopped at the intersection. The information conveyed by an indication can either be to proceed through the intersection ("go"), to exercise judgment on whether to go or stop because a change in right of way is about to occur or that the intersection needs to be cleared, or to come to a stop. The state of an indication can either be active or off. If the state is active, it can be either steady or flashing. A color is associated with the vehicle display and can be either green, yellow, or red.

Figure 10 shows an example of the response of the display to the traffic controller. Here, the green is displayed as long as both the minimum green timer and the passage timer are active. In this example, when the passage timer expires, the display changes to yellow and then to red. You will learn about these and other timing processes and how their status affects the displays later in this book.

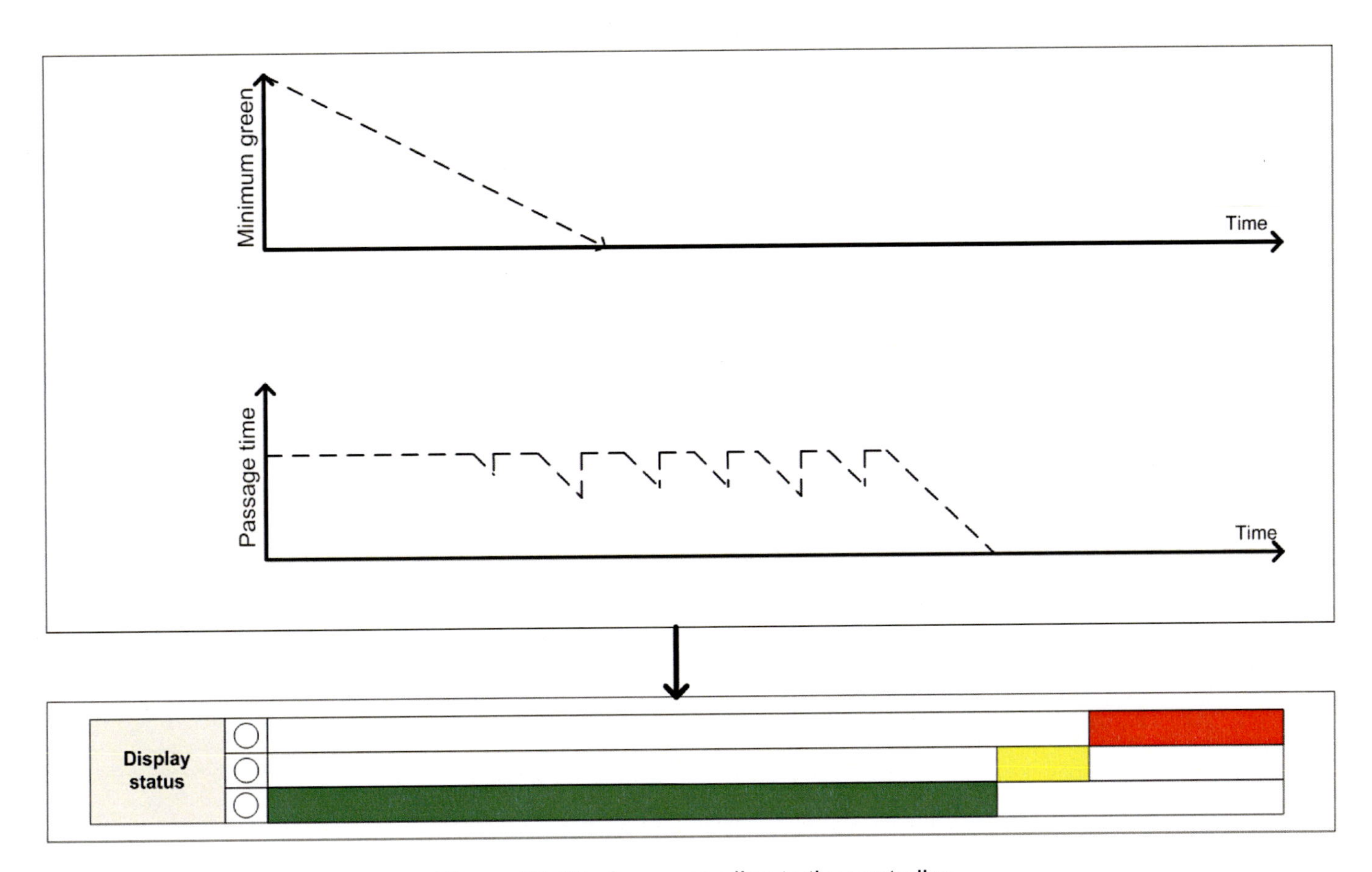

Figure 10. Display responding to the controller

We can also view these components together in one diagram, called a *traffic control process diagram*, as shown in Figure 11. You will use this diagram to follow the effects of a given traffic demand on the detectors, the traffic controller, and the display later in this book.

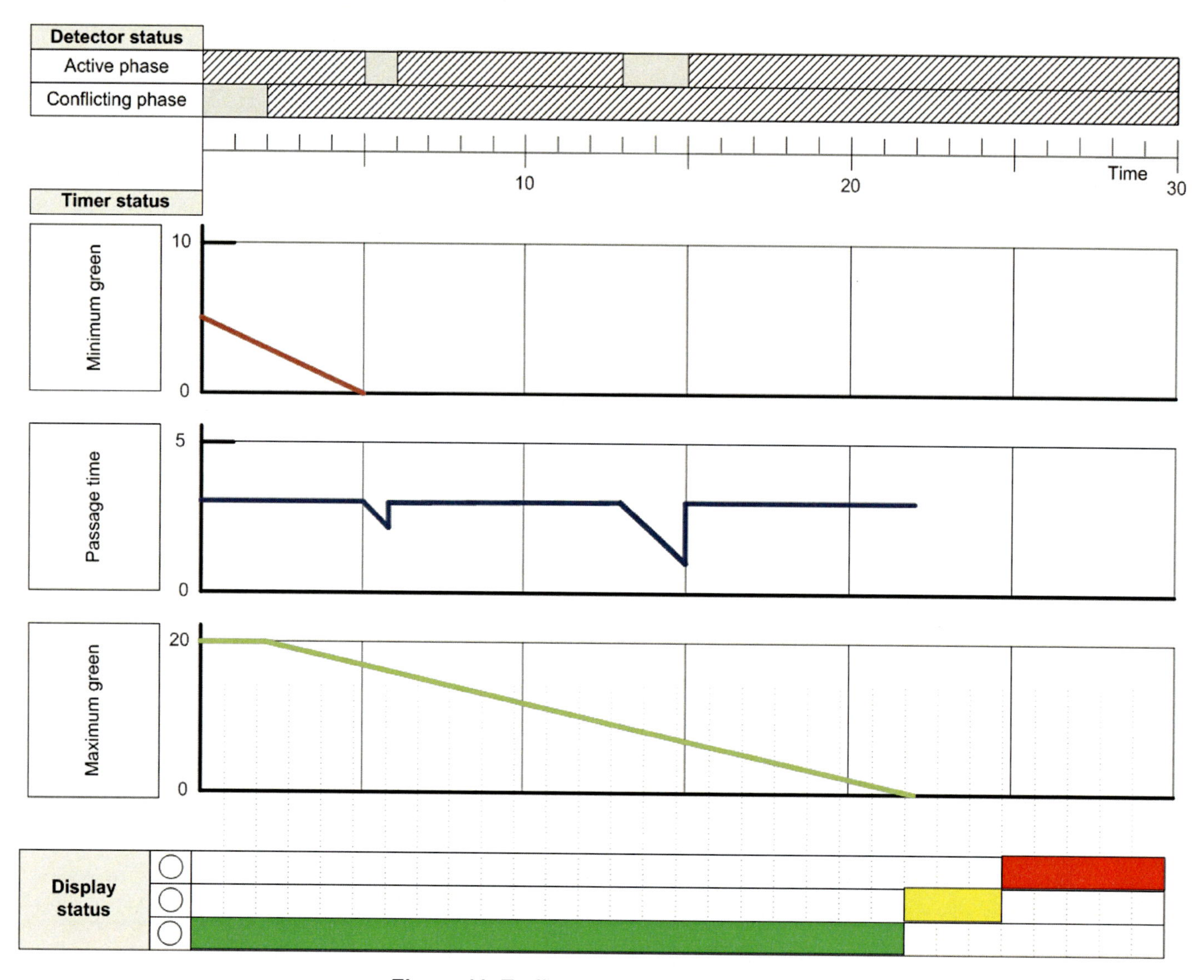

Figure 11. Traffic control process diagram

Field Observations

Let's now move to the field. You are at the intersection of State Highway 8 and Warbonnet Drive in Moscow, Idaho. What geometric information do you see in the photographs of this intersection presented Figure 12 and Figure 13?

Figure 12. Aerial view of SH 8 and Warbonnet Drive

Figure 13. Street level view of SH 8 and Warbonnet Drive

We note first the geometric layout and that it is a T-intersection. State Highway 8, the major arterial, has two through lanes on both approaches, a left turn lane, and a right turn lane on the westbound approach. Warbonnet Drive, the minor street, has two lanes on the southbound approach.

There are also several other physical components of the intersection. The vehicle displays are mounted on the mast arms above each approach as shown in Figure 14. Also visible are lighting and street signs, as well as pedestrian crosswalks and other lane striping. The cabinet, housing the traffic controller and other devices, is located on the northwest corner of the intersection and is shown in Figure 15.

Figure 14. Mast arm, lighting, signs, and signal displays

Figure 15. Cabinet housing traffic controller

Figure 16 shows the inside of a typical cabinet. The call for service from the detectors goes to the detector amplifiers, located at the top part of the cabinet. These calls are then routed to the traffic controller, where they are processed according the timing parameters and logic that have been set. Finally, the load switches set the proper display, based on the outputs from the controller.

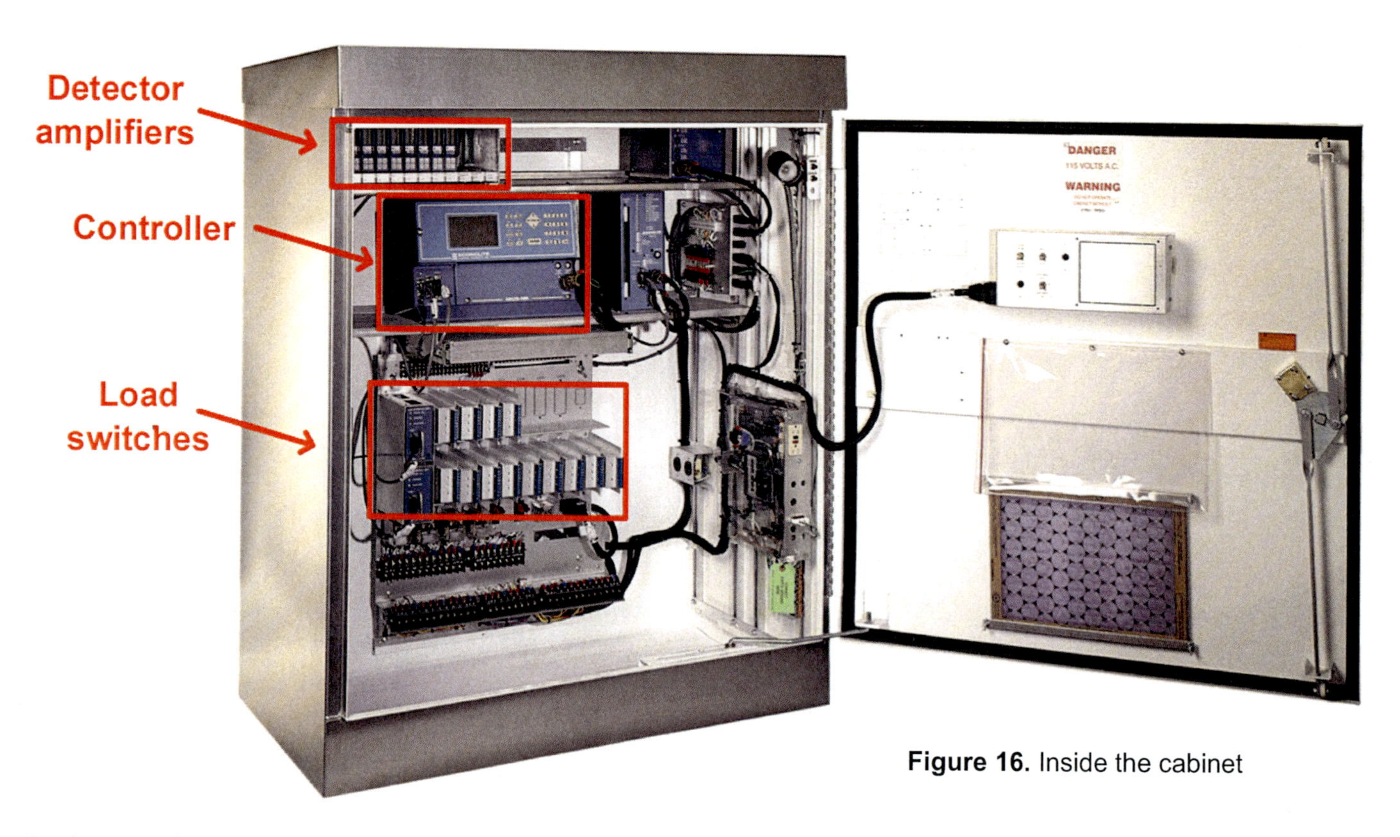

Figure 16. Inside the cabinet

Let's now focus on the flow of the vehicle users through the intersection, specifically on the westbound approach as shown in Figure 17.

Figure 17. Westbound approach and vehicle displays for eastbound lanes

Figure 18. Vehicle queue on westbound approach during red

We start observing the vehicle movements at the beginning of the red interval for the westbound approach. At the beginning of the red interval, as shown in Figure 18, vehicles respond to the red signal indication by slowing as they approach the intersection and coming to a complete stop either at the stop line or behind a vehicle in front of them. Vehicles queue up at the stop line with the queue growing as the red interval times.

Figure 19. Queue moving at beginning of green

Figure 20. Queue moving at beginning of green showing headways

When the displays change to green, vehicles begin to enter and cross the intersection (Figure 19). The driver in the first vehicle at the stop line sees the change in the signal and reacts to the change by accelerating his vehicle into and through the intersection. Drivers in the following vehicles begin the same process: they see the change in the signal, they see the vehicle in front of them begin to move, and then they begin to accelerate their vehicle into and through the intersection. As the drivers of the vehicles approach their desired speeds, the headways between the vehicles drop to a relatively constant or stable value, called the saturation flow rate. The queue that forms during red continues to be served (Figure 20).

After the queue clears, as shown in Figure 21, vehicles arrive at and leave the intersection without delay and without much change in their speeds. At some point, the signal changes from green to yellow (See Figure 22). A vehicle on the approach reacts to this change by deciding to stop. The display then changes to red so that the next set of movements can be served.

Figure 21. Vehicles arrive and leave without delay

Figure 22. Vehicle responds to yellow

In summary, the process that we've just observed through the photographs shown on the previous pages consists of four parts, each describing a response of the vehicle to the display:

- Drivers slow and stop as they arrive at the intersection during the red indication
- At the beginning of green, the queue begins to move as drivers respond to the green indication

- After the queue has cleared, vehicles respond to the green indication by arriving at the intersection and passing through without stopping
- When the yellow indication is displayed, vehicles either decide that they can safely travel through the intersection, or slow and stop in anticipation of the red indication

This flow process is one that you know well and have participated in as a driver many times. What is new is that you now have a context or framework for seeing and understanding this flow process and how it interacts with the other components of the traffic control system.

ACTIVITY
3 Exploring the System: Driving Along an Arterial and Noting What You See

Purpose

The purpose of this activity is to give you the opportunity to drive along an arterial and note what you see. The "seeing" and the "noting" are important, as it will help you to focus on the critical parts of the system, parts that we will document and include in the models that we will develop in the activities to follow.

Learning Objectives

- Describe how drivers respond to signal displays
- Identify and describe various physical components of a signalized arterial

Required Resource

- Movie file: A03.wmv

Deliverable

- Prepare a one page report summarizing your findings

Information

Learning to discern what is important is a critical skill for the transportation engineer. For example, when you are driving down an arterial to travel to a destination, what do you look for? You want to keep a safe distance from other vehicles. You want to watch for pedestrians and bicyclists. And, you want to watch for control devices such as stop signs or traffic signals.

As a traffic engineer, you will begin to watch for other things:

- Flow rates or traffic volumes
- How the intersection is laid out and the striping of the lanes
- Information, guide, and warning signs
- Signal displays
- The controller cabinet
- How the intersection is performing (do vehicles arrive primarily during red or during green, do queues clear before the end of green, is there a queue spillback from one intersection to another, and can pedestrians safely cross the street?)

This video takes you on a tour of Lankershim Blvd, an arterial located in Los Angeles, California. The four signalized intersections in the video are near Universal Studios. The tour begins with a trip northbound through the four intersections and then returning southbound. The tour lasts a little more than four and a half minutes. Your perspective is that of the driver of the car and his passenger.

Figure 24 shows an aerial photograph of one intersection along the driver's route. Figure 25 is a diagram, showing the location of the four intersections.

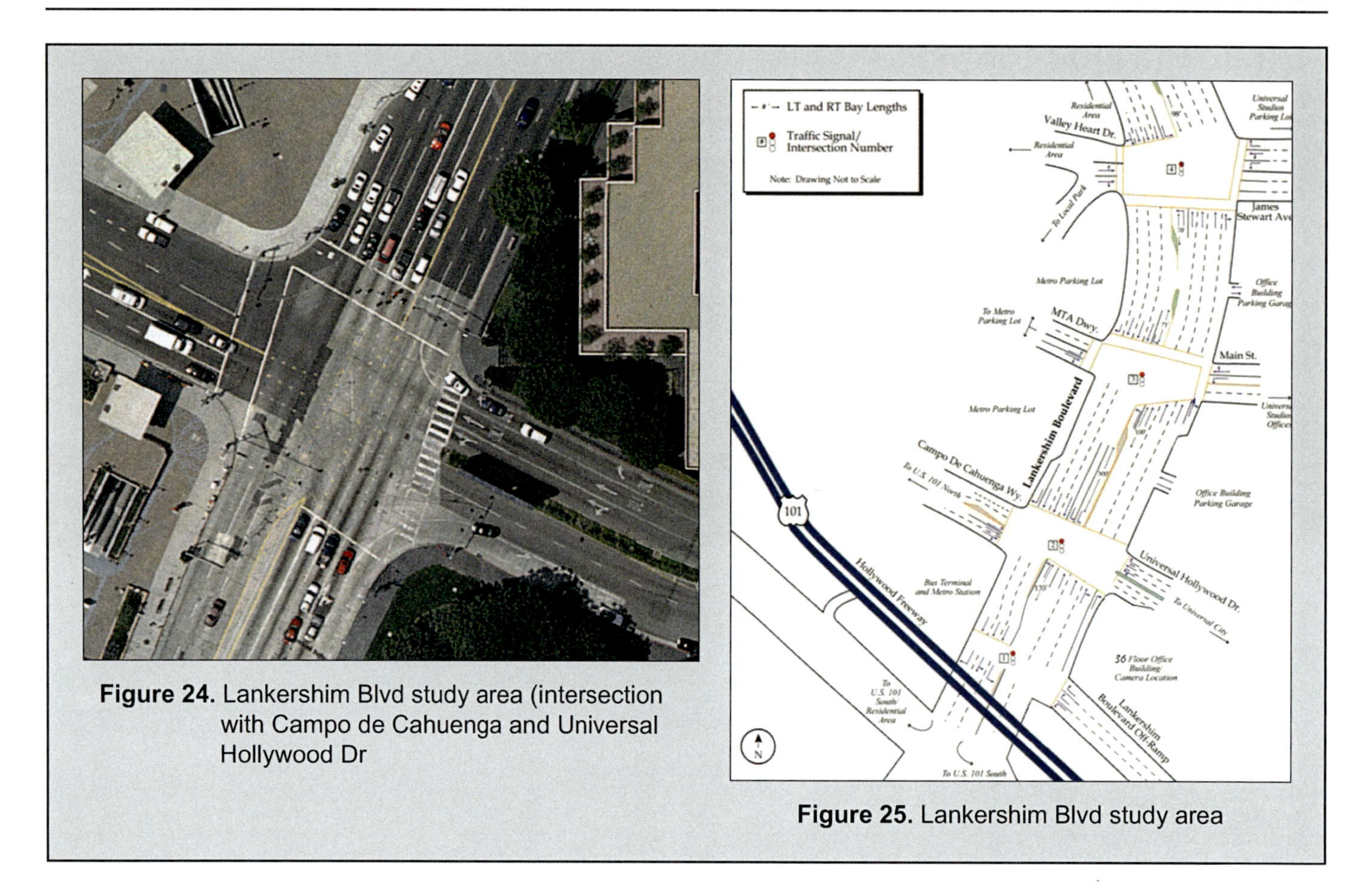

Figure 24. Lankershim Blvd study area (intersection with Campo de Cahuenga and Universal Hollywood Dr

Figure 25. Lankershim Blvd study area

Task 1

As you watch the video, remember the kinds of user responses that we discussed in Activity #1. For each intersection, record the following information:

1. The geometric layout of the approach, as seen from the perspective of the driver (user).
2. A description of each display for each of the four signalized intersections, as seen from the perspective of the driver.
3. The status of each vehicle display as the vehicle approaches and the response of the driver to the display.

Learning to See: The Simulation Environment in Which We Will Work

Purpose

The purpose of this activity is for you to appreciate how realistic a simulation model can be in replicating traffic flow at a real signalized intersection.

Learning Objectives

- Assess the realism of a simulation environment by comparing it with a video of actual field operations
- Develop your ability to "see" and "observe" video and animation of traffic flow at a signalized intersection and relate these observations to traffic flow theory and principles

Required Resource

- Movie file: A04.wmv

Deliverable

- Prepare a document with your answers to the Critical Thinking Questions

Critical Thinking Questions

As you begin this activity, consider the following questions. You will come back to these questions once you have completed the activity.

1. How realistic does the simulation appear to be? How realistic does a simulation model need to be? What is the basis of your conclusions?

2. Based on the information provided in the video, how do you know that a southbound vehicle has been detected?

3. Is the process of queue formation and clearance for the southbound approach similar or different to the description of traffic flow in Activity #1? Explain your answer.

4. Why does the phase end (or the display turn from green to yellow) for the southbound approach near the end of the video?

5. What other observations can you make that are relevant to the operation of the intersection?

INFORMATION

Not all models realistically duplicate traffic flow conditions found in the field. However, if a simulation model such as VISSIM is calibrated correctly, the results can closely approximate conditions that you would observe in the field. You will observe two videos, each of the intersection of State Highway 8 and Line Street in Moscow, Idaho. In this activity you will compare a field video with the simulation of the same intersection and conditions. When you start the movie file, your computer screen will look like Figure 26. The video on the left shows the VISSIM animation while the one on the right shows a video from the field showing traffic flow at the intersection.

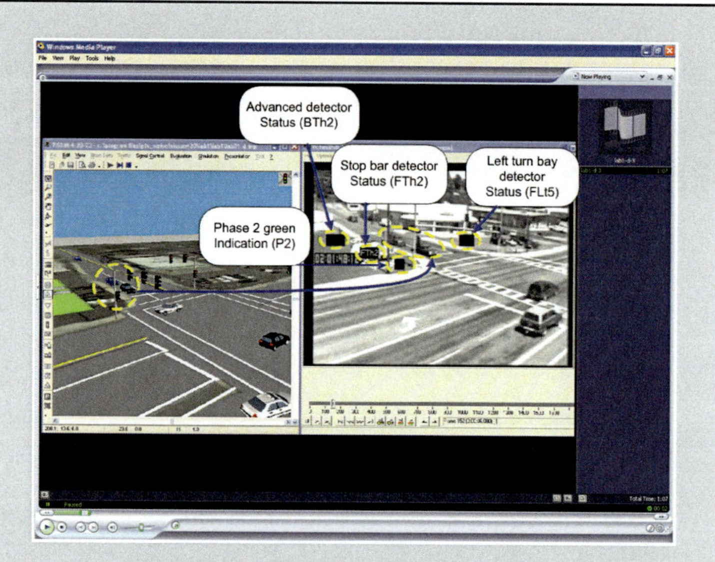

Figure 26. Field validation video

Three detector status indicators and the phase 2 green indication are imprinted in black boxes on the video frames. When a detector is "ON," the black text box corresponding to the detector is filled with the name of the detector. The detector names are indicated in parentheses in Figure 26. When the phase 2 green is "ON," the corresponding black text box is filled with "P2," for phase 2. In Figure 26, the two circled vehicles are queued in both fields of view and are calling for service on phase 2 as can be seen by the fact that the stop bar detector is on.

TASK 1

Open the movie file, A04.wmv.

TASK 2

Watch the video all the way through. Keeping in mind the Critical Thinking Questions for this activity, observe the traffic flow, detection information, and display status for both the southbound approach (upper left in both the video and animation) and the eastbound approach (bottom left). Phase 2 serves the southbound approach and phase 4 serves the eastbound approach. Make notes on your observations.

TASK 3

Based on the notes that you made during your observation of the video, prepare brief answers to the Critical Thinking Questions.

ACTIVITY 5

Working Together: Team Building for Effective Learning and Design

Purpose

In this activity, you will learn about effective team building and the factors that contribute to successful teams.

Learning Objectives

- Identify essential team behaviors that lead to successful completion of activities and the design project
- Explain how team behaviors support different team roles

Required Resources

- "Designing Teams and Assigning Roles," by Peter Smith
- "Teamwork Methodology," by Peter Smith, Marie Baehr, and Karl Krumsieg
- "Team Reflection," by Patricia Hare

Deliverable

- Prepare a one-page document that addresses your work in Tasks 1 and 4

Information

This activity includes three readings from the *Faculty Guidebook – A Comprehensive Tool for Improving Faculty Performance*, 4th Edition, edited by Steve Beyerlein, Carol Holmes, and Dan Apple and is used by permission. Each of the readings addresses an aspect of team building. More about the *Faculty Guidebook* is available at: www.pcrest.com

Task 1

Spend about ten minutes reading one of the three documents that has been assigned to you. Once you have completed the reading, spend another five minutes preparing a one page outline of the most important points covered in the document.

Task 2

Discuss your outline with your assigned partner. Your task is to prepare a three minute presentation on the paper that you read. The presentation should focus on what the audience should know and be able to do after hearing your presentation.

Task 3

Make your three minute presentation.

Task 4

As you listen to the other presentations, make a list of 2 to 3 key conclusions from the presentations. Confirm your conclusions with the presenters.

Student Notes:

3.4.2 Designing Teams and Assigning Roles

by Peter Smith (Mathematics & Computer Science, St. Mary's College, Emeritus)

For many faculty members, the issues surrounding team construction and management are significant. This module explores methods for implementing the use of roles in the classroom, including assigning students to teams and requiring team members to perform in roles. Because the workplace has become much more team-oriented over the past two decades, it is important that students learn to work well in teams; and students who participate in team environments are much better prepared to succeed on the job than are those without teaming experience. Although it is not yet common for business or industry to employ formal process-oriented roles for team members, graduates who have used roles frequently in undergraduate courses realize that the use of roles would dramatically improve team performance.

Why Roles are Important

Using roles helps team members to become interdependent (Johnson, Johnson, & Smith, 1991) and to be individually accountable for team success. It helps them to increase their learning skills (Apple, Duncan-Hewitt, Krumsieg, & Mount, 2000), and speed up the four stages of team development: forming (goal setting), storming (conflict resolution), norming (problem-solving), and performing (Tuckman, 1965). Roles should be rotated frequently so that each student has the opportunity to practice each role and to realize that effective learning requires that teams use all of the roles simultaneously. Rotating roles discourages dominance by one person and gives all students opportunities to practice social, communication, and leadership skills (Millis & Cottell, 1998). The roles introduced in this module are effective for enhancing team performance because each team member is empowered by his or her role to make a unique and significant contribution to the learning process.

Cooperative versus Collaborative Learning

The use of roles in learning activities is at the heart of the controversy between cooperative and collaborative learning. Although both approaches use small-group learning and encourage cooperative behavior, positive interdependence, and individual accountability, collaborative learning advocates hold that interdependence will occur naturally and that no attempt should be made to structure it. Therefore, the facilitator should not assign teams or roles, and should neither assess learning skills and performance, nor structure their development (Davidson, 1994). Cooperative learning is much more structured. The facilitator strives to ensure that the teams have diverse membership, he or she constantly assesses the skill level and performance of each student, and plans activities that will allow students to improve. When fulfilling the responsibilities of team roles, students must use many learning skills, so the facilitator has opportunities to intervene to help the individual student while ostensibly helping the team improve its performance (***3.2.3 Facilitation Methodology***).

Table 1 ***Various Team Roles and When to Use Them***

E = Essential O = Optional NA = Not Applicable

Learning Situation	Captain	Recorder	Reflector	Spokes-person	Technology Specialist	Planner	Time Keeper	Skeptic	Optimist	Spy
Cooperative Learning	E	E	E	O	O	O	O	O	O	O
Laboratory	E	E	E	NA	E	O	O	O	O	O
Project	E	E	E	O	O	E	O	O	O	O
Problem Solving	E	E	E	O	O	E	O	O	O	NA
Student Presentation	E	O	E	E	O	O	E	O	O	O
Student Teaching	E	O	E	E	O	O	E	O	O	O
Committee Work	E	E	E	E	O	E	O	O	O	NA
Department Business	E	E	O	NA	E	E	O	O	O	NA
Grant Writing	E	E	E	O	O	E	O	O	O	NA
Peer Assessment	O	E	E	E	NA	NA	O	O	O	NA

Performance Criteria for Team Roles

Captain

1. Facilitate the team process, keeping it enjoyable and rewarding for all team members.
2. Make sure each member has a role and is performing within that role.
3. Ensure that all team members can articulate and apply what has been learned.
4. Manage time, stress, and conflict.
5. Accept accountability for the overall performance of the team.
6. Contribute to the group as an active learner.

Recorder

1. Record group roles and instructions at the beginning of a task or activity.
2. During an activity, record and collect important information and data, integrating and synthesizing different points of view.
3. Document group decisions and discoveries legibly and accurately.
4. Accept accountability for the overall quality of the recorder's report.
5. Control information flow and articulate concepts in alternative forms if necessary.
6. Contribute to the group as an active learner.

Reflector

1. Assess performance, interactions, and the dynamics among team members, recording strengths, improvements, and insights ***(4.1.9 SII Method for Assessment Reporting)***.
2. Be a good listener and observer.
3. Accept accountability for the overall quality of the reflector's journal.
4. Present an oral reflector's report positively and constructively if asked to do so.
5. Intervene with suggestions and strategies for improving the team's processes.
6. Contribute to the group as an active learner.

Spokesperson

1. Speak for the team when called upon to do so.
2. Ask questions or request clarification for the team.
3. Make oral presentations to the class for the team.
4. Use the recorder's journal to share the team's discoveries and insights.
5. Collaborate periodically with the recorder.
6. Contribute to the group as an active learner.

Technology Specialist

1. Use the available technological tools for the team activity.
2. Listen, converse, and collaborate with team members; synthesize inputs, try suggestions and/or follow directions for the technology.
3. Retrieve information from various sources; manage the available resources and information.
4. Help team members understand the technology and its use.
5. Be willing to experiment, take risks, and try things.
6. Contribute to the group as an active learner.

Planner

1. Review the activity, develop a plan of action, and revise the plan to ensure task completion.
2. Monitor the team's performance against the plan and report deviations.
3. Contribute to the group as an active learner.

Timekeeper

1. Observe the time resource for the activity and/or record the time allocation announced by the facilitator.
2. Keep track of the elapsed time for various tasks and notify the captain when the agreed-upon time has expired.
3. Contribute to the group as an active learner.

Optimist

1. Focus on why things will work.
2. Keep the team in a positive frame of mind.
3. Look for ways in which team discoveries can be applied or used to the team's advantage.
4. Contribute to the group as an active learner.

Skeptic

1. Question and check the assumptions that are being made.
2. Determine the issues or reasons why quality is not being met at the expected level.
3. Be constructive in helping the team improve performance.
4. Contribute to the group as an active learner.

Spy

1. Eavesdrop on other teams during an activity to gather information and seek clarification of direction.
2. Relay information that can help the team perform better.
3. Contribute to the group as an active learner.
(Myrvaagnes, Brooks, Carroll, Smith, & Wolf, 1999)

Issues Surrounding Team Design

1. Although teams can contain any number of participants, most college and university level practitioners prefer groups of four for cooperative learning activities (Millis et al., 1998). Quads are small enough to engage each student, but large enough to provide a rich mix of ideas. Four-person teams can also be easily split into pairs for "think, pair, share" activities. Cooperative learning advocates David and Roger Johnson recommend three-person teams, and up to three of these may be necessary if the number of participants is not divisible by four. If absenteeism is a serious problem, five-person teams may be optimal, although regular attendance is vital because each student has a responsibility to contribute to the team's efforts. Sporadic attendance is a severe handicap to success with cooperative learning.

2. Project, problem-solving, committee-work, and grant-writing teams can be, and probably should be, larger than four, depending on the task to be accomplished and the number of available participants. It is best to assign permanent roles in these teams based on the strengths of the individual members because consistently high performance is more important than learning growth in these circumstances. The captain should be extremely well organized, self-confident, and able to inspire the team to excel. The recorder should be skilled at synthesizing the essential meaning from team discussion and keeping very organized records. The reflector should excel at multiprocessing and be confident enough to suggest improvements, even if they may imply substandard performance by one or more team members. The planner should be very creative and persistent, but flexible enough to accept changes in the plan as the project evolves.

3. Peer assessment and student teaching teams may well be smaller than four, perhaps as small as two, since they have sharply focused goals. All required role activities have to be accomplished, but formal role assignments may not be necessary. The work in these teams is usually divided fairly between the members.

4. Cooperative learning teams may be formed in a number of ways, such as random selection by counting off or drawing cards from a deck, student or participant selection, teacher or supervisor selection, or a combination of the last two. The goal is always to provide the greatest diversity within each team. Random teams often provide this diversity, but there is no way to ensure it. Research shows that participant-selected teams are not diverse and are unlikely to be successful (Fiechtner and Davis, 1985). Participants can suggest several people they would like to work with, and the facilitator can take these requests into account when assigning teams; the aim being to preserve diversity in gender, ethnic background, academic preparation and ability, and discipline or major. In order to gather the information needed to assign teams, many facilitators delay forming permanent teams until they can collect data sheets from students and observe them in learning situations.

5. Teams should be designed to accomplish the task for which they are formed. They can exist for short periods (e.g., formed to complete a five-minute in-class exercise), they can work together for several weeks to complete a project, or they can stay together for a whole semester or longer to provide long-term emotional and academic support (Duncan-Hewitt, 1995). Forming new groups midway through the semester gives students the chance to work with new individuals, thus providing a more realistic simulation of on-the-job teamwork. When deciding if and when to restructure the teams, it is important to carefully consider the learning needs of the participants and how well the current teams are functioning. Younger or more inexperienced students are more likely to need the support that long-term groups provide. Those close to graduation may profit from more frequent team membership changes.

6. One of the first team activities should encourage the team members to introduce themselves and learn about each others' learning styles. At this time, the team should agree to expectations or ground rules for all members.

   Suggested ground rules for team activities (Silberman, 1998):

   - Start on time with everyone present
   - Be prepared
   - Get to know members who are "different" from oneself
   - Be gender/race/ethnicity sensitive
   - Give everyone a chance to speak
   - Let others finish speaking without interrupting them
   - Be brief and to the point
   - Share the workload
   - Rotate group roles
   - Reach decisions by consensus
   - Assess team functioning periodically

Guidelines for Implementing Team Roles

1. The facilitator must check that students have assumed and rotated roles, must intervene to improve role performance, and give credit for conscientious role fulfillment via learning journal reports or other means (***4.1.4 Assessment Methodology***).
2. To ensure that reflectors improve their performance, the facilitator should take time for reflectors to share oral reports with the class, frequently in the beginning of the term and at regular intervals thereafter.
3. Students with roles other than captain, recorder, and reflector often fail to appreciate the importance of their roles. The facilitator should intervene to recognize team members who do well in these other roles or to ask the team if they would like to be informed of specific instances when using these roles would enhance performance.
4. Reflectors may withdraw from active participation in the group in order to observe and write down their assessments. The facilitator should encourage them to observe while fully participating and to take a minute at 15-20 minute intervals to jot down their assessment.
5. Recorders may complain that they are so busy writing that they have no time to think or to process what the team is doing. They need to be encouraged not to write everything down, but to synthesize the discussion in a few well-constructed sentences.
6. When the captain is very shy, introverted, or not confident, another team member is likely to take over that role. To fix this situation, the facilitator should address all team intervention questions to the captain, refer to the team using the name of the captain (unless the team has chosen another name), hold the captain responsible for time management, and attempt to make eye contact with the captain when giving positive nonverbal feedback to the team.
7. The nature of the team roles and the responsibilities of those fulfilling the different roles may well change as the team moves through stages of development. More permanent roles may be appropriate in the latter stages of team development.

Concluding Thoughts

The effort needed to establish team roles and train students in their use pays big dividends in increasing learning. Roles also help ensure fair participation in the group process by all the learners. Students who value and experience defined roles in the group process will be prepared to assume a variety of roles in the workplace and in community and extracurricular activities as well.

References

Apple, D. K., Duncan-Hewitt, W., Krumsieg, K., & Mount, D. (2000). *Handbook on cooperative learn*ing. Lisle, IL: Pacific Crest.

Davidson, N. (1994). Cooperative and collaborative learning: An integrative perspective. In J. Thousand, R. Villa, & A. Nevin, (Eds.). *Creativity and collaborative learning: A practical guide to empowering students and teachers*. Baltimore, MD: Paul H. Brooks.

Fiechtner, S. B., & Davis, E. A. (1985). Why groups fail: A survey of student experiences with learning groups. *The organizational behavior teaching review, 9 (4)*, 58-73.

Johnson, D. W., Johnson, R. T., & Smith, K. A. (1991). *Active learning: Cooperation in the college classroom*. Edina, MN: Interaction.

Millis, B. J., & Cottell, P. G. (1998). *Cooperative learning for higher education faculty*. Phoenix, AZ: Oryx Press.

Myrvaagnes, E., with Brooks, P., Carroll, S., Smith, P. D., & Wolf, P. (1999). *Foundations of problem solving*. Lisle, IL: Pacific Crest.

Secretary's Commission on Achieving Necessary Skills (SCANS). (1991). *What work requires of schools: A SCANS report for America 2000*. Washington, DC: Department of Labor.

Silbernam, M. (1998). Building cooperative learning teams. *Cooperative learning and college teaching, 8 (3)*, 16-17. Stillwater, OK: New Forums Press.

Tuckman, B. W. (1965). Developmental sequence in small groups. *Psychological Bulletin, 63 (6)*, 384-399.

3.4.3 Teamwork Methodology

by Peter Smith (Mathematics & Computer Science, St. Mary's College, Emeritus), Marie Baehr (Vice President for Academic Affairs, Coe College), and Karl Krumsieg (Vice President of Operations, Pacific Crest)

Teamwork is essential when a group of people strive to achieve a common goal. Because teamwork is a process, a methodology is needed to better understand and master performance in this area. This module presents a methodology that can benefit team performance by experts as well as novices. To demonstrate how the methodology might be practiced by people with different levels of experience, two examples are presented: one that involves a team of students, and one that involves a team of faculty members.

Need for a Teamwork Methodology

The dictionary defines *teamwork* as the joint action by a group of people in which individual interests be-come secondary to the achievement of group goals, unity, and efficiency. In other words, teamwork involves a group of people actively cooperating in an organized way to achieve a goal. The process of teamwork has become commonplace throughout organizations in all areas of society, including business and industry, health care, public service, government, and education (Commission on Accountability in Higher Education, 2005). The benefits of teamwork are numerous. When an effective teamwork process is employed, more can be accomplished with better results. Individuals working within teams also develop many beneficial skills: teaching new skills to others, learning to negotiate, exercising leadership, and working with diverse people in diverse situations; being part of a team effort in which individual members are held accountable (Millis & Cottell, 1998).

Teamwork is an individual skill: everyone in an organization must take responsibility for the performance of each team to which they are assigned; otherwise individuals can justify non-performance by blaming the team (Avery, 2001). Teamwork differs from project management in that it focuses on team formation and the behaviors and attitudes of the team members; not just the successful accomplishment of the project, goal, or product (Smith & Imbrie, 2005). Team membership calls on a participant's personal development skills (***4.2.3 Personal Development Methodology***), but it also requires individuals to establish relationships and interact with other team members. Teamwork requires leadership skills, but it is just as important for team members to be able to support the leadership of another. In fact, teams are seldom allowed to reach their full potential when they adhere to the traditional concept of a leader who makes a majority of the decisions (Maxwell, 2001). Team activities can employ a number of methodologies, such as communication, learning, problem solving, research, and design; but none of these focus on team formation and group processing. The reader is encouraged to compare the guidebook modules describing these methodologies with the Teamwork Methodology listed in Table 1 and to note differences in emphasis.

Table 1 ***Teamwork Methodology***

1. **Define the Mission:** Establish a common vision and goals for the team.
2. **Recruit Members**: Assemble the individuals to meet the needs of the team.
3. **Collect Resources**: Identify and collect resources available to the team.
4. **Build the Team**: Assign members to appropriate roles.
5. **Create and Implement Plan:** Schedule the resources for identified tasks and perform the plan.
6. **Assess Performance**: Assess the performance of the team against the plan
7. **Modify the Plan**: Make periodic improvements to the plan.
8. **Provide Closure**: Provide a final point or end; celebrate accomplishments.

A Simple Example of the Methodology

Scenario: The instructor for a capstone course in an engineering program has decided to use teams to more realistically simulate the process of project design and development. One of the project options is to design a high-impact tester to shock test circuit boards for naval warships.

1. **Define the Mission**

   Student teams must complete a high-quality drawings package for a high-impact tester that can be used by an electronics company for the eventual fabrication of this device.

2. **Recruit Members**

   The course instructor composes teams using information gained from an email questionnaire completed by students shortly before the first class session. The aim is to preserve diversity in gender, ethnic background, academic preparation, and discipline or major. The instructor must staff multiple projects, and all must have an equal likelihood for success.

3. **Collect Resources**

   The instructor identifies software tools, fabrication facilities, and working space required by each team. In this case, solid modeling software is installed on the lab computers. The lab also has meeting areas for pri-vate team meetings. The flawed and incomplete blueprints from the Navy are also available.

4. **Build the Team**

   On project assignment day, team members interview each other to discover their interests and personal goals for the course. Each team is asked to develop a shared vision of project success and to determine relevant roles that motivate each member. The team working on the impact tester sets a goal of having their design selected by the client and having their device installed within two months after the finalization of the drawing package. They select roles of project manager, document clerk, reflector, and client com-munications coordinator.

5. **Create and Implement the Plan**

   Each team creates milestones to ensure that the project is completed on time. They use insights gained from a client interview with a lead engineer at the electronics company to divide the work as fairly as possible according to ability and interest. Each team draws up a detailed plan to meet their first milestone and all team members commit to it.

6. **Assess Performance**

   The project manager periodically checks to ensure that tasks are being satisfactorily accomplished in a timely manner. He finds that two tasks require more work than was anticipated. The reflector also touches base with each team member regularly and finds that those assigned to the difficult tasks are becoming frustrated.

7. **Modify the Plan**

   At the team's next meeting they confront the problem of task difficulty. Two members who have com-pleted their tasks volunteer to help with the tasks that are falling behind schedule. They agree to distribute tasks more equitably when they plan for the next milestone.

8. **Provide Closure**

   After presenting the results of the first milestone to their client, the team goes out for pizza to celebrate a job well done and to get revved up for the next milestone.

Discussion of the Teamwork Methodology

1. **Define the Mission**

   The first step to building a team involves identifying and defining the purpose and objectives for the team. The mission influences who is recruited, what resources are needed, and what main tasks need to be performed. In some cases, teams are formed to accomplish a specific goal; in other cases, teams maintain their structure but may change the people involved.

2. **Recruit Members**

   One should identify and recruit people who believe in, and are committed to, the stated mission. These individuals should define their goals and objectives, share their reasons for involvement, and indicate how their participation can strengthen the performance of the team.

3. **Collect Resources**

   The mission statement influences what resources are required to meet the team's goals and objectives. One should identify the available resources and determine what additional resources need be obtained. Examples of a team's resources include the team members and their skills, financial assets, information, computers, physical equipment and facilities, time, and the team members' individual resources that they are willing to contribute for the team's use.

4. **Build the Team**

   It is important that team building occurs at this point and continues throughout the process. Participants must build shared ownership of the team's goals and objectives, and all must believe that these are worthwhile and attainable. By assigning roles with job descriptions, one enhances the team's accountability, performance, and unity while helping to facilitate team goals. Depending on the purpose and length of the team's mission, roles should be periodically rotated so that everyone can gain experience and improve skills in different areas.

5. **Create and Implement the Plan**

   The process of creating the plan need not be democratic; however, it is important that all members accept responsibility for implementing it. Successful completion of the plan depends on "buy in," or acceptance, by all team members. It is important that as the team implements the plan, all members perform according to their roles. The team captain is responsible for team's overall performance.

6. **Assess Performance**

   Each member's performance should be regularly assessed according to the criteria set for each role. The team as a whole should also be regularly assessed as it works toward meeting its goals and objectives. By assessing during the early stages of the plan (as well as on a regular basis), it is possible to determine what is working and what needs to be changed.

7. **Modify the Plan**

   The plan of action should be updated as dictated by the situation and/or by the team's performance. Changes and modifications can be made for both the short term and long term. In addition to modifying the plan, one may also change the situation by shifting roles within the team, adding new team members, obtaining additional resources, or by changing the goals and objectives.

8. **Provide Closure**

   All team members should know when the plan is completed or the objectives have been met. Both individual and team accomplishments should be acknowledged and celebrated.

Another Example of the Methodology

Scenario: A college needs to replace its Director of Information Technology. The Academic Dean assigns a search committee to this task.

1. **Define the Mission**

   The mission is to find an outstanding candidate for the position of IT director and to prepare a report for the dean.

2. **Recruit Members**

   The dean consults the department heads of those areas that interact most closely with IT and asks them to recommend persons under them who will be competent and committed to the search com-mittee process. She suggests that potential search committee members submit written statements de-scribing what they hope to contribute to the search process. The dean consults the department heads of Business Affairs, Human Resources, Registrar, Admissions, and the Library; and faculty from Math, Science, and Business. She asks the math faculty representative to chair the committee, and together they select the other members. The dean sends appointment letters to each.

3. **Collect Resources**

   The Educause organization has a national clearinghouse to advertise for IT directors. There are some other technical journals and trade publications like Compuworld that accept advertising. The college will pay for these ads as well as travel and housing expenses to bring three candidates to campus for interviews. The Business Affairs Office offers their conference room for meetings and the services of their secretary to take minutes.

4. **Build the Team**

   At the first team meeting the members share their goals and accept the following roles and tasks which they all agree will accomplish the search committee's mission. Peter represents the math department. He will act as team leader. He facilitates team meetings and manages time resources. He checks to make sure that each team member accomplishes his or her job, and he helps the team stay focused on its objectives. Debby is the Human Resources Director and Researcher. She agrees to schedule candidates' interviews, and to arrange their travel and lodging. She identifies the references provided on each candidate's application, and sets up phone contacts with them. She also arranges phone interviews with each candidate, and discusses the salary and benefits packages with them. Mary, the Admissions Director, ensures that the candidates are committed to the support of student recruitment. She also coordinates campus tours and student lunches with the on-campus interviews. Alice, the Registrar, makes sure that the candidates are aware of the difficulties presented by in-house registration software and the lack of a system of computerized class scheduling. She also informs the candidates of the IT concerns of the administrative and reports their responses back to the team. Jill, represents the business department, and acts as the team's reflector. She assesses the process of the team, offering regular feedback about the team's performance, including their strengths, areas in need of improvement, and insights. She also serves as mediator to help resolve conflicts that arise among team members. Les, the Business Affairs Manager, assesses the candidates' fiscal knowledge and ability to work with Financial Aid as well as their commitment to include business functions, especially billing and payroll, into an integrated networked IT system. John, the Development Office Representative, observes the candidates during their on-site interview visits, watching how they deal with college personnel who have little IT savvy but significant IT needs. He is particularly interested in the candidates' creativity in problem solving. Dick represents the science department faculty. He as-sesses the candidates' level of commitment to working towards IT literacy among students, and their interest in utilizing work study students in the IT department. Dick also constructs an efficient process for narrowing the candidates.

5. **Create and Implement the Plan**

   The team meets and, after brainstorming goals and objectives, agrees on several project goals, and objectives having to do with their overall work process. Their goals are to select the three best candidates to bring to campus for interviews, to choose the best of the three to recommend to the Academic Dean, and to complete this process within two months. They agree on a project implementation plan. They begin by reading the written applications and rejecting the candidates who are clearly unsuitable. They interview each remaining candidate by phone with at least two committee members participating in each call. They agree to use Dick's plan for narrowing the pool of candidates. They check the references of those candidates who made the cut, and, if they still have more than three, select the best three for on-campus interviews. Debbie arranges travel and lodging for each finalist, and sets up a full-day interview schedule span-ning a two-week period, during which all constituencies will be included in the interview process. They schedule a final meeting to identify the best candidate and prepare a report for the dean, followed by a ca-tered lunch to celebrate a job well done.

6. **Assess Performance**

   After the application deadline, the team meets weekly to monitor progress on the project as well as the team's process. At the first meeting they eliminate five candidates who clearly do not have the qualifications for the job. The team completes the phone interviews in record time. By the second meeting they are ready to apply Dick's weighted voting process to the remaining candidates. There is a clear separation between the top five candidates and the others. Debby is swamped trying to check references for these five.

7. **Modify the Plan**

   Peter asks different members of the team to call the references, three members to each call. Also, a different group from the committee talks to each of the remaining candidates until each member has talked to each candidate. Even though this takes an additional two weeks, the search process is still on schedule. After all of the interviews have taken place, the committee is able to eliminate two candidates from consid-eration and everyone believes that the decision process has been fair.

8. **Provide Closure**

   The on-campus interviews are very revealing. In the opinion of all participating constituencies one candidate is clearly superior (Debby gathers this information using a standard questionnaire to solicit feedback from everyone involved). The committee decides that if the dean rejects this choice or if the desired candidate turns down the offer, the search should be reopened. After completing the report to the dean well within the two-month time frame, the committee celebrates its work with catered lunch, and then disbands. Fortunately, the dean agrees with the committee's choice and the candidate accepts the position and does an outstanding job for the college for the next ten years.

Concluding Thoughts

Teamwork is a process that challenges each team member to accept accountability for accomplishing the team's goals and for actively contributing his or her utmost to enhance team synergy; potential team members should not accept membership on a team if they cannot make this commitment. This methodology provides a blueprint for bringing every team performance to the highest level so that participation becomes a growth experience for all. It is important to remember that, in many cases, team performance will be degraded if team members focus only on the product the team is expected to produce without paying attention to the process of team formation, interaction, and closure. Try implementing this methodology during your next teamwork experience. While your teammates may be skeptical initially, they will likely appreciate the improved teamwork that this structure will produce.

References

Avery, C. (2001). *Teamwork is an individual skill.* San Francisco: Berrett-Koehler.

Commission on Accountability in Higher Education. (2005). *Accountability for better results: A national imperative for higher education.* Boulder, CO: State Higher Education Executive Officers.

Fiechtner, S. B., & Davis, E. A. (1985). Why groups fail: A survey of student experiences with learning groups. *The organizational behavior teaching review, 9 (4),* 58-73.

Maxwell, J. (2001). *The 17 indisputable laws of teamwork.* Nashville, TN: Thomas Nelson.

Millis, B. J., & Cottell, P. G. (1998). *Cooperative learning for higher education faculty.* Phoenix, AZ: Oryx Press.

Smith, K. (2004). *Teamwork and project management.* (2nd ed.). New York. McGraw Hill.

3.4.4 Team Reflection

by *Patricia Hare (Dean of Developmental Programs, Brevard Community College)*

To attain optimum productivity in team projects and cooperative learning situations, it is critical that team members spend some portion of their time thinking critically about the effectiveness of their work as a team. *Team reflection* is a process in which team members bring closure to their work or learning experience, and focus on ways to increase future learning and performance. Ideally this should use no more than 5% of a group's actual performance time. Ideally team reflection uses an assessment-oriented approach, analyzing personal and team happenings against important criteria, and producing action plans that can add value to future performances.

Nature of Reflective Practice

What separates humans from animals is our ability to examine our world carefully, to think about our surroundings, and to think about our own thinking. Thinking about what we were thinking, doing, or feeling is known as *critical thinking* or *reflection* (Chaffee, 2004). Experts tell us that when we reflect, we must allow space (with no distractions), silence, and time to ponder and to self-assess (Ferrett, 2006). Reflection is a cornerstone of purposeful learning and of critical thought.

Donald Schön distinguishes between two different types of reflection: reflection-in-action and reflection-on-action. *Reflection-in-action* is "thinking on our feet." We observe our experience, paying attention to what might be unfolding, connecting with our feelings, and building new understandings to inform our actions in that experience. *Reflection-on-action* involves thinking about our experience after it has happened, to think about why we acted as we did during the experience, to consider what was happening individually or in a group, and to explore circumstances that might have been present.

Reflective practice, whether it is in action or on action, is a habit, structure, or routine for examining individual and group experiences. It can vary in depth, frequency, and length depending on its purpose (Amulya). Reflective practice can be based on finding solutions (***3.3.4 Problem-Based Learning***), habitually journaling personal experiences (***4.2.3 Personal Development Methodology***), or making deliberate improvements in quality (***4.1.4 Assessment Methodology***).

Role for Assessment

When one practices reflection with a mindset toward assessment, one focuses on helping performers improve the quality of their future performances rather than simply analyzing and evaluating past events (***4.1.2 Distinctions Between Assessment and Evaluation***). Assessment is assessee-centered and is guided by appropriate performance criteria (***4.1.7 Writing Performance Criteria for Individuals and Teams***). For the assessment to be effective, the assessor and the assessee must trust and respect each other, and the assessee must be prepared to act on the assessment feedback (***4.1.8 Issues in Choosing Performance Criteria***). Models such as the ***SII Method for Assessment Reporting (4.1.9)*** provide a structure for identifying strengths in performance, including explanations for why they were strengths; for prioritizing improvements, including descriptions of how they might be implemented; and for generating insights about knowledge construction, problem solving, or personal development that have value in other contexts. Formal aids such as a reflection journal, periodic reflector reports, and team worksheets can recover important data associated with an individual or team performance that makes assessment feedback more specific and therefore more useful (Apple, 2000).

Team reflection is an excellent process to start developing team assessment skills. Use of peer reporting leads to dialogue between peers and others involved in the learning process. Each member takes a turn recounting a key event, accepting feedback and analyzing it, making assumptions and connections, and formulating questions that emerge in the process. This practice allows the group to explore assumptions and connections across multiple perspectives. One advantage of team reflection is that reflections emerge from collective work that is frequently connected to or aligned with team values. Based on what the team learns through reflective thinking and sharing, the team can assess whether they have met their own performance criteria and can generate action plans to improve future performances.

Team-based reflection forces students or team members to think at higher levels in Bloom's taxonomy (***2.2.1 Bloom's Taxonomy—Expanding its Meaning***). Reflective practice is enhanced by active listening, questioning, discussing, and storytelling. Team reflection that results in high-quality assessment feedback can be promoted by assigning and using the role of team reflector (***3.4.2***

Table 1 **_Criteria for an Oral Reflector's Report_**

The report should
• Be loud and clear enough for all to hear
• Be concluded within 30 seconds (unless specific otherwise)
• Identify one strength of the team's performance and explain why it is a strength
• Identify one area for improvement on which the team can focus, and explain how the team can make this improvement
• Provide one insight gained about the learning process, and explain the significance of the insight

Table 2 **_Criteria for a Written Reflector's Report_**

The report should
• Be concise
• Prioritize information
• Relate to the focus area of performance
• Refer to key skills used by the team
• Address affective issues
• Be clear
• Be accurate
• Cite specific examples to support assessment results
• Provide supporting documentation in the *Learning Assessment Journal*

Designing Teams and Assigning Roles). The reflector should keep a journal in which to record team strengths, improvements, and insights. This person should report his or her findings in a positive and constructive manner (reflection-on-action), and also intervene during teamwork sessions with suggestions and strategies for improving the teams' processes (reflection-in-action). The reflector role should be rotated among team members.

Tools for Team Reflection

A number of tools for stimulating team reflection are available. One tool is the reflector's journal mentioned above. Entries are made in real time and are grounded in the knowledge gained through the experience. The *Learning Assessment Journal* contains reflector report forms and weekly reflector report forms which provide excellent prompts for reflection-in-action and reflection-on-action (Apple, 2000). Findings can be summarized and acknowledged in two ways: in oral reflectors' reports and in written reflectors' reports. Criteria for these reports are given in Tables 1 and 2.

Reflective journaling can be used in online discussions about an event or an experience that is shared by the online team. Members engage in discussions, reflecting on what they have discovered in the experience or the event. Problem-based reflective practices and assessment can also be conducted online, with reflections written in discussion boards. Members can then assess what they have learned during these sessions.

Concluding Thoughts

Reflective practices can add significant value to cooperative learning as well as student and faculty projects (Rodrique-Dehmer, 2007). Implementing reflective practices in a team environment will certainly take more time initially. However, faculty who make the commitment to use reflection on a formal and regular basis, both in their classes and in their committee work, find that the benefits of team learning, productivity, and participant satisfaction significantly outweigh the initial time investment.

References

Amulaya, J. *What is reflective practice?* Cambridge, MA: Massachusets Institute of Technology Center for Reflective Practice.

Apple, D. K. (2000). *Learning assessment journal.* Lisle, IL: Pacific Crest.

Chaffee, J. (2004). *Thinking critically*. Boston: Houghton-Mifflin.

Ferrett, S. (2006). *Peak performance: Success in college and beyond* (6th ed.). New York: McGraw-Hill.

Rodriguez-Dehmer, I. (2007, Winter). What is the role of responsive and reflective instructor? Florida *Developmental Education Association Newsletter.*

Schön, D.A. (1983). *The reflective practitioner: How professionals think in action.* New York: Basic Books.

Schön, D. A. (1990). *Educating the reflective practitioner: Toward a new design for teaching and learning in the professions.* San Francisco: Jossey-Bass.

ACTIVITY 6 Design Team Agreement

Purpose

Forming a team is a non-trivial task. Each member of a team has different expectations, experiences, and ways of working that have to be blended and negotiated with other team members. One of the readings from Activity #5 defines teamwork as "... the joint action by a group of people in which individual interests become secondary to the achievement of group goals, unity, and efficiency." Since much of the work done in engineering practice today is based on working effectively in teams, the experiences that you gain in learning how to do this will be valuable ones.

Learning Objective

- Develop group consensus on how the team will work with, treat, and communicate with each other.

Required Resources

- "Designing Teams and Assigning Role", by Peter Smith
- "Teamwork Methodology", by Peter Smith, Marie Baehr, and Karl Krumsieg
- "Team Reflection", by Patricia Hare

Deliverables

- Prepare a document with your team's answers to the Critical Thinking Questions
- Prepare and sign the team agreement

Critical Thinking Questions

1. How do you want to be treated by your group?

2. What do you expect from each group member during a group meeting?

3. How do you intend to communicate with each other and by what means?

4. How do you intend to keep records of work conducted by the group and by individuals within the group?

5. How do you intend to resolve disputes or conflicts?

6. When do you intend to regularly meet outside of normal class hours?

7. How do you intend to divide work tasks among group members?

8. How do you intend to assess the performance of the group and its members?

Task 1

Assemble into your assigned group. Reflect on the reading that you completed in Activity #5 and the two presentations that you listened to. Answer the Critical Thinking Questions and discuss them with your group.

Task 2

Prepare a one-page document that addresses the issues listed above. The document should be in the form of a "team agreement" that starts with "We agree to …" and that is signed by each team member.

Activity 7 Introduction to the Traffic Signal Timing Manual

Purpose

The purpose of this activity is to give you the opportunity to explore the basic reference used by practitioners in this field, the *Traffic Signal Timing Manual*.

Learning Objective

- Describe the content, scope, and organization of the *Traffic Signal Timing Manual*

Required Resource

- *Traffic Signal Timing Manual*

Deliverables

Prepare a document that includes

- Answers to the Critical Thinking Questions
- Completed Concept Map

Link to Practice

Your instructor will assign a reading from the *Traffic Signal Timing Manual*.

Critical Thinking Questions

When you have completed the reading, prepare answers to the following questions:

1. What is the purpose of the *Traffic Signal Timing Manual*?

2. List each of the chapters in the manual and briefly describe the purpose of each.

3. Which of the chapters do you think are most pertinent to the work that you will be doing during this class?

4. What is the difference between a policy and a standard?

5. What defines a signal timing policy?

6. Find the section of the *Traffic Signal Timing Manual* that deals with the signal timing design process. Target two aspects of the signal timing design process where your understanding could be strengthened. Write a critical thinking question for each of these two aspects. Provide answers to these questions.

7. What are some of the interesting or important findings of the National Signal Timing study and why do you find them interesting or important?

8. How does policy support the design of the traffic control system?

9. In addition to the examples of signal timing policy application described in your reading of the *Traffic Signal Timing Manual*, find one other example based on a search of the Internet. Briefly describe it and provide the URL.

10. What are some advanced traffic control concepts that are described in the *Traffic Signal Timing Manual*? List and define three of these concepts.

IN MY PRACTICE...

by Tom Urbanik

Traffic signal timing has several aspects ranging from a simple signalized intersection to traffic signals adjacent to railroad grade crossings. While the timing engineer may work regularly on isolated and coordinated intersections, occasionally a special problem may arise that the engineer has not worked on such as railroad preemption of a traffic signal. The *Traffic Signal Timing Manual* has practical guidance which the signal timing engineer can use to determine the requirements for operating the traffic signal. It should be noted that there may be situations that are too complex for the timing engineer to comfortably address. By referring to the *Traffic Signal Timing Manual*, the engineer may conclude that they are not comfortable with tackling the problem, but can become knowledgeable enough to seek the appropriate assistance to create a safe and efficient solution.

Concept Map

Terms and variables that should appear in your map are listed below.

actuated control	detector	fixed time control	queue
controller	display	movement	user

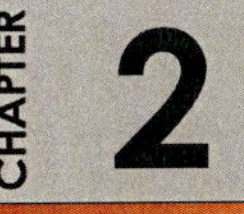

Modeling What We've Observed: Queuing Systems

Purpose

The purpose of this chapter is for you to learn how to apply queuing theory to the operation of a signalized intersection.

Learning Objectives

When you have completed the activities in this chapter, you will be able to

- Describe the components of a queuing model and how a queuing model can represent a signalized intersection
- Connect your observation of traffic flow at a signalized intersection with a model framework
- Represent and interpret queuing diagrams for a range of traffic flow and control conditions
- Connect a theoretical queuing model with real world conditions
- Represent and interpret queue accumulation polygons for a range of traffic flow and control conditions
- Compare and contrast the traffic flow representations used in the *Traffic Signal Timing Manual* with those that you studied in the activities in this chapter

Chapter Overview

A model is a representation of reality. It includes only those elements or features of the real world ("reality") that are important for understanding a system or a process, in this case the traffic signal control system. In Chapter 2, you will learn about queuing models and how to connect the theory included in these models with what you observe in the real world. A queue is a line waiting for service: people in a check-out line at a grocery store, vehicles at a stop sign waiting to safely enter the intersection, or pedestrians waiting for the Walk indication at a signalized intersection. In each case, we need to specify the pattern of arrivals, the manner in which service is provided, and the discipline within the queue. For example, vehicles may arrive at a traffic signal in a random manner (arrival pattern), be delayed during red and be served during green (service pattern), in a first-come, first-served manner (queue discipline).

This chapter begins with a *Reading* activity (Activity #8) on the modeling of traffic flow at a signalized intersection using queuing theory. In Activity #9, you will be asked to complete several tasks to validate your understanding of queuing models by answering questions about and preparing sketches of queuing models that represent various traffic flow conditions. In Activity #10 you will link this queuing model with the real world, using a high resolution data set of field observations from a signalized intersection. Your ability to link theory with what you observe in the field is an important skill. In Activity #11, you will enhance your understanding of queuing models by observing traffic flow in the field and collecting data that you will use to link to the models. Finally, in Activity #12, you will learn how the *Traffic Signal Timing Manual* represents traffic flow at a signalized intersection and compare this representation with the descriptions in this chapter.

Activity List

Number and Title		Type
8	Modeling Traffic Flow at Signalized Intersections	*Reading*
9	What Do You Know About Queuing Systems?	*Assessment*
10	Using High Resolution Field Data to Visualize Traffic Flow	*Discovery*
11	From Model to the Real World: Field Observations	*Field*
12	Basic Operational Principles	*In Practice*

Modeling Traffic Flow at Signalized Intersections

PURPOSE

The purpose of this activity is to introduce you to queuing systems as they apply to the operation of signalized intersections.

LEARNING OBJECTIVE

- Describe the basic components of a queuing model representing traffic flow at a signalized intersection

DELIVERABLES

- Define the terms and variables in the Glossary
- Prepare a document that includes answers to the Critical Thinking Questions

GLOSSARY

Provide a definition for each of the following terms and variables. Paraphrasing a formal definition (as provided by your text, instructor, or another resource) demonstrates that you understand the meaning of the term or phrase.

average delay	
cumulative vehicle diagram	
platoon	
queue accumulation polygon	
queuing system	
saturation flow rate	
time space diagram	

v	
t	
r	
g	
g_q	
s	
d_i	
t_d	
t_a	
d_a	
c (lowercase)	
C (uppercase)	
X	

Critical Thinking Questions

When you have completed the reading, prepare answers to the following questions.

1. What is a queuing system?

2. Which elements of a traffic control system are included in the queuing system?

3. Which elements of traffic flow can you represent in a time space diagram?

4. What is a *queue accumulation polygon* and what information does it show about intersection operation and performance?

5. How realistic is the uniform delay equation or model?

6. What performance measures can the cumulative vehicle diagram show?

7. What are the elements common to a flow profile diagram, a cumulative vehicle diagram, and a queue accumulation polygon?

Information

In the previous chapter, we described the process of vehicles arriving at a signalized intersection and the response of the drivers of these vehicles to the various vehicle displays. We considered driver responses during the red interval, the beginning of the green interval when the queue is clearing, during the green interval after the queue clears and entering the intersection without delay, and during the yellow and red intervals. In some ways, this is the most basic kind of intersection operation, the ebb and flow of vehicles arriving at and traveling through the intersection, as time progresses from the red interval to the green interval to the yellow interval, and back again to red.

In this chapter, we will consider an idealized representation, or model, of vehicular traffic arriving at and departing from the intersection. It is useful to abstract or model this narrative description into more mathematical terms that we can use and extend to more complex conditions. As we proceed to these idealized representations, which are important as learning and visualization tools, we shouldn't forget that the real world isn't quite this clean and sharp, something that we shall see in later in this chapter!

Representing traffic flow using a time-space diagram

Let's first consider vehicles arriving at the intersection. Figure 27 shows three vehicles representing the ideal condition of uniform or constant headways, as they approach an intersection. The "space" of the intersection approach is shown on the y-axis, while time is shown on the x-axis. The vehicle display is shown at the bottom of the figure, with each of the three horizontal spaces available to show the time histories of the red, yellow, and green displays. This representation is known as a time-space diagram.

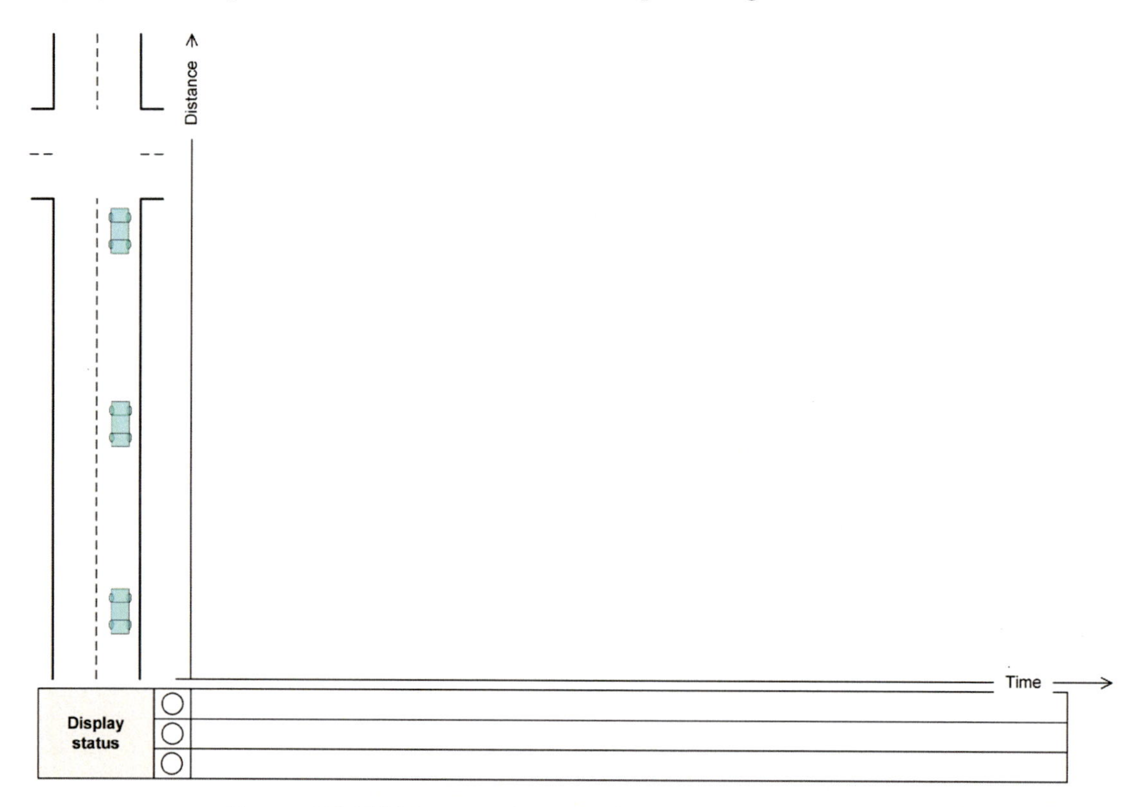

Figure 27. Vehicles evenly spaced approaching intersection

Figure 28 shows the trajectories for the three vehicles. The time space diagram shows that the vehicles had been traveling at constant (and equal) speeds as the slopes of the trajectories are parallel and linear. The display status is red.

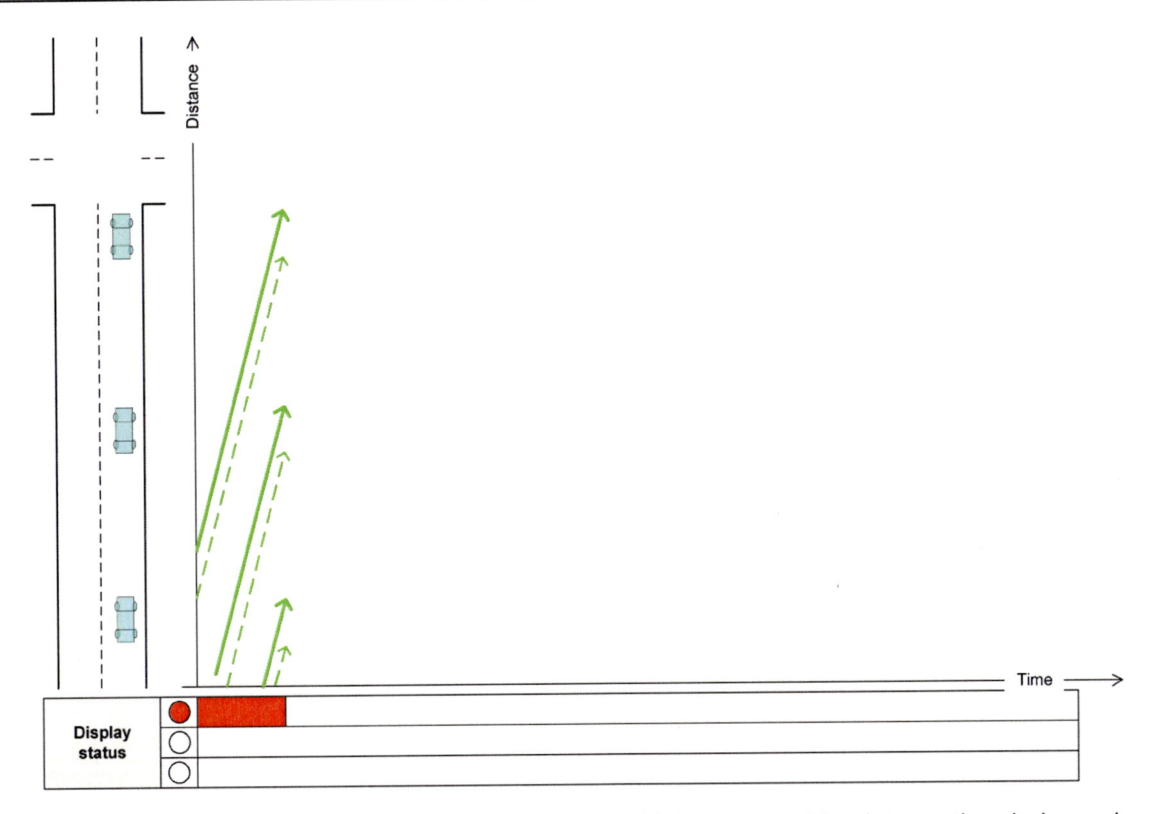

Figure 28. Vehicle trajectories, for evenly spaced vehicles approaching intersection during red

The response of the vehicles to the red indication is shown in Figure 29. The trajectories of the vehicles show this response as vehicles decelerate and their speeds go to zero. There is a queue of five vehicles at the end of the red interval.

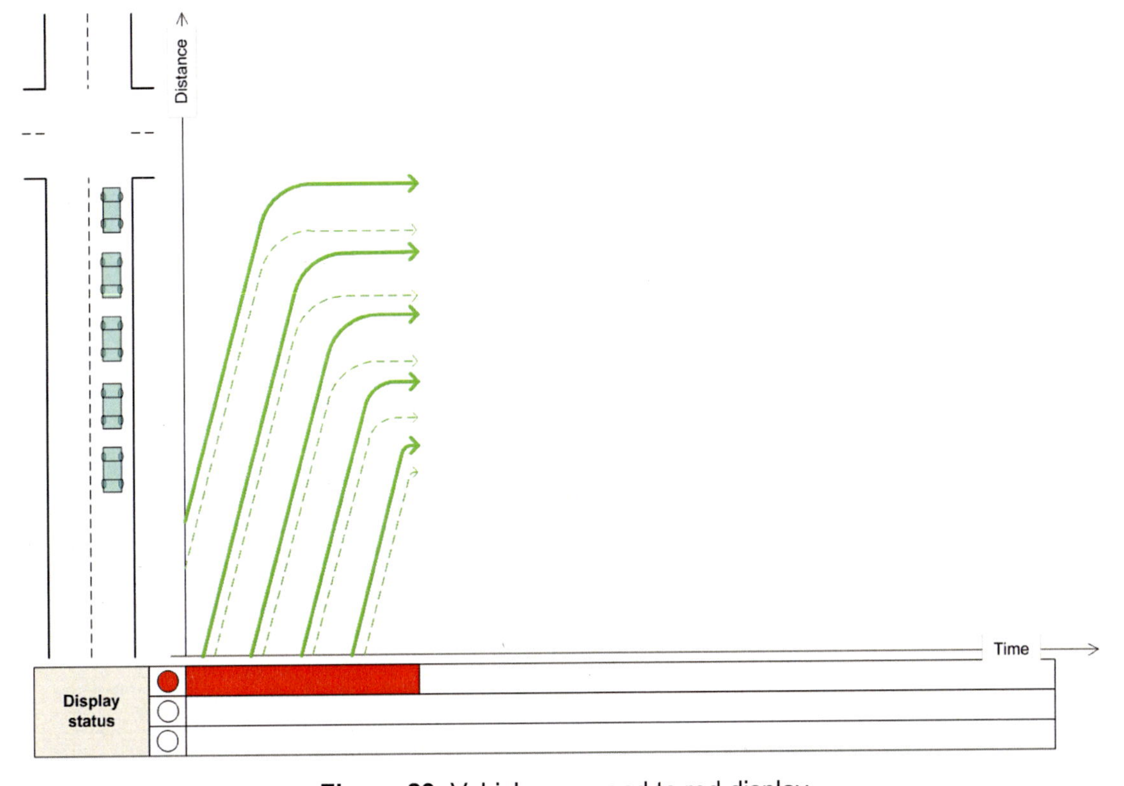

Figure 29. Vehicles respond to red display

Figure 30 shows the response of the queued vehicles to the change in display to green, as they begin to accelerate and move into the intersection.

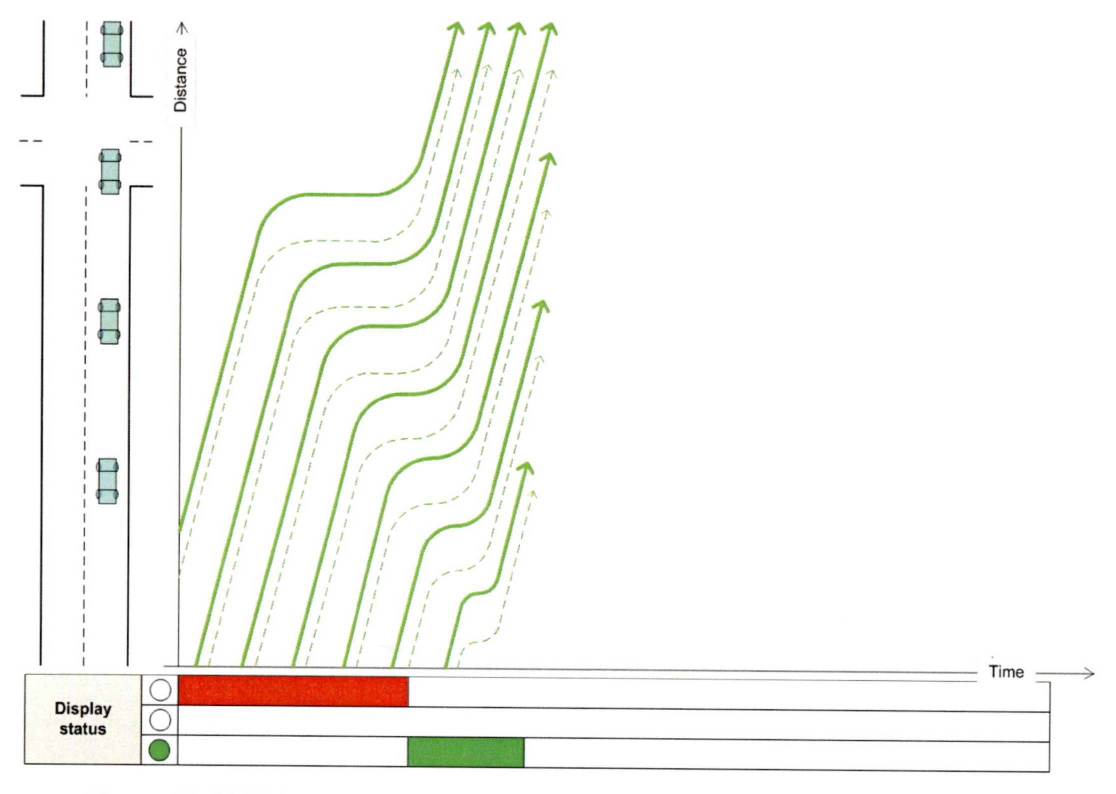

Figure 30. Vehicles respond to green display as queue begins to move and clear.

Figure 31 shows the vehicles that arrive after the queue has cleared. They arrive and leave with no delay, as shown by the constant slope of the vehicle trajectories.

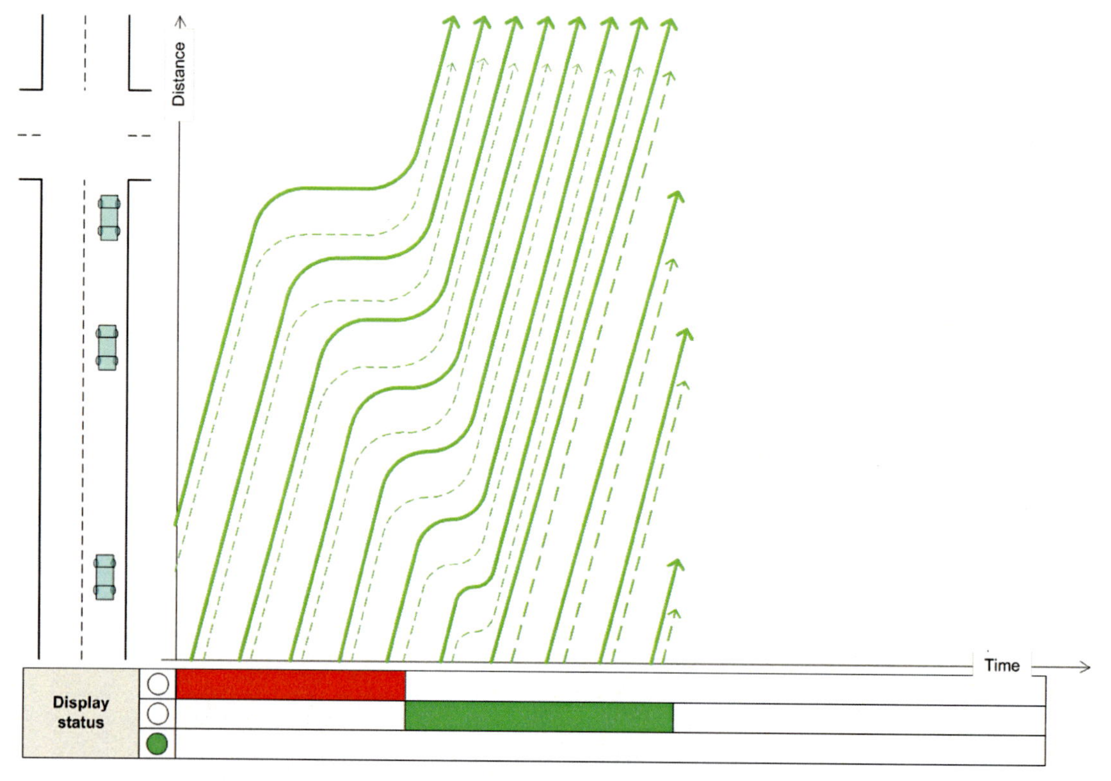

Figure 31. Vehicles responding to the green display after the clearance of the queue

Figure 32 shows vehicles approaching the intersection at the onset of the yellow interval. The last two vehicles shown will not enter the intersection but will begin to stop.

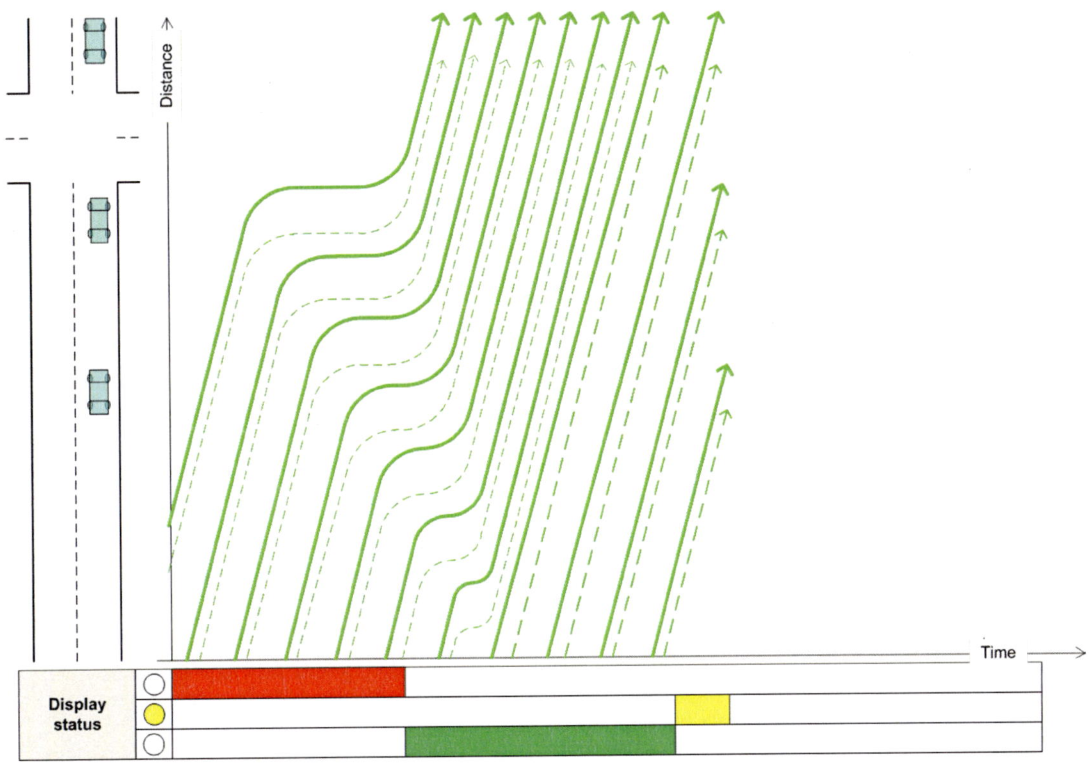

Figure 32. Vehicles responding to yellow display

Figure 33 shows the last two vehicles responding to the red display, stopping at the intersection and forming a new queue.

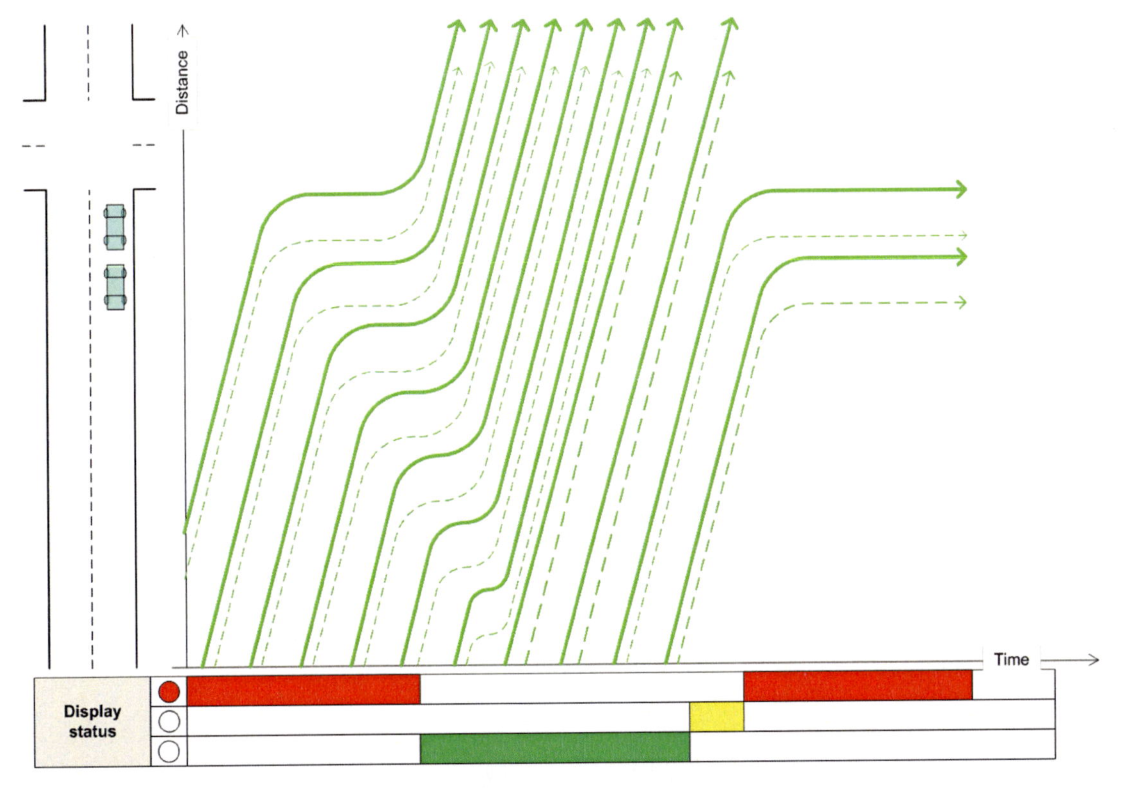

Figure 33. Vehicles responding to red display

While the time-space diagram is an approximation or model of the vehicles traveling through the intersection, particularly with our assumption of uniform headways, we can learn a lot from this diagram. The following six bullets are illustrated in Figure 34 (as adapted from May, 1989) for the vehicles that we've just observed.

1. Shock wave of the queue forming during red.
2. Shock wave of the queue clearing during green.
3. Delay for each vehicle, the horizontal line between the time that the vehicle arrives and the time that it leaves.
4. The saturation headway, the headway between two vehicles departing as part of the clearing queue at the beginning of green, measured at the stop line.
5. The slope of the vehicle trajectory, the vehicle speed.
6. The time between the passage of the end of the leaving vehicle and the front of the following vehicle by a given point, the time gap.

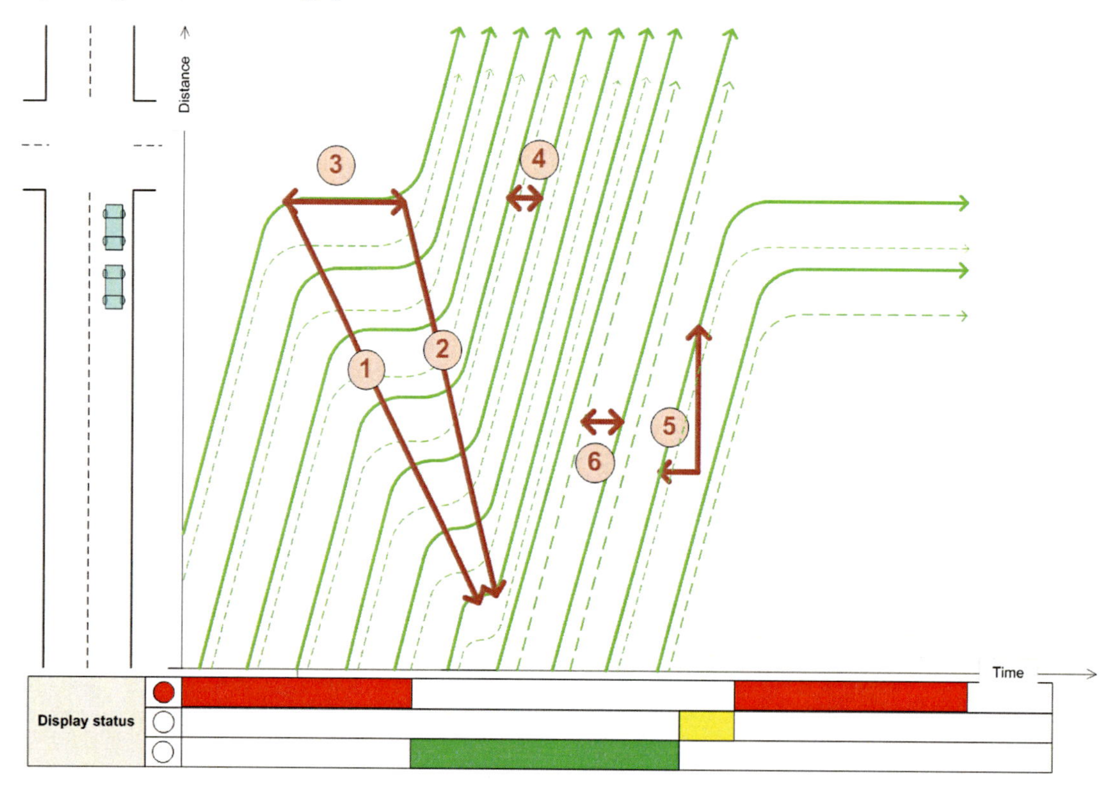

Figure 34. Examples of information available from time-space diagram

We can summarize the categories of driver response in Figure 35, showing:

1. Drivers responding to the red indication by stopping.
2. Drivers responding to the green indication by beginning to move through the intersection.
3. Drivers responding to the green indication (after the queue is cleared) by traveling through the intersection without stopping (without delay).
4. Drivers responding to the yellow indication by stopping.

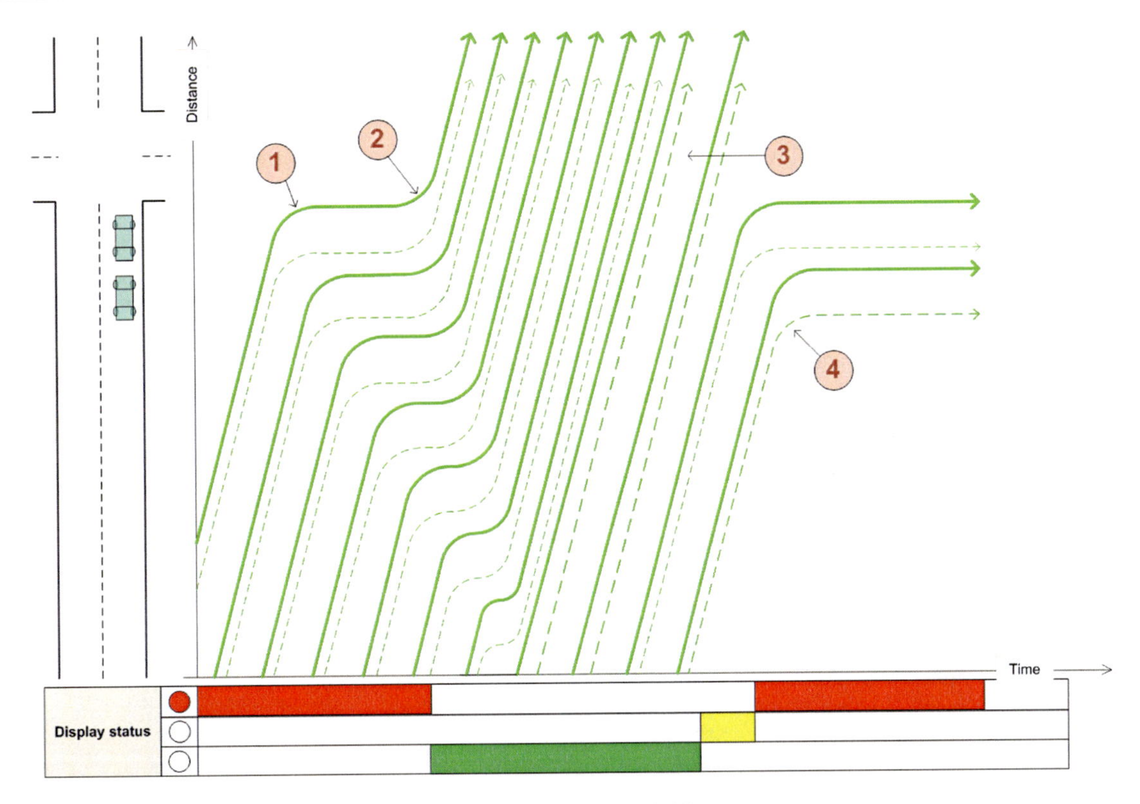

Figure 35. Four categories of driver response

Queuing System Representation

Another way of modeling the traffic flow at one approach of a signalized intersection is to use queuing theory. Queuing theory, originally developed to model and help design the nation's telephone communications system, is based on a system that includes users who desire to be served in some way, a server, and a process for serving these users. For example, consider shoppers who are being checked out at a supermarket in Figure 36. They arrive at the check stand with the shopping carts, they wait in line, and they are checked out or served by the checker at the check stand. Formal queuing theory includes specifying the arrival pattern, the service pattern, and the number of service channels, among other factors. The model that we will consider here is called the D/D/1 model, for deterministic (D) arrivals, a deterministic (D) service pattern, and one service channel.

Figure 36. The supermarket checkout line as a queuing system

Figure 37 shows the elements of a queuing system for vehicles traveling on the through lane of a signalized intersection. A vehicle at the stop line, waiting to travel through the intersection, is said to occupy the server as it is waiting to be served. The act of service for a vehicle involves having a green indication and responding to it. Any vehicles waiting behind the server position are said to be waiting in queue. It should be noted that the term "queue" as used by traffic engineers includes vehicles in both the server and the queue. In queuing theory terminology, the queue does not include the first-in-line position (the server). In addition, in queuing theory the queue is assumed to be stacked vertically at the intersection stop bar.

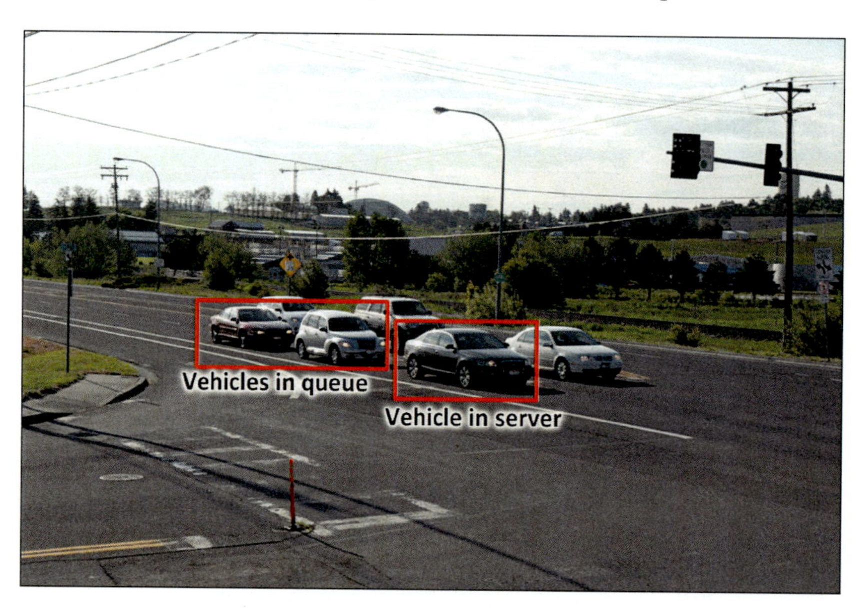

Figure 37. Elements of queuing system

Since one vehicle can be served at a time, the number of service channels is said to be one. And, since vehicles are served in the order that they arrive at the intersection, the queue discipline is "first in, first out."

The flow rate as measured upstream of the approach stop line is called the demand or arrival rate. The flow rate measured at the stop bar, as vehicles are being served, is called the service rate. As we will see in the next section, the service rate varies by time during the signal cycle and can be divided into three segments:

- zero, during the red interval
- the saturation flow rate, during the initial period of green when the queue is clearing
- the arrival rate, during green after the queue has cleared

While the time space diagram shows the response of vehicles to the signal display, we can use other graphical representations of the process, each related to the arrival and departure patterns of vehicles at the intersection. We can represent the flow patterns for this queuing system in three different ways, using a flow profile diagram, a cumulative vehicle diagram, and a queue accumulation polygon. Each diagram will be described in the following sections of this chapter.

Flow Profile Diagram

The flow profile diagram represents the flow rates of vehicles arriving at and departing from the intersection over time. Figure 38 shows a graph of the rates that vehicles arrive at the intersection and then depart from it during one cycle. During both the red and green intervals, vehicles arrive at a constant or uniform rate that we will call the arrival rate or approach demand. This rate is shown in Figure 38 as a solid dark line. The departure, or service, rate (shown as the lighter dashed line) depends on the status of the signal. During the red interval, the service rate is zero. During the initial part of the green interval (which we will call the queue service time, g_q), the service rate is equal to the saturation flow rate (shown below as *s*). After the queue has dissipated, the service rate equals the arrival rate for the remainder of the green interval, which we will call g_u, or the unsaturated portion of the green interval. Note that the total green interval (*g*) is what is sometimes called the "effective green", or the total time available to and usable by vehicles traveling through the intersection.

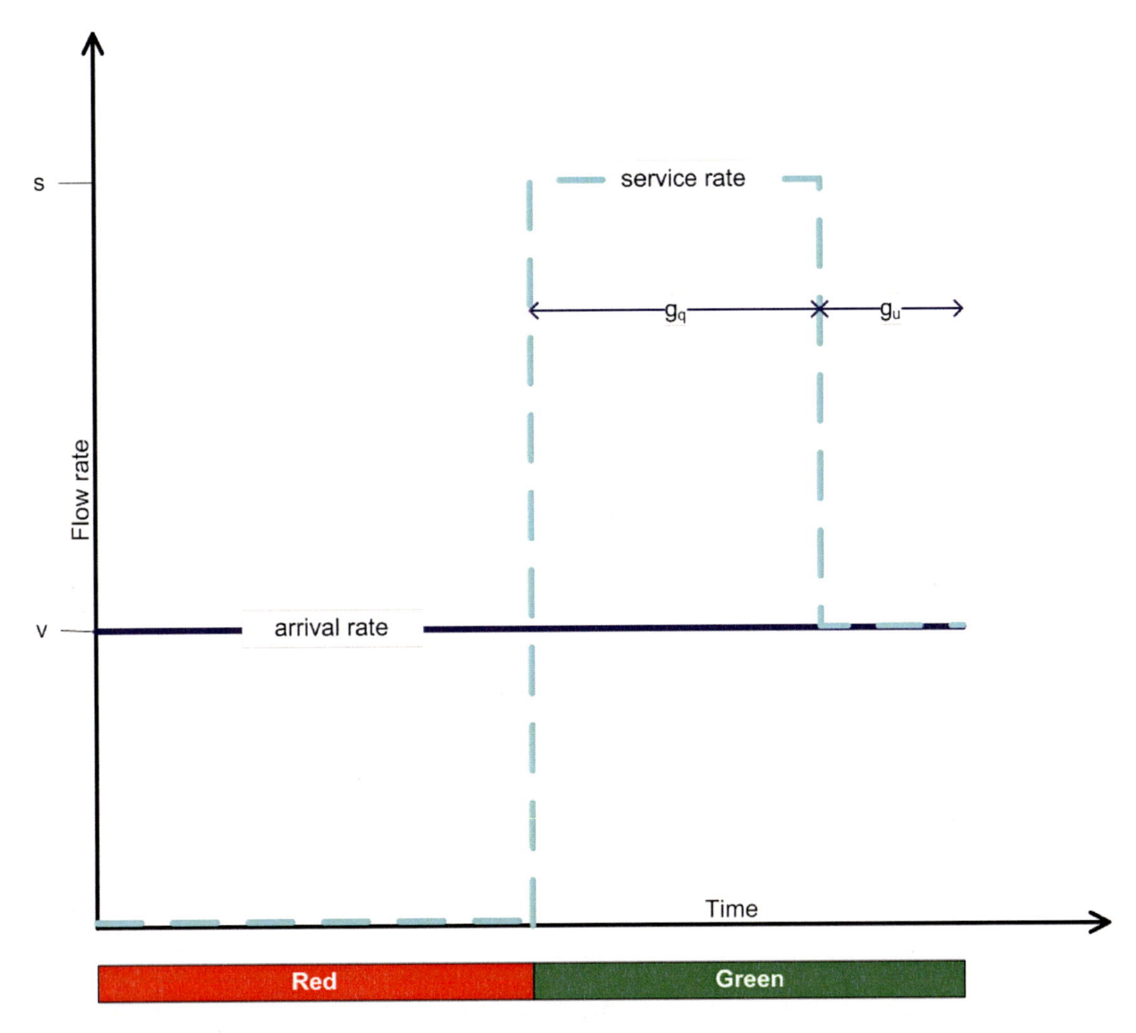

Figure 38. Flow profile diagram

Cumulative Vehicle Diagram

Another representation of this same process is the cumulative vehicle diagram, or the number of vehicles that have arrived at and departed from the intersection at any point in time during the cycle. We call these functions the cumulative arrivals and cumulative departures and we can graph them as shown in Figure 39. The slope of the arrival curve is equal to the arrival rate, while the slope of the second segment of the departure curve (its steepest portion) is equal to the saturation flow rate.

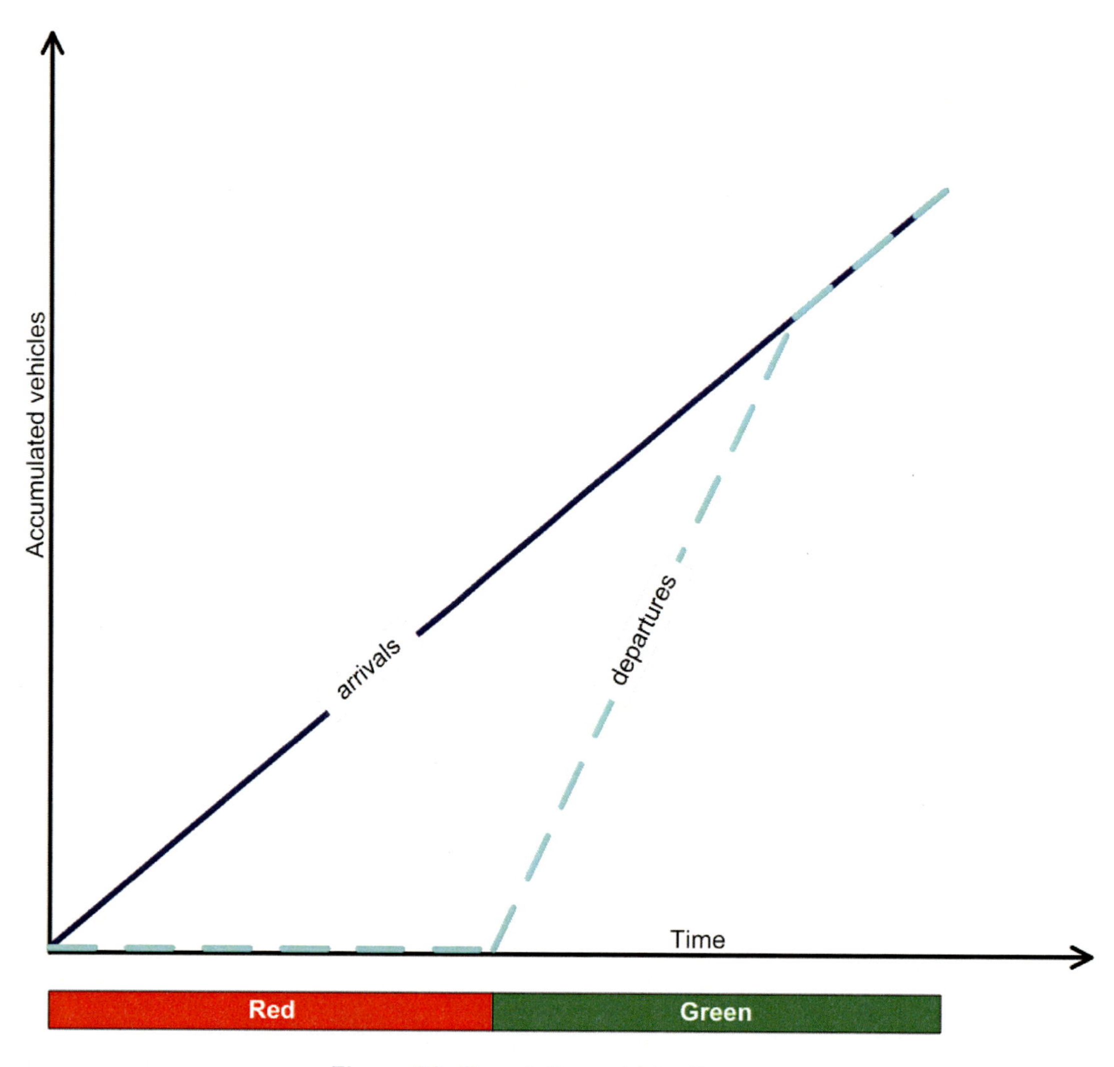

Figure 39. Cumulative vehicle diagram

We will now look at several important mathematical representations that can be derived from the graphical representation of the cumulative arrivals and departures shown in Figure 39. First consider the duration of time from the beginning of the red interval to the time during the green interval that the queue has just cleared. This duration is equal to $r + g_q$, and is shown in Figure 40. We are particularly interested in the time duration g_q, or the time that it takes for the queue to clear after the beginning of green.

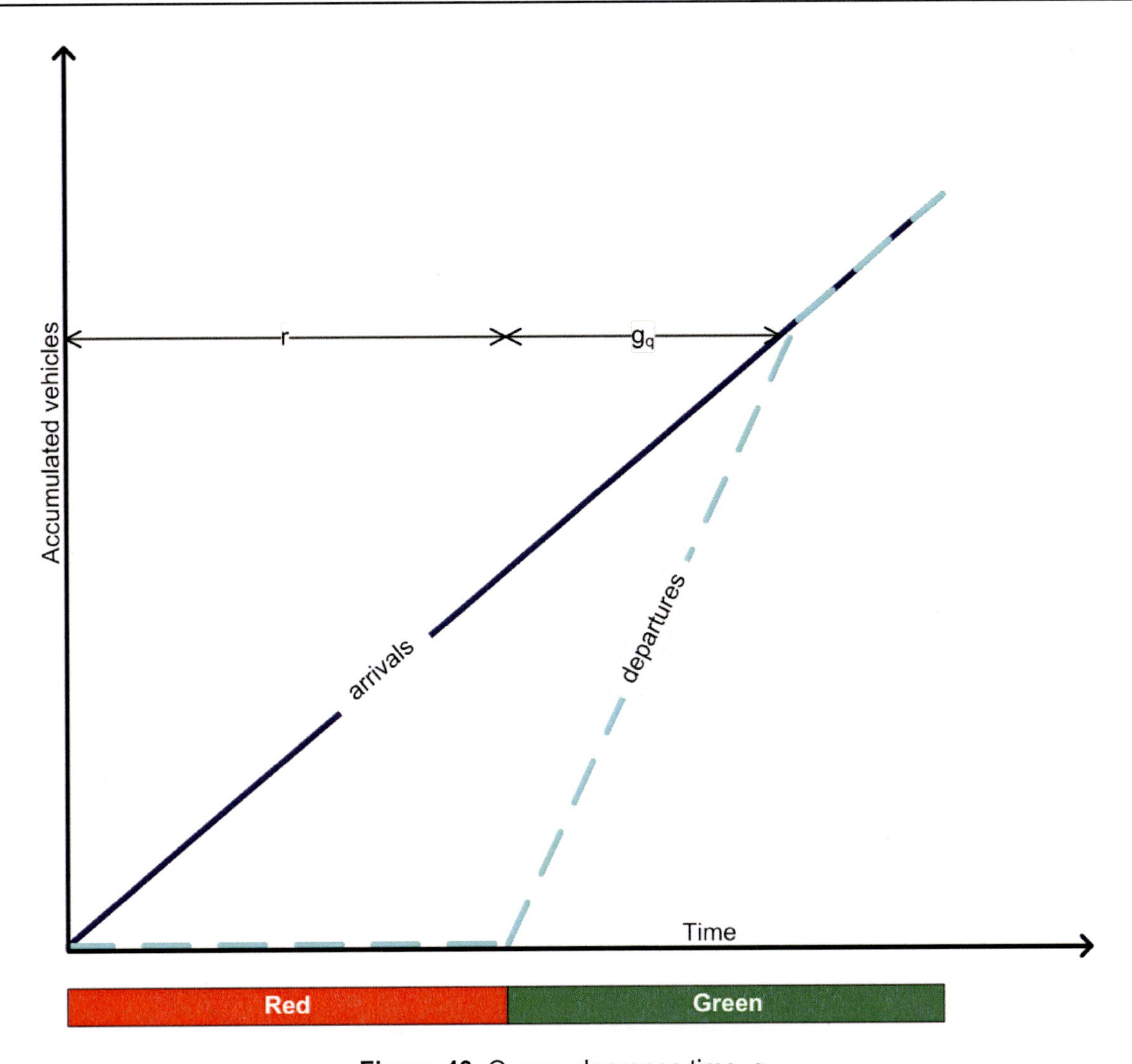

Figure 40. Queue clearance time, g_q

A derivation of a method to compute g_q follows. Assuming a uniform arrival rate, the number of vehicles that have arrived at the intersection at any time t after the start of the red interval is the product of the arrival flow rate v and the time t after the start of red, or vt. The number of vehicles that have arrived at the intersection at the point that the queue clears can be written as:

$$vt = v(r + g_q)$$

The rate at which vehicles depart from the intersection depends on the time after the start of red. During the red interval, the rate is zero. During the time that the queue is clearing, the departure rate is s, or the saturation flow rate. The number of vehicles that depart from the intersection during this same time interval $(r + g_q)$ can be written as the product of the saturation flow rate, s, and the duration of the queue clearance time, g_q, since the departure flow rate during red is zero.

Since the number of vehicles that arrive during this time interval must equal the number of vehicles that depart during this time:

$$v\left(r + g_q\right) = sg_q \qquad \text{Solving for } g_q, \qquad g_q = \frac{vr}{s - v}$$

In words, the duration of time for the queue to dissipate after the beginning of the green interval, g_q, is the number of vehicles that are in the queue at the end of the red interval (or vr) divided by the net rate that the queue dissipates, or $s - v$.

Clearly, the longer the length of the red interval, the longer it will take for the queue to dissipate after the beginning of the green interval. This conclusion has a very important implication for signal timing design, a point to which we will return periodically in this book: how long should we remain in the green interval? (And, the longer the green, the longer the cycle, and the longer the resulting delay.) While this question could generate complex responses, the simple answer is just enough time to serve the standing queue of vehicles that is present at the beginning of the green interval. And, we've seen how to compute this time, which we've called the queue service time or g_q. Once the standing queue has been served, and the flow rate drops below the saturation flow rate, the green interval should end so that the traffic on the other approaches can be served. Again, we want to keep the cycle length as short as possible to keep the delay as low as possible. However, it must also be noted that cycle lengths that are too short could lead to oversaturation.

The cumulative vehicle diagram also tells us some important information about the performance of the intersection, as shown in Figure 41. We can note the arrival of the ith vehicle at time t_a and its departure at t_d. The horizontal line connecting these two time points represents the delay d_i experienced by this vehicle.

$d_i = t_d - t_a$ where d_i is the delay experienced by the ith vehicle

t_d is the time that the ith vehicle departs the intersection

t_a is the time the that i^{th} vehicle arrives at the intersection

Here, the delay experienced by the vehicle is simply its time in the system, from its arrival at t_a to its departure at t_d.

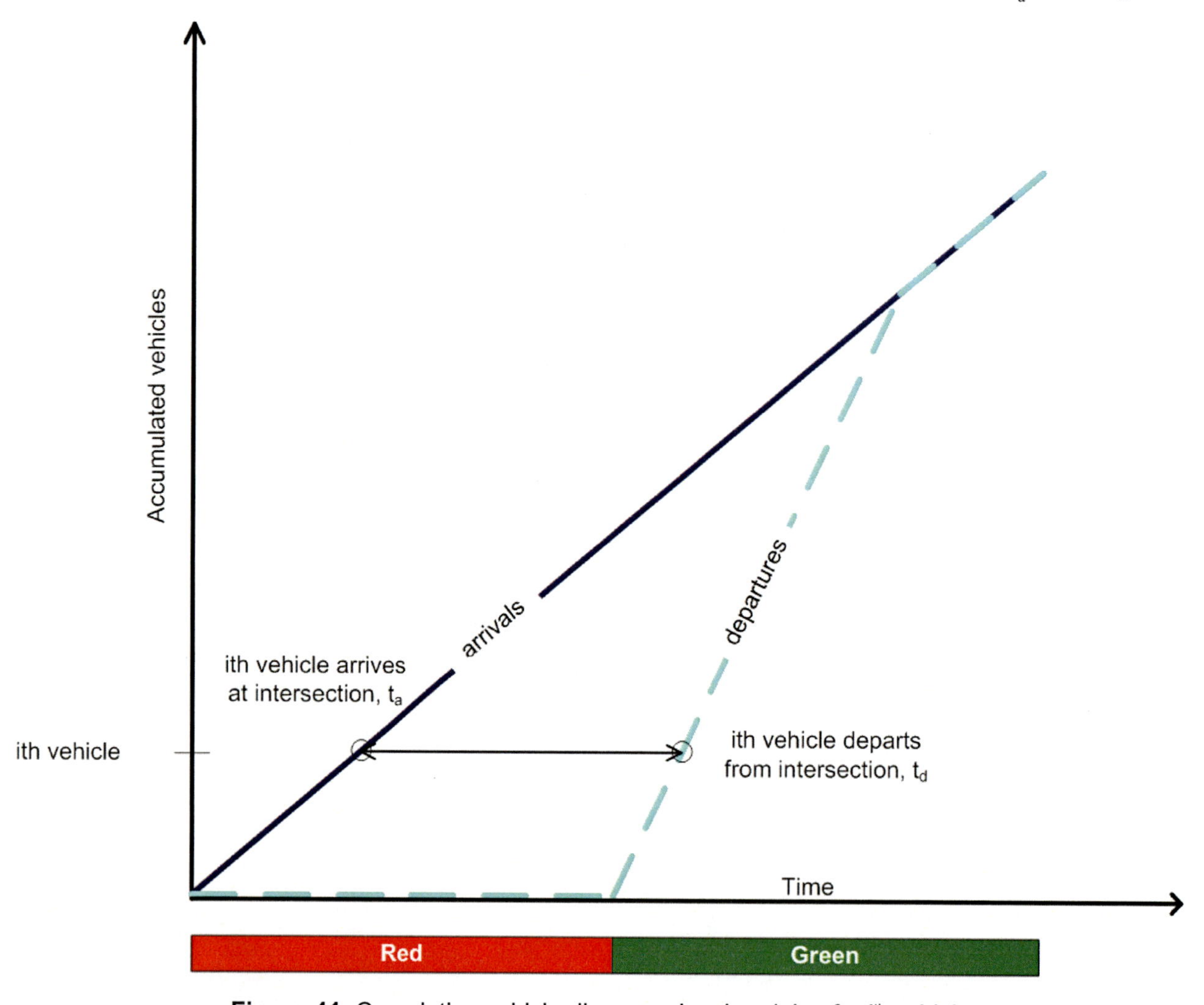

Figure 41. Cumulative vehicle diagram showing delay for i^{th} vehicle

If we consider each vehicle that arrives and departs from the intersection, we can represent the delay that each vehicle experiences as the horizontal line that connects the arrival curve with the departure curve. Examples of four such vehicles, and their individual delays, are shown in Figure 42. Further, if we integrate the area between the arrival and departure curves or add these individual delays, the total area (shown as shaded in the figure) is the total delay experienced by all vehicles that arrive at the intersection during the red and green intervals.

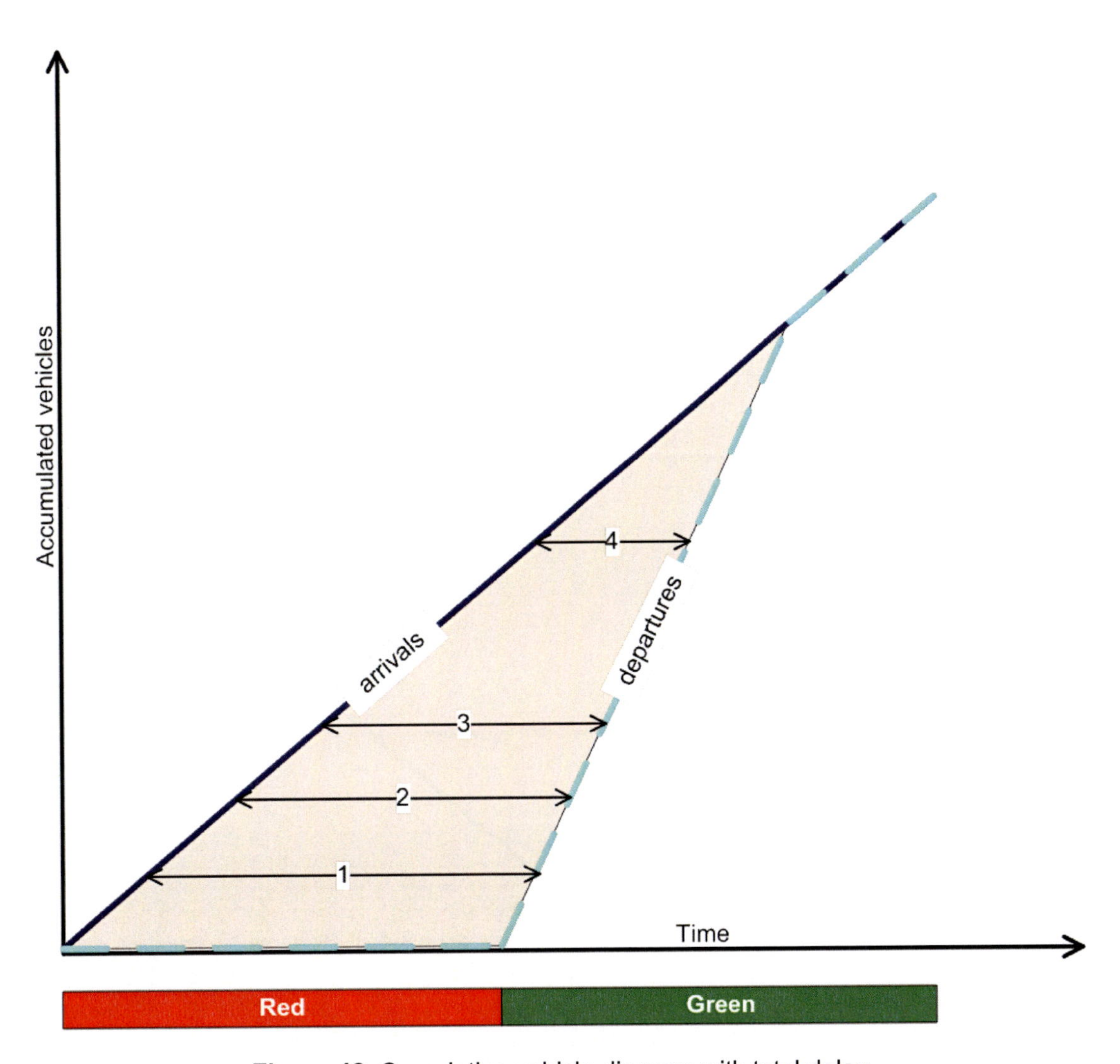

Figure 42. Cumulative vehicle diagram with total delay

Mathematically, we can write an expression for this total delay, as the area of the triangle: one half the product of the base and height of the triangle. The base is the length of the red interval and the height is the number of vehicles that have arrived between the start of red and the time that the queue clears:

$$d_t = (0.5)(base)(height)$$
$$d_t = (0.5)(r)(sg_q)$$
$$d_t = (0.5)(r)(s)\left(\frac{vr}{s-v}\right)$$

Rearranging terms, we can write the total delay experienced by all vehicles arriving during the red and green intervals as:

$$d_t = \frac{0.5r^2v}{1-v/s}$$

We can also compute the average delay experienced by all vehicles by dividing this expression by the total number of vehicles that arrive during the cycle, vC, where C is the cycle length:

$$d_a = \frac{0.5r^2v}{(vC)(1-v/s)} = \frac{0.5r^2}{(C)(1-v/s)} = \frac{0.5r(1-g/C)}{1-v/s} = \frac{0.5C(1-g/C)^2}{1-X(g/C)}$$

This delay is sometimes called the uniform delay since this model assumes that vehicles arrive at the intersection at a uniform rate. X is the degree of saturation or volume/capacity (v/c) ratio.

Note also that a vertical line drawn between the arrival curve and the departure curve at any point in time shows the number of vehicles in the queue on the intersection approach at that point in time. An example of this is shown in Figure 43. If we draw this line continuously from left to right, it would start at zero, grow to its maximum value at the end of the red interval, and then decrease until the arrival and departure curves join together at g_q, after the beginning of the green interval.

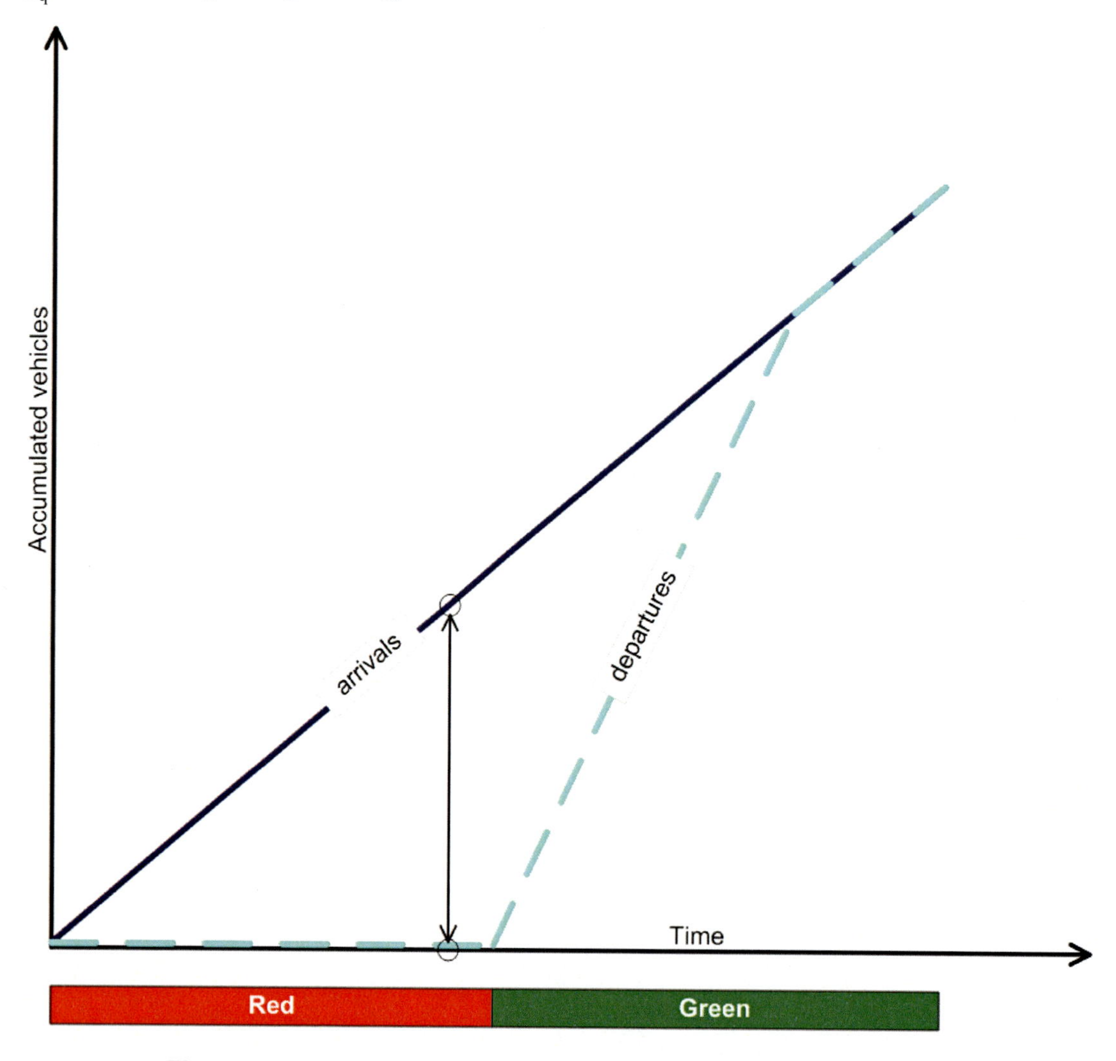

Figure 43. Cumulative vehicle model showing instantaneous queue

Queue Accumulation Polygon

A third representation is the queue accumulation polygon (QAP), the queue length at any point in time or the vertical distance between the arrival and departure curves shown in Figure 43. The QAP is shown in Figure 44. Here the queue grows from zero at the beginning of the red interval reaching a maximum at the end of the red interval. The queue begins to dissipate at the beginning of the green interval and finally clears at a point g_q after the beginning of the green interval. The queue is zero after this time and continues to be zero until the end of the green interval.

The area of the QAP is the total delay experienced by all vehicles arriving at the intersection during both the red and green intervals. This area is equal to the area between the arrival and departure curves presented in the cumulative vehicle diagram in Figure 42.

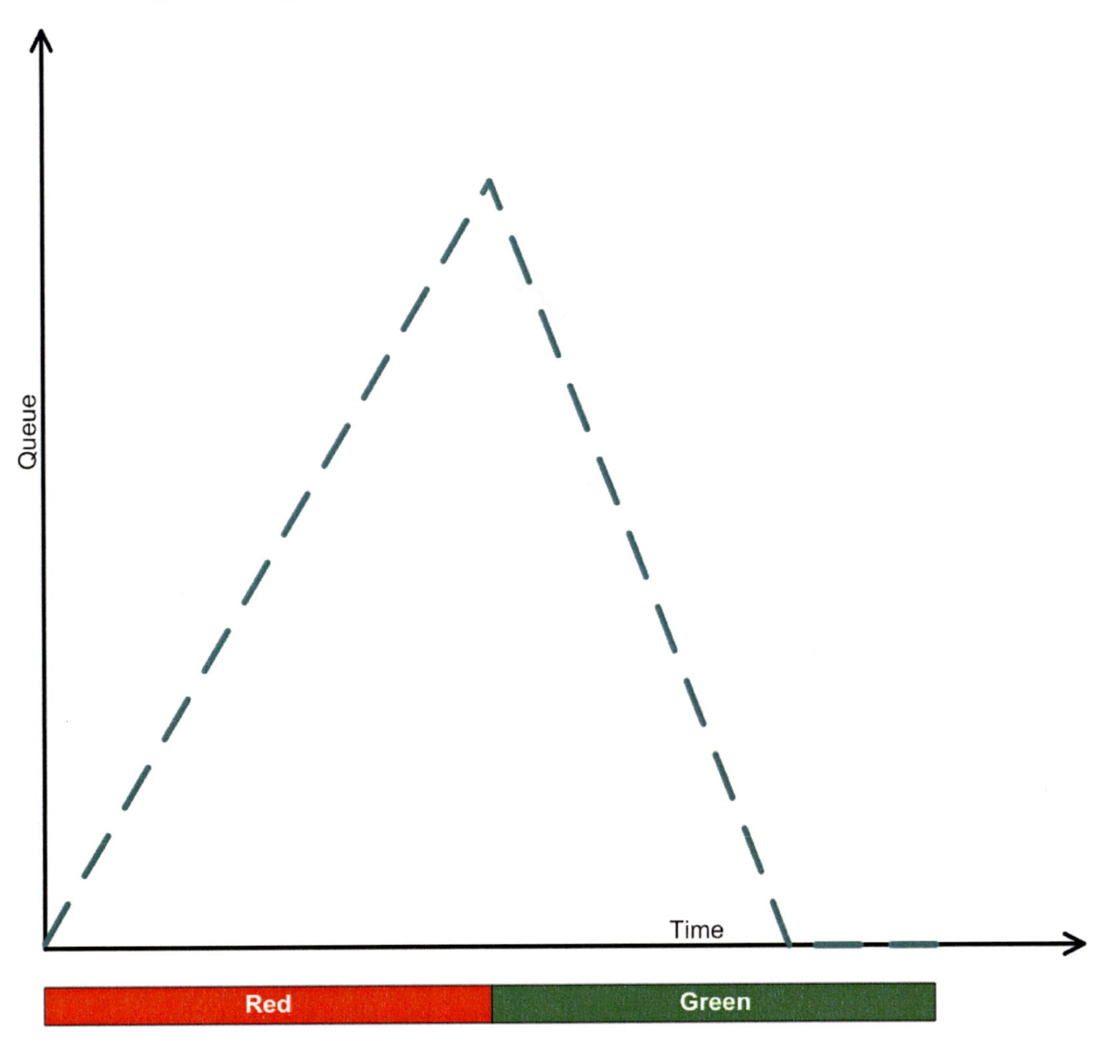

Figure 44. Queue accumulation polygon

Example Calculation of Delay

Let's consider an example using the queuing diagrams to illustrate a point that we discussed earlier: longer cycle lengths result in longer delay. Here we will assume the following input data for two cycle lengths, $C = 60$ seconds and $C = 120$ seconds, for one approach of a signalized intersection in which the arrival flow is uniform throughout the cycle:

- green ratio, g/C, = 0.5
- volume, v, = 800 vehicles per hour
- saturation flow rate, s, = 1900 vehicles per hour of green

The capacity (c) is calculated as the product of the saturation flow rate and the green ratio:

$$c = (g/C)s = (0.5)(1900) = 950\ veh/hr$$

The average delay for the two cases is calculated below:

$$d_{a1} = \frac{0.5C(1-g/C)^2}{1-X(g/C)} \qquad d_{a2} = \frac{0.5C(1-g/C)^2}{1-X(g/C)}$$

$$d_{a1} = \frac{0.5(60)(1-0.5)^2}{1-(800/950)(0.5)} \qquad d_{a2} = \frac{0.5(120)(1-0.5)^2}{1-(800/950)(0.5)}$$

$$d_{a1} = 13.0\ sec/veh \qquad d_{a2} = 25.9\ sec/veh$$

We can also observe the increase in average delay per vehicle for a range of cycle lengths, from 40 seconds to 150 seconds, as shown in Figure 45, using these same conditions. This finding will be important later as you determine the maximum green time parameter, as this parameter directly affects the cycle length. Keeping the maximum green time (and thus the cycle length) low, will keep delay lower.

We should also note that when the cycle length is decreased to much smaller values, the delay begins to increase. This concept will be discussed in Chapter 7.

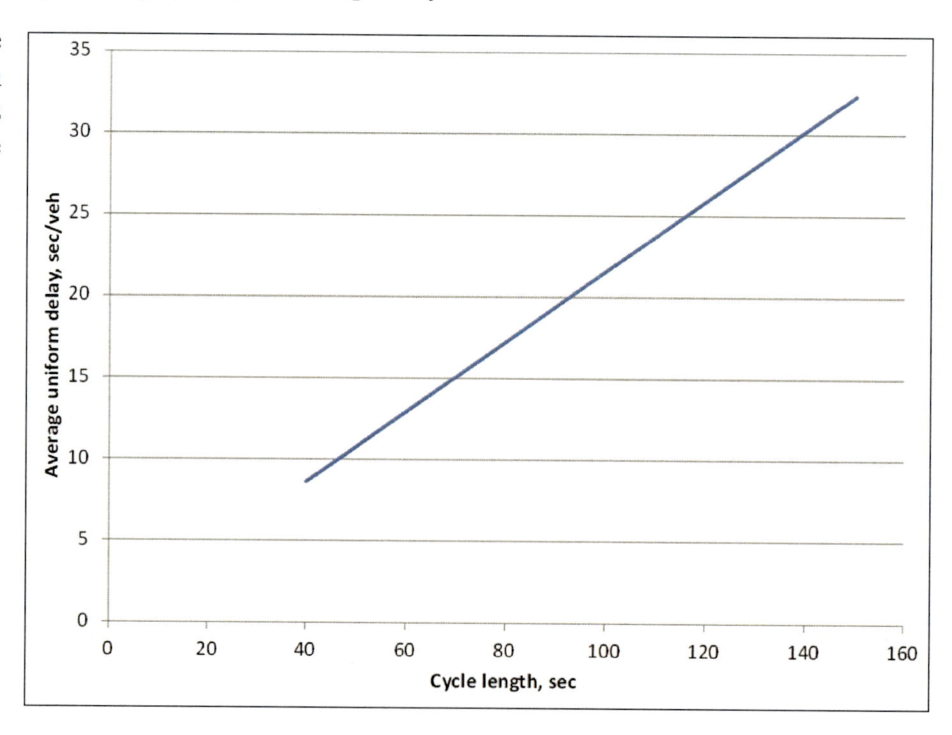

Figure 45. Uniform delay as a function of cycle length

What Do You Know About Queuing Systems?

PURPOSE

The purpose of this activity is to provide a framework for you to think about traffic flow at signalized intersections. In this activity, you will build a base of knowledge of modeling traffic flow at a signalized intersection using queuing theory as your model framework. You will learn to recognize patterns through visualizing arrivals and departures at a signalized intersection. You will also make connections between words and charts, finding alignment between alternative ways of representing traffic flow patterns.

LEARNING OBJECTIVES

- Connect your observation of traffic flow at a signalized intersection with a model framework
- Represent and interpret queuing diagrams for a range of traffic flow and control conditions

REQUIRED RESOURCE

- Activity #8: "Modeling Traffic Flow at Signalized Intersections"

DELIVERABLE

- A document with the required sketches from Tasks 1 and 2, plus your answer to the Critical Thinking Question

CRITICAL THINKING QUESTION

1. What insights did you gain about intersection operation or performance from these cases?

Task 1

Complete the following sketches.

Make a sketch showing the flow profile of the arrival flow and departure flow at a signalized intersection for a period of one cycle. Assume that the arrival flow is uniform. Label the axes and the important parameter values on the sketch.

Make a sketch that shows the cumulative vehicle arrivals and departures during one cycle. Again, assume that the arrival flow is uniform.

Based on the two sketches that you made above, sketch the queue accumulation polygon for these same conditions.

Task 2

Draw a flow profile diagram, a cumulative vehicle diagram, and a queue accumulation polygon for the following three cases and describe how each of these cases differs from the original case that you drew in Task 1.

Case 1: Uniform vehicle arrivals throughout the cycle with the queue clearing just at the end of green.
Flow profile diagram
Cumulative vehicle diagram
Queue accumulation polygon

Case 2: Uniform vehicle arrivals throughout the cycle but the queue does not clear before the end of green.

Flow profile diagram

Cumulative vehicle diagram

Queue accumulation polygon

Case 3: No vehicle arrivals during red; a platoon (or group of vehicles) arrives during the first half of the green interval only, with no arrivals during the second half of the green interval.
Flow profile diagram
Cumulative vehicle diagram
Queue accumulation polygon

Student Notes:

ACTIVITY

10 Using High Resolution Field Data to Visualize Traffic Flow

PURPOSE

The purpose of this activity is to give you the opportunity to see how queuing system models relate to high resolution field data.

LEARNING OBJECTIVES

- Connect your observation of traffic flow at a signalized intersection with a model framework
- Represent and interpret queuing diagrams for a range of traffic flow and control conditions

REQUIRED RESOURCE

- Spreadsheet file: A10.xlsx

DELIVERABLES

- Prepare an Excel worksheet that includes the following information:

  Tab 1: Title page with activity number and title, authors, and date completed

  Tab 2: Original data

  Tab 3: Time-space diagram plot with answers to questions from Task 2

  Tab 4: Cumulative vehicle diagram plot

  Tab 5: Uniform delay calculation

  Tab 6: Summary and answers to Critical Thinking Questions

CRITICAL THINKING QUESTIONS

1. Consider the two estimates of delay from Tasks 4 and 5. Why are they different? How would you refine your calculation method from Task 4 to reduce this difference in the two delay estimates?

2. Compare the cumulative vehicle diagram that you prepared in Task 3 with the one that you sketched in Task 1 of Activity #9. Describe and explain any differences in these two diagrams.

INFORMATION

While the models that we considered in the reading (Activity #8) provide an excellent framework for understanding traffic flow at a signalized intersection, they lack an important ingredient that we observe in the real world. The models are deterministic and the real world is stochastic. In this activity we consider a very high resolution data set that was collected in Los Angeles that will allow us to consider the messiness, or stochasticity, that is ever present in the real world.

In 2006, the Federal Highway Administration published the results of a study of traffic flow characteristics along a four-block section of Lankershim Blvd. in Los Angeles. The study was based on very high quality video that was taken from a 30-story building located adjacent to Lankershim Blvd. Video image processing software extracted data on position, velocity, and acceleration for vehicles traveling along the arterial for a 30 minute period at time intervals of 0.1 second. This is by far the most detailed study of vehicle trajectories ever compiled.

Figure 46 shows an aerial view of one of the four signalized intersections included in this study. It is the intersection of Lankershim Blvd., Campo de Cahuenga Way and Universal Hollywood Drive, located near Universal Studios. Figure 47 shows the entire arterial. You took a video tour of this arterial using the file a03.wmv (See Activity #3 in Chapter 1).

Figure 46. Aerial view of Lankershim Blvd. intersection

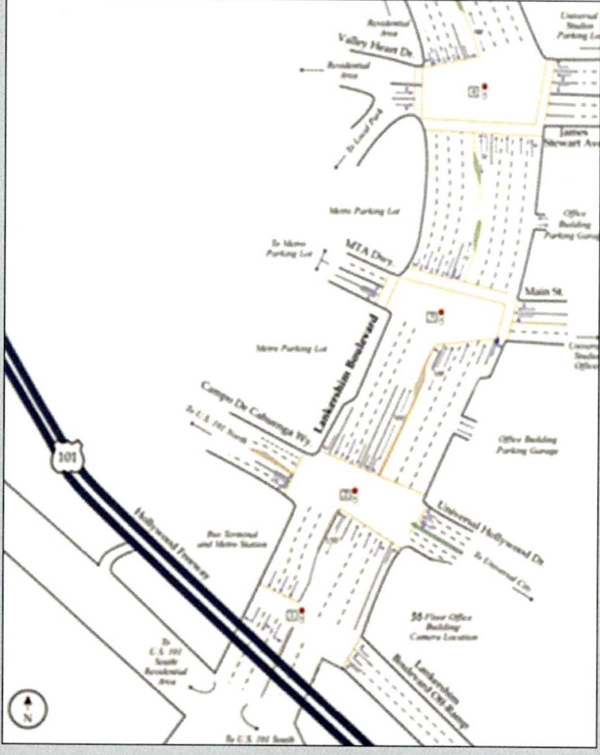

Figure 47. Lankershim Blvd. study area

You are given field data for one lane of a signalized intersection in the Excel spreadsheet. The data in the "field data tab" give the location of eight vehicles over a period of three minutes at one foot resolution, and the time that each vehicle passes each one foot point. The data in the "arrival-departure data" tab includes the arrival time at a given point and the departure time from a given point.

Task 1

Using the field data, prepare a time-space plot for the eight vehicles, placing distance on the y-axis and time on the x-axis. Note that the location of the stop bar for the subject intersection is at a distance of $y = 346$ feet. The stop bar should be shown on your plot.

Task 2

Change the chart settings to show the range $y = 200$ feet to 400 feet and $x = 20$ seconds to 120 seconds. Answer the following questions:

1. Is there movement in the queue while the vehicles are supposedly stopped?

2. Which vehicles are directly affected by the red display?

3. Which vehicles are affected only by the behavior of their leading vehicles?

4. Which vehicles are not affected by either the red display or their leading vehicles?

5. How far upstream does the queue extend?

Task 3

Review the data on the "arrival-departure tab." Using the maximum extent of the queue upstream from the stop bar as the system entry point to your queuing system, prepare a cumulative vehicle diagram showing the arrival time into the system and the departure time from the system.

Task 4

Using the cumulative vehicle diagram that you prepared in Task 3, show on the diagram the time that each vehicle is in the system (delay time). Compute the average delay (average time in system) per vehicle. Remember that this delay does not consider free flow travel time.

Task 5

Using the uniform delay equation from Activity #8, compute the average delay per vehicle for this system. For the uniform delay calculation, make the following assumptions: $C = 102$ seconds, $g = 35$ seconds, and $s = 1681$ vehicles per hour of green. Use your diagram to determine any other data needed for this calculation.

Student Notes:

ACTIVITY

11 From Model to the Real World: Field Observations

PURPOSE

The purpose of this activity is to provide a framework for you to think about traffic flow at signalized intersections, both in a model representation and in connecting this model to what you observe in the field.

LEARNING OBJECTIVE

- Represent and interpret queue accumulation polygons for a range of traffic flow and control conditions

REQUIRED RESOURCES

- Activity #8: "Modeling Traffic Flow at Signalized Intersections"

DELIVERABLE

- Using an Excel spreadsheet, summarize your field observations, including your field notes and the data that you've collected. The spreadsheet should include the following sections, integrating field data and answers to the questions from Tasks 1 through 4.

  Tab 1: Title page with activity number and title, authors, and date completed

  Tab 2: Summary of your general observations and sketch

  Tab 3: Table 4 data and a description of the sequence of movements that you observed

  Tab 4: Table 5 data and a discussion of the queue pattern that you observed, including a chart of the queue accumulation polygon that results from your data and a description of how you would compute total delay using this information

  Tab 5: Table 6 data and a summary of the headway data that you've observed. Description of the efficiency of the intersection timing for the lane that you observed, including an estimate of the green utilization time and the number of cycle failures. Note that (1) green utilization time is the ratio of the duration of the green interval during which the queue is clearing to the total duration of the green interval, and (2) cycle failure occurs when the queue fails to clear during the green interval.

- Summary of traffic flow problems that you observed

EQUIPMENT NEEDED

- Phone that records time to the nearest second

TASKS

These tasks should be completed during both the morning and afternoon peak periods.

TASK 1

Walk or drive to your assigned intersection. Spend 15 minutes observing the operation of the intersection. Record the physical elements of the intersection, including the intersection geometry, lane striping, the location of the cabinet and other signal furniture, and other features that you consider to be important. Prepare a sketch of the intersection and note each of these items on the sketch.

Task 2

Observe the operation of the intersection for three full cycles. Record the duration of the green, yellow, and red clearance intervals for each movement served at the intersection to the nearest second for each of these cycles. Prepare a chart summarizing the sequence of the movements served (in order) and the mean duration of each of these sequences. The following table shows an example of data collected for this task.

Table 1. Example phase durations

Cycle number	Movement (direction)	Duration (sec)		
		Green	Yellow	Red Clearance
1	NBLT, SBLT NBTH, SBTH EBLT, WBLT EBTH, WBTH	5 25 11 30	3 3 3 3	2 2 2 2
2	NBLT, SBLT NBTH, SBTH EBLT, WBLT EBTH, WBTH	4 20 5 25	3 3 3 3	2 2 2 2
3	NB LT, SB LT NBTH, SBTH EBLT, WBLT EBTH, WBTH	5 20 10 23	3 3 3 3	2 2 2 2

Task 3

Continue to observe the operation. For one through lane (the most heavily traveled lane based on your earlier observations), record the length of the standing queue for **five cycles**. The length of the queue should be recorded **every ten seconds**. During green, the queue should only be considered non-zero if it is stopped or is beginning to move at the beginning of the green. Once the initial queue that has formed during red has cleared, the queue will be zero during the remainder of the green interval. Table 2 shows an example of data collected during a two minute period for one lane of an intersection approach. Using these data, prepare a queue accumulation polygon for each of the five cycles that you observed.

Table 2. Example queue evolution

Beginning of time interval (hh:mm:ss)	Number of vehicles in standing queue	Display status
2:00:00 pm	3	Red
2:00:10 pm	5	Red
2:00:20 pm	7	Red
2:00:30 pm	7	Red
2:00:40 pm	7	Red
2:00:50 pm	7	Red
2:01:00 pm	5	Green
2:01:10 pm	2	Green
2:01:20 pm	1	Green
2:01:30 pm	0	Green
2:01:40 pm	0	Green
2:01:50 pm	0	Green

Task 4

Observe the operation of the same heavily traveled lane for another five cycles. Record the beginning and ending clock time for the green interval for each of the five cycles. During the green period, record the clock time that each vehicle in this lane passes by the stop bar. The following table shows an example of data recorded for one cycle when six vehicles entered the intersection during green. Estimate the green utilization and determine the number of cycle failures (when the queue doesn't clear before the end of green). An example of the headway data is shown in Table 3.

Table 3. Example headway data

Cycle	Clock time (hh:mm:ss)	Event
1	2:20:30	Beginning of green interval
	2:20:33	Passage of vehicle 1
	2:20:35	Passage of vehicle 2
	2:20:38	Passage of vehicle 3
	2:20:41	Passage of vehicle 4
	2:20:43	Passage of vehicle 5
	2:20:47	Passage of vehicle 6
	2:20:59	Beginning of yellow interval

Table 4. Data collection sheet for phase durations

Cycle number	Movement (direction)	Duration (sec)		
		Green	Yellow	Red Clearance

Table 5. Data collection sheet for queue evolution

Beginning of time interval (hh:mm:ss)	Number of vehicles in standing queue	Display status

Table 6. Data collection sheet for headway data

Cycle	Clock time (hh:mm:ss)	Event

Student Notes:

ACTIVITY

12 Basic Operational Principles

PURPOSE

The purpose of this activity is to provide you with the opportunity to learn how the *Traffic Signal Timing Manual* addresses basic operational principles of traffic flow at signalized intersections.

LEARNING OBJECTIVE

- Compare and contrast the traffic flow representations used in the *Traffic Signal Timing Manual* with those that you studied in the activities in this chapter

REQUIRED RESOURCE

- *Traffic Signal Timing Manual*

DELIVERABLES

Prepare a document that includes

- Answers to the Critical Thinking Questions
- Completed Concept Map

LINK TO PRACTICE

Read the section of the *Traffic Signal Timing Manual* that covers the basic principles of traffic flow as assigned by your instructor.

CRITICAL THINKING QUESTIONS

When you have completed the reading, prepare answers to the following questions:

1. Describe how the *Traffic Signal Timing Manual* represents traffic flow arriving at and departing from a signalized intersection.

2. Contrast these representations with those that you studied in the activities of this chapter. Identify the differences in these representations.

3. Are there any basic traffic flow concepts presented in the *Traffic Signal Timing Manual* which you have not encountered before? Briefly describe them.

In My Practice...

by Tom Urbanik

As you work through problems as an engineer, there is significant focus on the mathematics of traffic flow. It is important to understand that traffic theory is a tool to facilitate understanding. It should be understood that vehicles are driven by individuals with different characteristics. These human factors issues result in performance that is different than the orderly movement assumed in traffic models. In addition, drivers may not line up in lanes to facilitate overall orderly movement. They are making a trip which has an origin and a destination (which may change when they get a call to pick up some milk on the way home). These realities make traffic messier than our tools and models can replicate.

As a result of the realities of traffic flow, successful operations necessitates that the engineer journey out to the field to see if his or her analysis is functioning as designed. As the engineer acquires more field knowledge, future analysis will become better and "field tuning" modifications will, although always necessary, be more minor in their extent.

CONCEPT MAP

Terms and variables that should appear in your map are listed below.

cumulative vehicle diagram

flow profile diagram

saturation flow rate

D/D/1 model

queue accumulation polygon

time space diagram

Student Notes:

Whose Turn is it? Phasing, Rings, and Barriers

Purpose

One of the primary functions of a traffic signal is to provide a time separation between the conflicting movements at an intersection. A stop sign accomplishes the same result but leaves much of the discretion to the driver to decide "whose turn it is." A traffic signal provides clear instructions to drivers traveling through the intersection that it is their "turn to go" using an appropriate signal or display.

In Chapter 3, you will complete a set of activities to learn about the most commonly used method of separating conflicting movements at an intersection. Known as NEMA (National Electrical Manufacturing Association) phasing, this method assigns numbers to each of the four left turn and four through movements, and provides a logical process through which each of the movements is served in turn. Each movement is controlled by a phase (or "timing unit"), and eight phase numbers account for the basis of a NEMA phasing plan. Left turn phasing is of particular importance, particularly the determination of whether a left turn should be protected (indicated by a green arrow) or permitted (indicated by a flashing yellow arrow or a green ball). Left turn phasing is covered in Chapter 8 of this book.

Learning Objectives

When you have completed the activities in this chapter, you will be able to

- Describe NEMA phasing and the concept of rings and barriers
- List the phase numbers at a standard intersection with eight movements
- Draw and describe a ring barrier diagram in which there are two rings and eight phases
- Determine the phasing pattern and sequence for a signalized intersection in the field
- Begin to appreciate how theoretical information about phasing, rings, and barriers is used and applied by professionals

Chapter Overview

This chapter begins with a *Reading* activity and ends with an *In Practice* activity. The reading content describes the concept of the ring barrier diagram and how phases are sequenced. The In Practice activity provides further information on these concepts. The chapter also includes two other activities. Activity #14 is an exercise to practice what you've learned from the reading. In Activity #15, you will document the phasing pattern for an intersection in the field using the NEMA phasing scheme.

Activity List

Number and Title	Type
13 Phasing, Rings, and Barriers	*Reading*
14 What Do You Know About Phasing and Ring Barrier Diagrams?	*Assessment*
15 Verifying Ring Barrier Operation in the Field	*Field*
16 Phasing, Rings, and Barriers in Practice	*In Practice*

Student Notes:

Activity 13 Phasing, Rings, and Barriers

Purpose

This activity will give you the opportunity to learn how movements and phases are safely sequenced.

Learning Objective

- Describe NEMA phasing and the concept of rings and barriers
- List the NEMA ring operational rules

Deliverables

- Define the terms and variables in the Glossary
- Prepare a document that includes answers to the Critical Thinking Questions

Glossary

Provide a definition for each of the following terms and variables. Paraphrasing a formal definition (as provided by your text, instructor, or another resource) demonstrates that you understand the meaning of the term or phrase.

concurrency group	
movement	
NEMA phase numbering	
overlap	
phase	
ring	
ring barrier diagram	

Critical Thinking Questions

When you have completed the reading, prepare answers to the following questions.

1. What is the purpose of the ring barrier diagram?

2. How is timing represented in a ring barrier diagram?

3. Why use a ring barrier diagram instead of a conflict matrix to describe the sequencing of phases?

4. What is the difference between a *movement* and a *phase*?

Information

Separating and Sequencing Movements

The primary function of a traffic signal is to safely separate and sequence the movement of vehicles traveling through an intersection so as to minimize the probability of crashes and to maximize the flow of vehicles. The "separation" of movements in time is accomplished by inserting the change and clearance timing intervals (displayed to the vehicle user as yellow and red clearance indications) between conflicting movements. The change and clearance intervals will be discussed later in this book, in Chapter 9. The "sequencing" of vehicle movements is the subject of this current chapter.

The sequencing of movements is accomplished through the sequencing of displays or indications to which users respond. The four-component traffic control process model, introduced in Chapter 1 and shown in Figure 48, shows the interdependencies of the user, the detector, the controller, and the display at a conceptual level. Here the dependence of the user on the display is clear: the user responds in a safe and appropriate way to the information conveyed in the display. But if we take a closer look at the actual hardware used in the field, we can better see the detector-controller-display linkages as well.

User → Detector
Display ← Controller

Figure 48. Traffic control process model

The Cabinet

The controller is housed in a box called the cabinet, which is located at the intersection. See Figure 49. Modern cabinets have 16 detector amplifiers, a controller that can accommodate up to 16 phases and 4 rings, and 16 load switches. Figure 50 shows an example of the location of the detector amplifiers, the controller, and the load switches inside a cabinet.

Figure 49. Cabinet at intersection

- A detector amplifier accepts calls from one or more loop detectors; each loop detector monitors traffic on one or more lanes. Each loop sets up an inductance field which is used to sense vehicles and generates an output signal to the traffic controller. The sensitivity of the loop is based on how well the metal interrupts the inductance field and the sensitivity of the amplifier settings that is needed to generate an output. The output signal from the amplifier effectively closes the detection circuit to the controller.
- The controller interprets this signal (on or off) as either a call or no call. The call is registered when an algorithm in the controller has processed the input.
- The controller provides the logic for determining when and how long each phase will time, based on the calls received from the detectors and the timing plan that has been programmed. A phase is a timing unit that controls the operation of one or more movements. One phase generally drives one load switch, although through the use of overlaps it can drive multiple outputs (load switches). Overlaps can also combine phases to meet special needs.
- Based on outputs from the controller, the load switches drive the displays that users respond to by converting the low voltage controller output to a higher voltage that drives the traffic signal display. Each display typically has up to three indications, the maximum number of indications driven by a load switch. The indications include a "go" indication that indicates that a user has the right-of-way and can safely proceed through the intersection and a "stop" indication that tells the user not to enter the intersection.

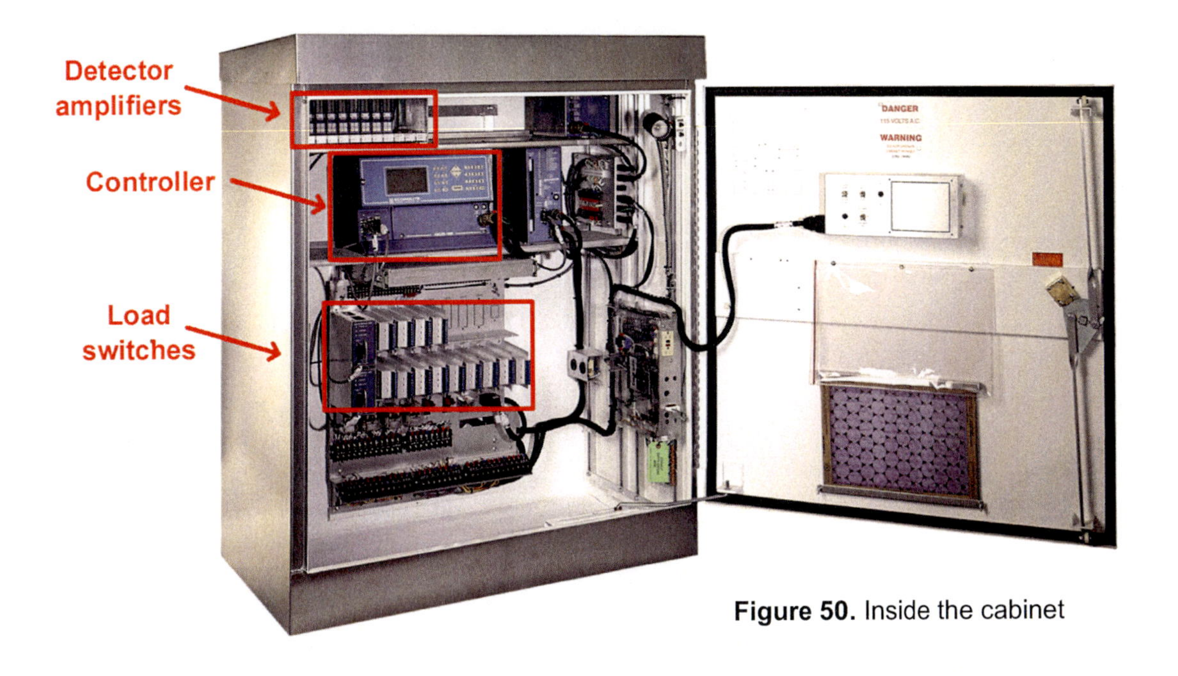

Figure 50. Inside the cabinet

Movements and Phases

Two of the concepts introduced in the previous section, the movement and the phase, are often used interchangeably (incorrectly so), but are distinct and mean different things. Let's start with the term "movement." We will define a movement as the response of the user to a "go" indication. This response has two attributes: the compass direction from which the user originates and the turn which the user makes. The compass direction normally includes one of the following designations: eastbound, westbound, northbound, or southbound. The turn designation includes left turn, through, or right turn.

We can also say whether a movement is opposed or not by another movement by using the terms protected, permitted, or not opposed. A protected movement has the unambiguous right-of-way, while a permitted movement may continue through the intersection only when there is a large enough (and safe) gap in the opposing traffic movement.

Movements are commonly numbered as shown in Figure 51. Left turn movements are designated with odd numbers (1, 3, 5, and 7) and through movements are designated as even numbers (2, 4, 6, and 8). While right turn movements are often given the same number as their compatible through movements, they are sometimes given the same right digit with a leading "1."

A phase is a timing unit that controls one or more movements. The timing unit includes three timing intervals, one each for the indications of green, yellow, and red. Phases are typically numbered in the same way as movements, with the exception that the even numbered phases normally control both the through and right-turn movements. Figure 52 shows the standard NEMA phasing convention (so-named since the National Electrical Manufacturers Association has established standards for the operation of an actuated traffic controller). For consistency, phases 2 and 6 should correspond to the through movements on the major street, while phases 4 and 8 should correspond to the through movements on the minor street. Similarly, phases 1 and 5 should correspond to the left turning movements on the major street and phases 3 and 7 should correspond to the left turning movements on the minor street.

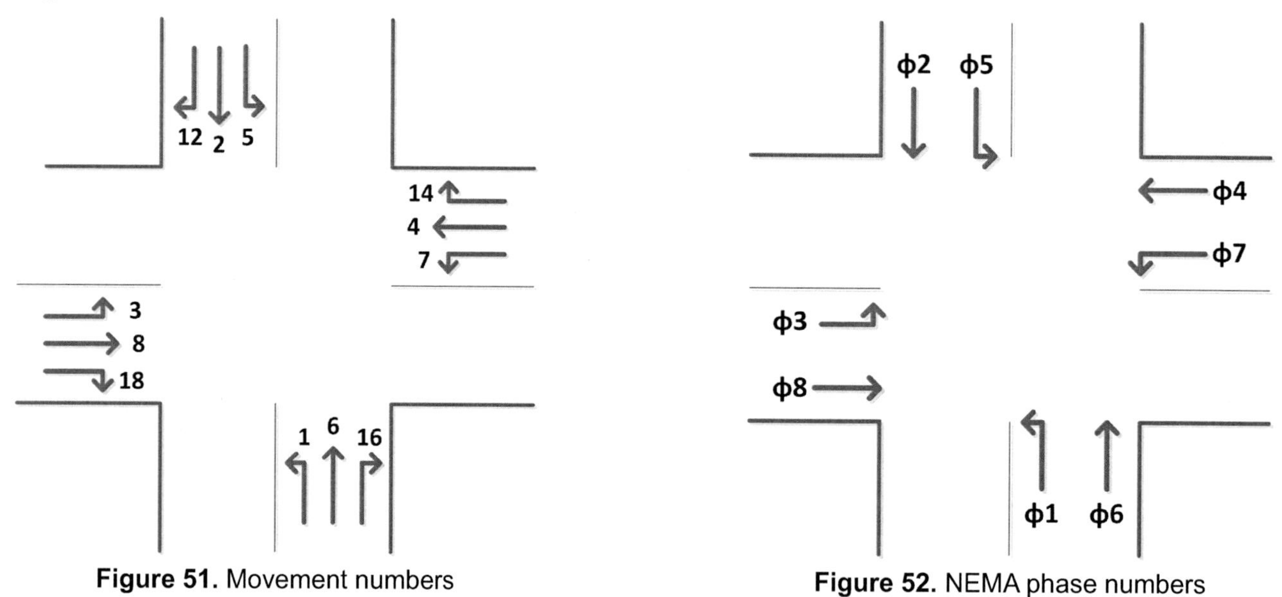

Figure 51. Movement numbers

Figure 52. NEMA phase numbers

The Conflict Matrix

The conflict matrix defines the phases that are compatible and can time concurrently. It also defines phases that are not compatible and thus cannot time concurrently. The safety task of preventing non-compatible phases from timing concurrently is managed by the malfunction monitoring unit (MMU) . The MMU monitors load switch outputs and ensures that conflicting green indications are not displayed at the same time.

For example, suppose that the phase 1 display indicates green. Either phase 5 or phase 6 can also display green concurrently with phase 1 green as shown in Figure 53 and Figure 54. This means that either movements 1 and 5 or movements 1 and 6 can be served concurrently.

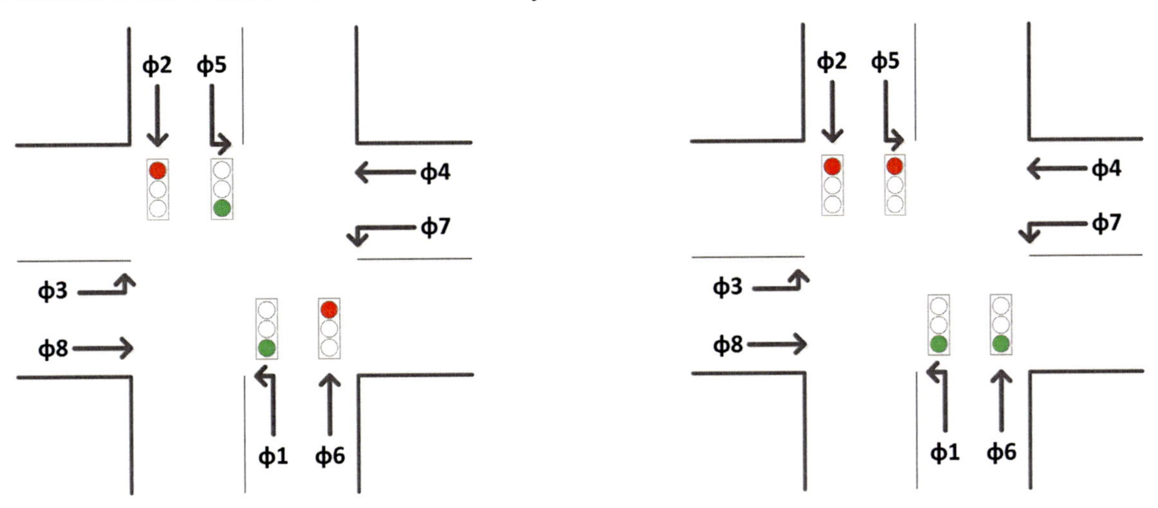

Figure 53. Phase 1 green and phase 5 green can be displayed concurrently

Figure 54. Phase 1 green and phase 6 green can be displayed concurrently

Figure 55 shows the phases that are both compatible with and in conflict with phase 1 in the form of a conflict matrix. Phases that conflict with phase 1 (phases 2, 3, 4, 7, and 8) are marked with an "X" in the column under Subject Phase 1. The empty squares next to phases 5 and 6 indicate that they are compatible with phase 1.

Compatible or Conflicting Phases	Subject Phase 1	2	3	4	5	6	7	8
1								
2	X							
3	X							
4	X							
5								
6								
7	X							
8	X							

Figure 55. Conflict Matrix showing conflicting and compatible phases for phase 1

Compatible or Conflicting Phases	Subject Phase 1	2	3	4	5	6	7	8
1		X	X	X			X	X
2	X		X	X			X	X
3	X	X		X	X	X		
4	X	X	X		X	X		
5			X	X		X	X	X
6			X	X	X		X	X
7	X	X			X	X		X
8	X	X			X	X	X	

Figure 56. Example conflict matrix

Now let's fill in the Conflict Matrix for all phases at a standard four leg intersection, as shown in Figure 56. Considering each subject phase in turn, the empty squares in each column show the phase pairs that are compatible, in which the movements controlled by these phases can move at the same time. The "X" squares show incompatible phases, in which the movements controlled by these phases cannot operate at the same time. The conflict matrix will be the basis for constructing a ring barrier diagram, which we will do in the next section of this reading.

Ring Barrier Diagram

The ring barrier diagram specifies the safe sequencing of phases (and thus the movements that they control) at a signalized intersection. We can construct the ring barrier diagram as follows. First, we separate the phases into two "concurrency groups", one group for the phases controlling the east-west movements and a second group for phases controlling the north-south movements. We separate the concurrency groups by a barrier. Figure 57 shows the beginning structure of the ring barrier diagram.

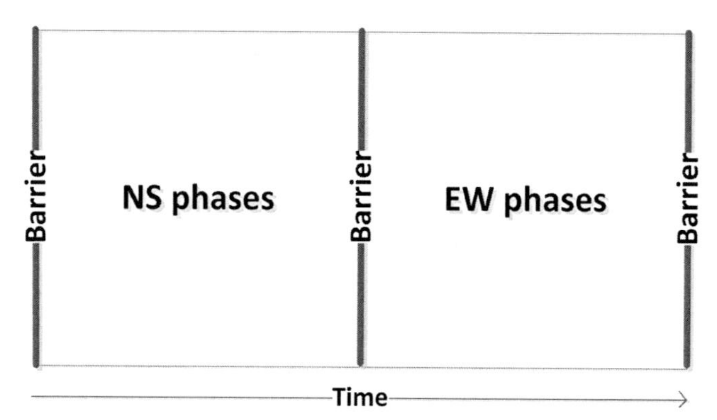

Figure 57. Beginning structure of ring barrier diagram

Next we consider, for the phases in each concurrency group, the order in which the phases must operate or time. We define a ring as a sequence of phases that are not compatible and that must time sequentially. For the north-south concurrency group, the movements controlled by phases 1 and 2 (the northbound left turn and the southbound through) must occur sequentially. Similarly, the southbound left turn and the northbound through movements, controlled by phases 5 and 6, must also occur sequentially. We can note these sequences in an expanded view of the ring barrier diagram (see Figure 58), where we have two rings, one for each sequence. Phases 1 and 2 are placed in ring 1, while phases 5 and 6 are in ring 2. The dashed lines indicate that the movement is "permitted." Here movements 12 and 16 are right turns (permitted) that must yield to pedestrians.

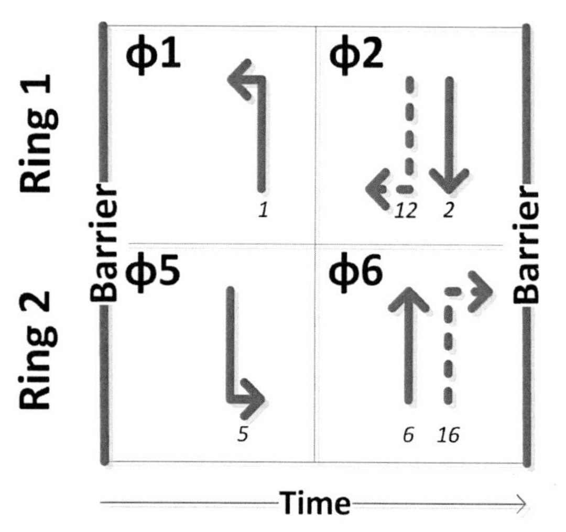

Figure 58. Partial ring barrier diagram for phases controlling EW movements

We are led to two rules that generally describe the standard ring barrier diagram:

1. The phases in a ring must be served sequentially and cannot be served concurrently because they are not compatible with each other.
2. A phase in one ring may be served concurrently with phases in the other ring in the same concurrency group.

A complete ring barrier diagram for the phases that control the east-west and north-south movements is shown in Figure 59. Similar to the north-south movements, the east-west movements are controlled in two rings or sequences: phase 4 follows phase 3 in ring 1, and phase 8 follows phase 7 in ring 2. Overall, ring 1 includes phases 1, 2, 3, and 4, while ring 2 includes phases 5, 6, 7, and 8.

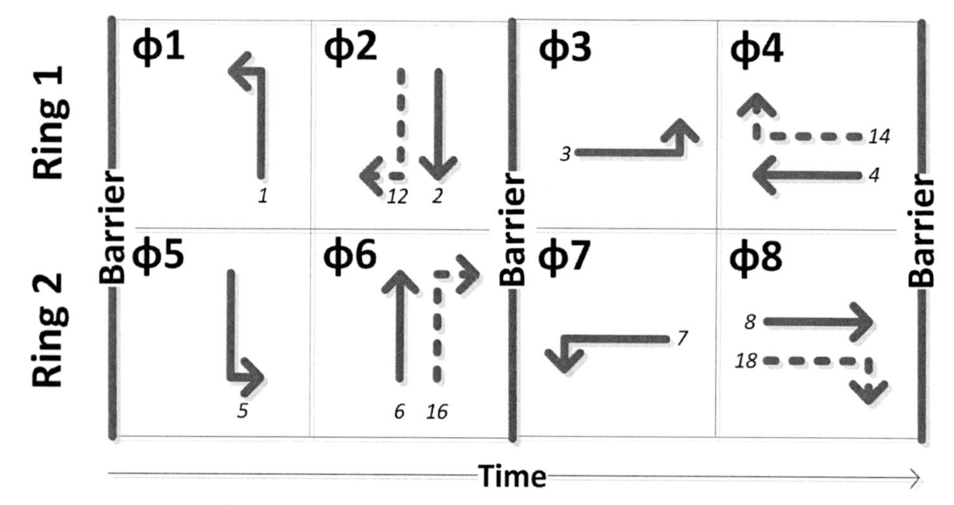

Figure 59. Complete ring barrier diagram

This depiction of the sequencing of the eight phases is now complete. However, there is an important feature of this sequencing that provides for more efficient operation when the timing of the phases is tied to the level of demand for each movement during a given signal cycle, as occurs with actuated signal control. Suppose, for example, that phase 1 times for 10 seconds, while the demand for the movement served by phase 5 requires 20 seconds to be served. We can represent this timing pattern in a partial ring barrier diagram as shown in Figure 60.

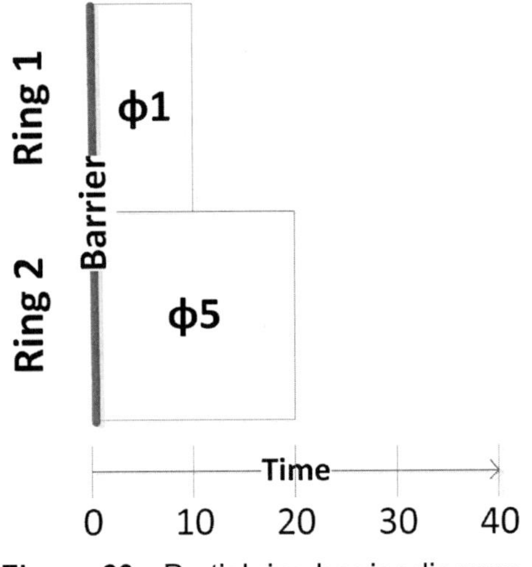

Figure 60. Partial ring barrier diagram showing phases 1 and 5

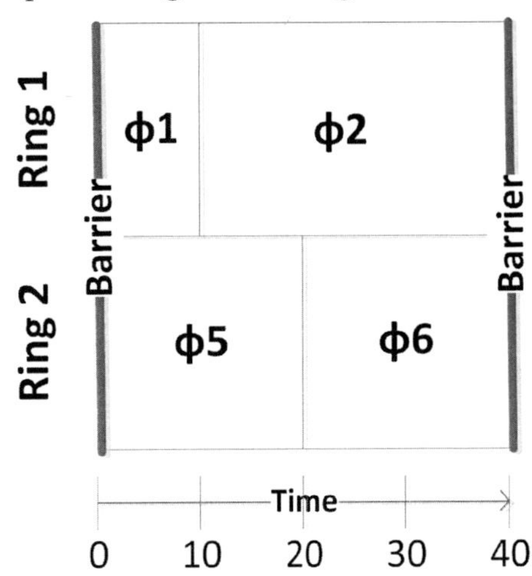

Figure 61. Example partial ring barrier diagram for north-south concurrency group

When phase 1 ends, phase 2 can begin timing since it can run concurrently with phase 5. If phase 2 then times for 30 seconds, and phase 6 times for 20 seconds, we can represent the partial ring barrier diagram for the phases serving the north-south movements as shown in Figure 61. Note that both rings 1 and 2 must "cross the barrier" at the same time, meaning that the active phases in these rings (in this case, phases 2 and 6) must end at the same time.

Consider another situation in which the demand for phase 1 requires 15 seconds, while the demand for phase 5 requires 10 seconds. Figure 62 shows the partial ring barrier diagram for this condition.

Now, what if the demand for phases 2 and 6 each requires 25 seconds to serve? What does the ring barrier diagram look like? Even though phase 6 requires 25 seconds, it will actually time for 30 seconds (with five seconds of slack time, or time that is not needed to service vehicle demand) because phase 2 and 6 must cross the barrier at the same time, as shown in Figure 63. During this "slack time" period, phase 6 is not timing but "resting in green."

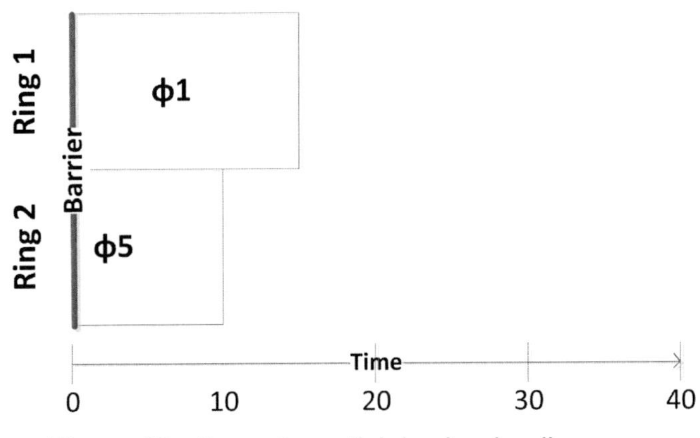

Figure 62. Example partial ring barrier diagram for north-south concurrency group

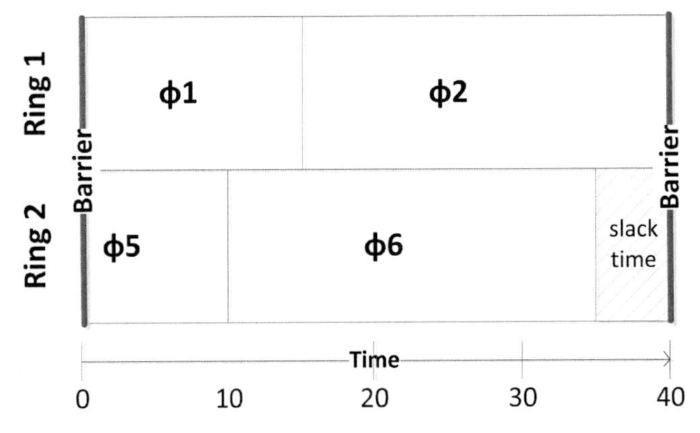

Figure 63. Example partial ring barrier diagram for north-south concurrency group with slack time

Other Issues

Phasing is a complex topic, and this reading covers only some of the basic concepts related to phasing. However, it is worth mentioning four other topics, including left turn phasing, split phasing, single ring operations and overlaps, and phasing for T-intersections.

Left turns

The previous discussion assumed leading protected left turns for the phasing sequence. This means that left turn movements are served first, before the compatible through movements, and that the left turn movements are protected. It is also common for left turns to be permitted, in which case left turning drivers must wait for a safe gap in the opposing traffic stream to complete their maneuver. An example ring barrier diagram for permitted left turn phasing is shown in Figure 64. Here all six movements in the north-south concurrency group are served at the same time by phases 2 and 6, while phases 4 and 8 simultaneously serve the six movements in the east-west concurrency group. The left turn movements (1, 3, 5, and 7) are shown as a dashed line since they are permitted and must yield to opposing through traffic. Various left turn phasing options will be discussed in more detail in Chapter 8 of this book.

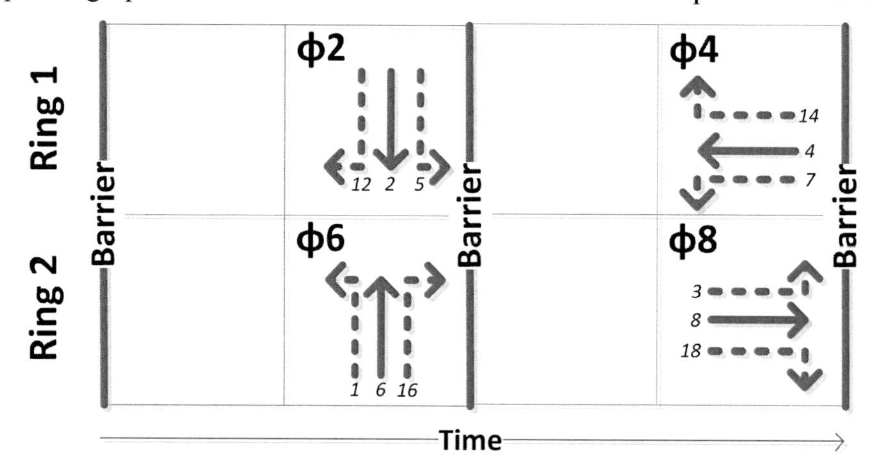

Figure 64. Ring barrier diagram for permitted left turn phasing

Split Phasing

Another phasing scheme is called split phasing. Split phasing is when each of the four intersection approaches is served in turn, sequentially, as shown in the ring barrier diagram in Figure 65. Split phasing is often used when the intersection geometrics limit opposing left turn movements from traveling at the same time. Split phasing can be accomplished with a single ring, a concept that will be discussed further in the next section.

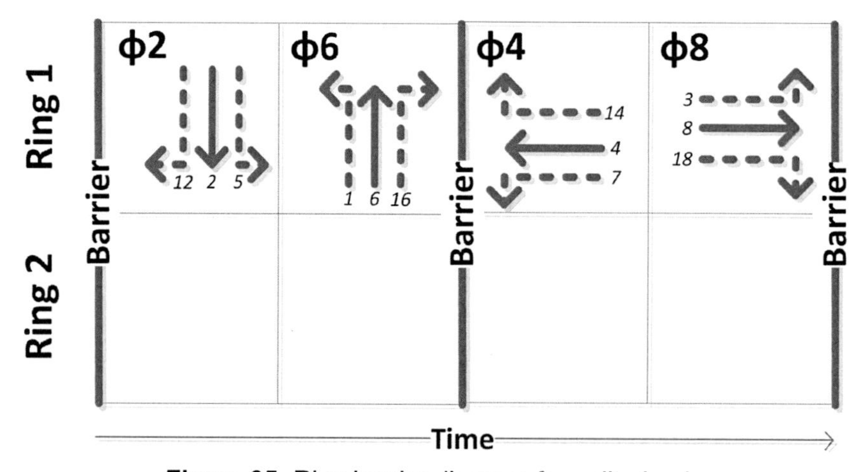

Figure 65. Ring barrier diagram for split phasing

Single Ring Operation and Overlaps

While two, or dual, ring operation is common, it is sometimes feasible for a single ring operation to be used. Consider the movement diagram shown in Figure 66. While this is a non-standard intersection (only three approaches with non-90 degree legs), it is not uncommon to find designs that are similarly unusual. One method of phase sequencing for this intersection is to have three phases: 8, 6, and 7. These phases would control movements 3 and 8, movements 6 and 16, and movement 7, respectively, as shown in Figure 67. Because the geometry of the intersection allows movement 18 to operate at the same time as movements 3 and 8 and movements 6 and 16, movement 18 can be controlled by an overlap, or sometimes called a child phase (overlap A) to the parent phase (phases 8 and 6). And, because movement 14 can operate compatibly with all of the vehicle movements, it can be operated as overlap B, with parent phases 8, 6, and 7.

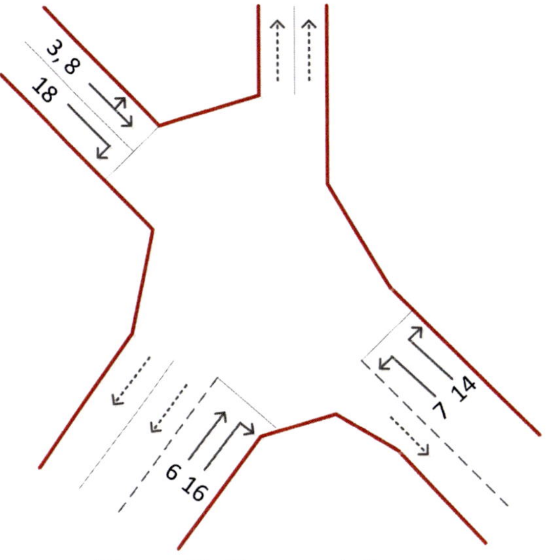

Figure 66. Movement diagram

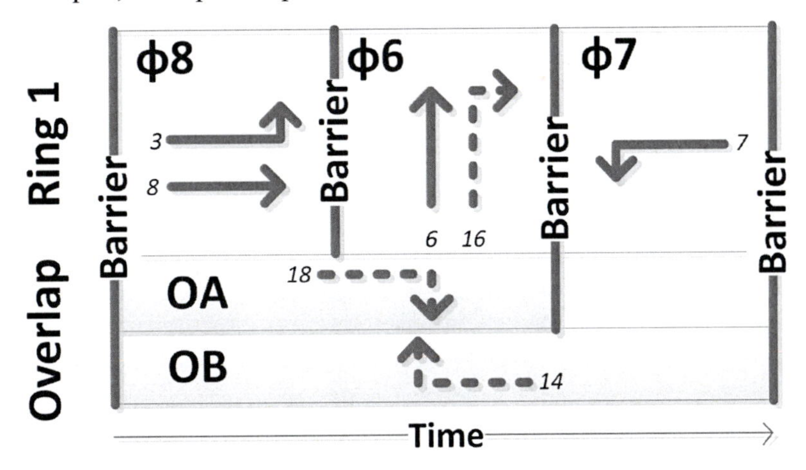

Figure 67. Ring barrier diagram for single ring and overlaps

A more standard overlap operation is shown in Figure 68. Here, the overlaps are each right turns that shadow what might be generally thought of as a conflicting left turn movement. For example, overlap A would "shadow" the parent phase 1. Standard cabinets can accommodate up to four overlaps, each one driving a load switch, which in turn drives a display and the included indications.

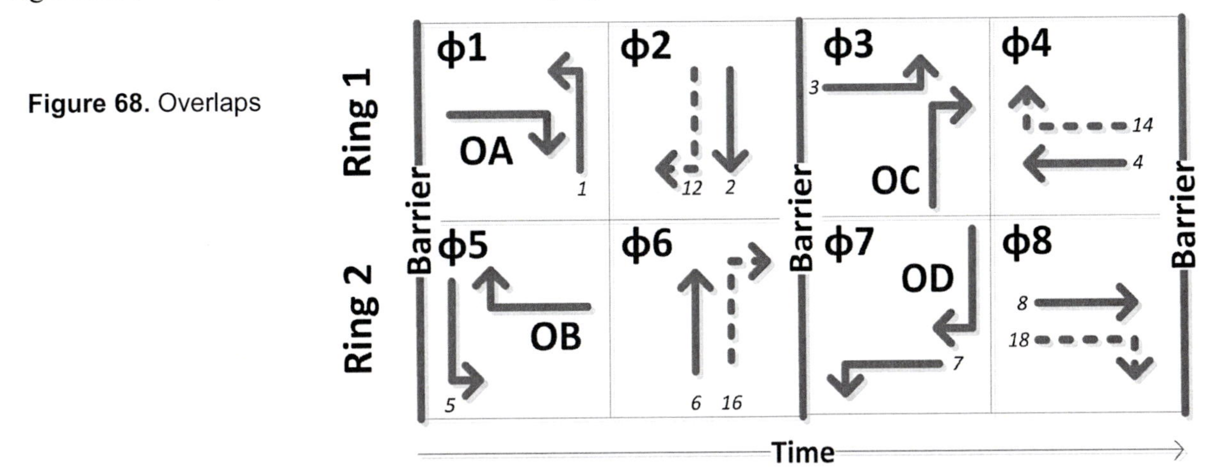

Figure 68. Overlaps

Phasing for T-Intersections

Suppose we had to develop a phasing plan for a T-intersection, in which there are only six movements, and not the twelve movements that are present at the standard four-leg intersection. The movement diagram is shown in Figure 69, while the phases that would control these movements are shown in the diagram in Figure 70. The ring barrier diagram is shown in Figure 71.

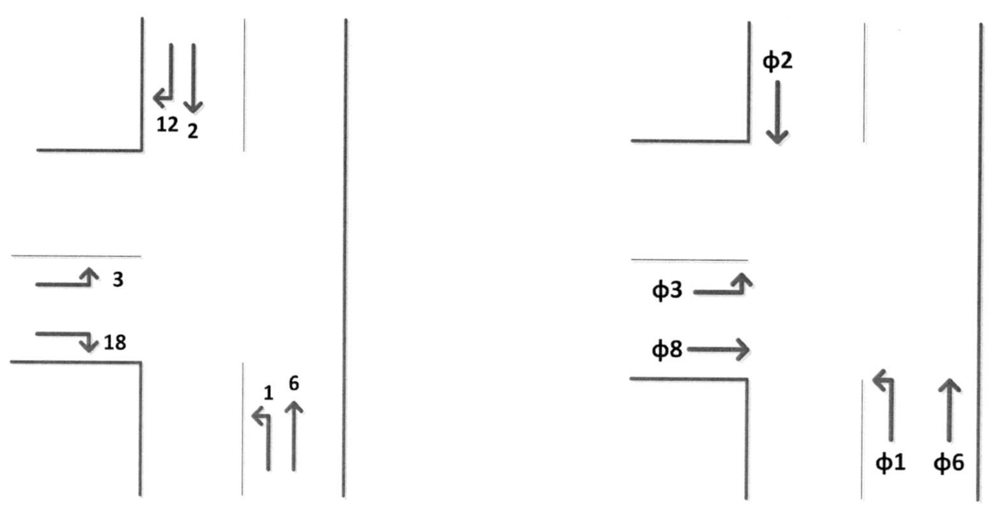

Figure 69. Movement diagram for T-intersection

Figure 70. Phase diagram for T-intersection

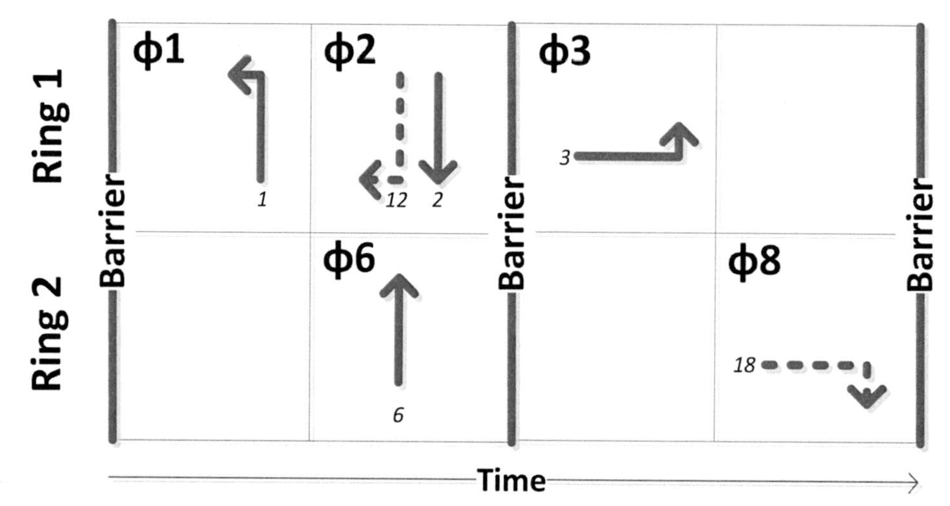

Figure 71. Ring barrier diagram for T-intersection

ACTIVITY

14 What Do You Know About Phasing and Ring Barrier Diagrams?

Purpose

The purpose of this activity is to test your understanding of the material covered in the Reading on ring barrier diagrams. This includes gaining a basic level of understanding of the concept of rings and barriers, and how they are used to safely separate the operation of conflicting movements at a signaled intersection.

Learning Objectives

- Describe NEMA phasing and the concept of rings and barriers
- List the phase numbers at a standard intersection with eight movements
- Draw and describe a ring barrier diagram in which there are two rings and eight phases

Required Resources

- Activity #13: "Phasing, Rings, and Barriers"

Deliverables

- Prepare a one page document that includes your results from Tasks 1 through 5

Task 1

Sketch a four leg intersection, showing an exclusive left turn lane and one through lane on each approach. Number each movement and list the phase number corresponding to the standard NEMA numbering scheme that would control each movement.

Task 2

Prepare a sketch of a ring barrier diagram that represents the condition described in Task 1.

Task 3

Prepare a brief description of the timing process for this eight phase operation by describing the order and manner in which each phase is served. Consider the various sequences that may occur depending on traffic flow volumes.

Task 4

Suppose the traffic demand for the east-west movements at a signalized intersection requires the following times for each movement to be served. Draw a partial ring barrier diagram showing the sequence and the timing of the phases controlling these movements. (See Table 7, following page.)

Table 7. Phases and required time to serve movements

Movement	Phase controlling movement	Required time (sec)
EBLT	5	5
EBTH	2	25
WBLT	1	10
WBTH	6	15

TASK 5

Figure 72 shows an intersection with five approaches. The movement are shown and numbered. Prepare a conflict matrix and a ring barrier diagram that would provide safe operation for this intersection.

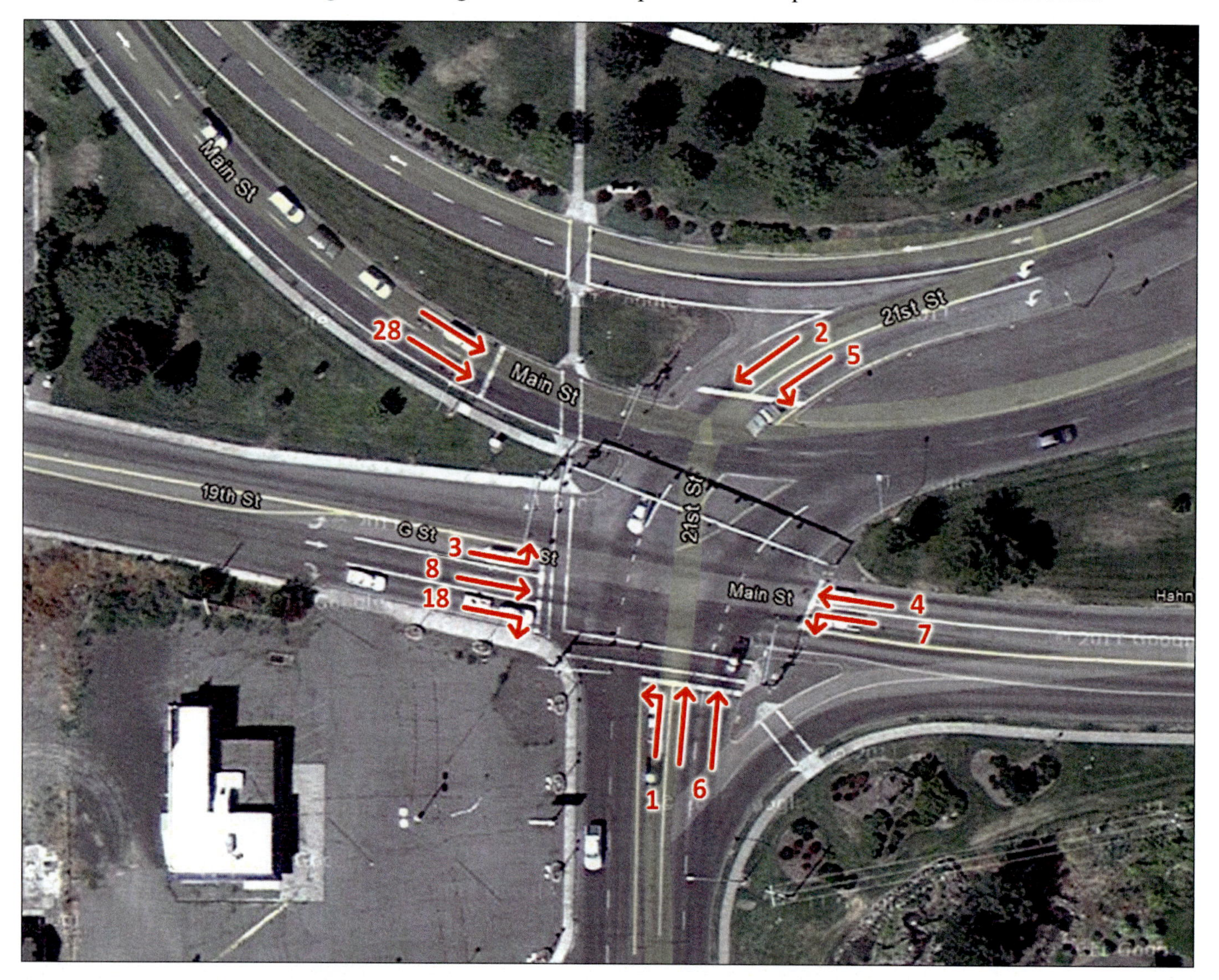

Figure 72. Intersection with five approaches, Lewiston, Idaho

ACTIVITY

15 Verifying Ring Barrier Operation in the Field

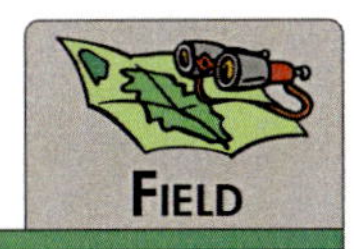

Purpose

The purpose of this activity is to give you the experience of observing the operation of an actuated controlled intersection in the field and to document the phasing sequence that you observe.

Learning Objective

- Determine the phasing pattern and sequence for a signalized intersection in the field

Deliverable

- Prepare a one page document including your field observations and the resulting ring barrier diagram

Information

You will be assigned an intersection in the field from which you will observe and record phasing information.

Task 1

Prepare a sketch of the intersection to which you have been assigned, including the geometry and the movements that you observe at the intersection.

Task 2

Based on standard NEMA phasing, add the phase numbers to the sketch of the movements that you prepared in Task 1.

Task 3

Observe the intersection for a period of 15 minutes. Record the sequence in which each movement is served during this period in Table 8.

Task 4

Prepare a ring barrier diagram showing the sequence of phases that you believe exist at this intersection. Document any differences between the normal sequence (Figure 59) and any special phasing sequences that you observe.

Table 8. Data collection form for sequence of phases

Cycle	Sequence of movements for phase

ACTIVITY

16 Phasing, Rings, and Barriers in Practice

PURPOSE

The purpose of this activity is to extend your understanding of the topics already encountered in this chapter.

LEARNING OBJECTIVE

- Appreciate how theoretical information about phasing, rings, and barriers is used and applied by professionals

REQUIRED RESOURCE

- *Traffic Signal Timing Manual*

DELIVERABLES

Prepare a document that includes

- Answers to the Critical Thinking Questions
- Completed Concept Map

LINK TO PRACTICE

Read the section of the *Traffic Signal Timing Manual* assigned by your instructor.

CRITICAL THINKING QUESTIONS

When you have completed the reading, prepare answers to the following questions:

1. What is the logic for the sequence of phases that are included in a ring?

2. What impact did your field experience have on your understanding of phasing, rings, and barriers?

In My Practice...

by Peter Koonce

The standard ring barrier configuration has been helpful to standardize design, operations, and maintenance functions of signalized intersections. Yet, there are some cases where the strength of the ring barrier diagram may limit our willingness to think outside the box. The intersection of NE 82nd & Airport Way in Portland, Oregon is a heavily travelled intersection with light rail in close proximity. Medians separating the movements provided an opportunity to allow pedestrians to cross in three stages during non-conflicting vehicle phases. This was done to insure safe operation of the light rail preemption.

The sketch of the intersection on the left shows the vehicle phases numbered 2, 4, 5, and 6. The pedestrian phases, numbered P4 and P6, operate at the same time as the vehicle phases 4 and 6, respectively. A pedestrian crossing Airport Way would be served by three sequential pedestrian phases: P4, P6, and P4. The sketch on the right shows a design that used overlaps to cross the barriers for pedestrians allowing the signal timing to permit crossings over the entire length of a cycle, reducing the pedestrian crossing time by up to 75 seconds. It also improved crossings for people on bicycles, addressing a complaint from a cyclist who correctly noted that the ring barrier structure prioritized vehicle movements with little concern for efficient pedestrian operations. A simple modification to the allowable phases and the order of the movements allowed a person on a bicycle to cross on the pedestrian movement more efficiently by assigning multiple phases to the various overlaps at the intersection.

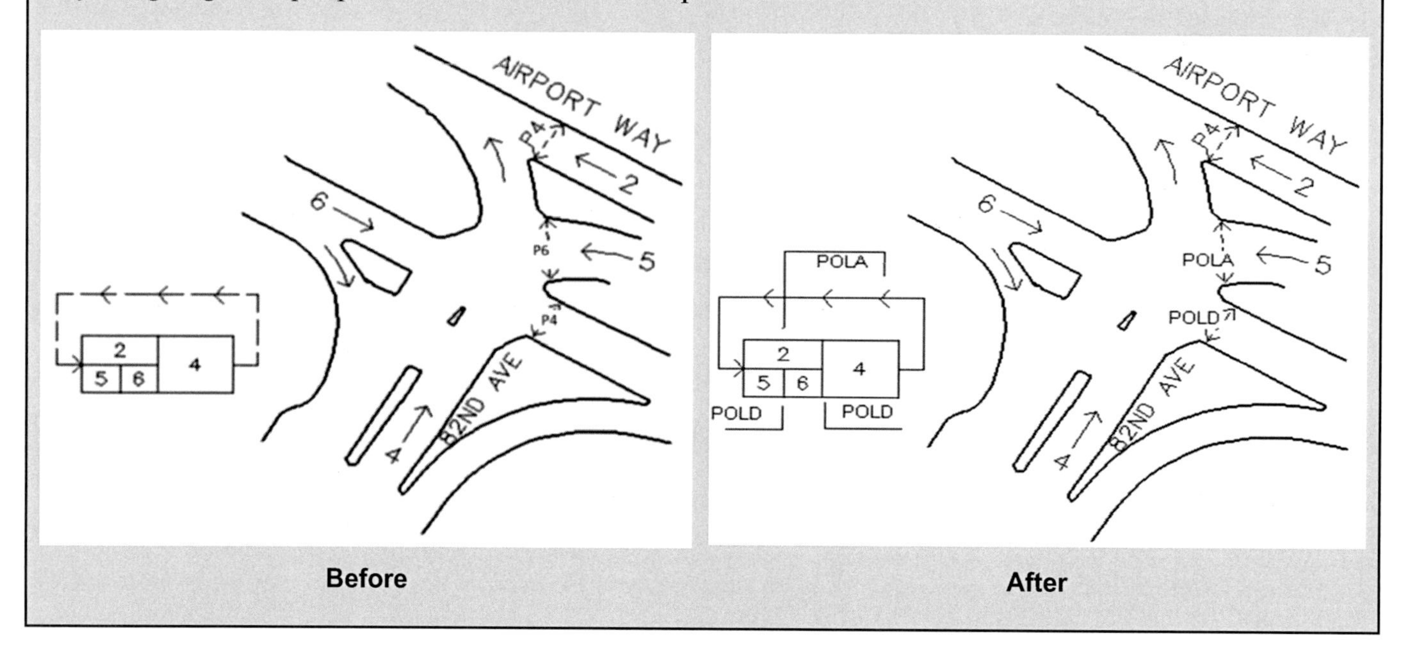

Concept Map

Terms and variables that should appear in your map are listed below.

concurrency group
NEMA phase numbering
phase
ring barrier diagram
movement
overlap
ring

Student Notes:

Actuated Traffic Controller Timing Processes

Purpose

In Chapter 4, you will learn about the timing processes that run an actuated traffic controller. Many transportation engineers begin their study of signalized intersections by making assumptions on the length of the cycle and the duration of the green intervals for each phase. However, the reality is the cycle length and the green interval durations are the result of the interaction of the traffic demand and a set of actuated controller timing processes. If the demand is low, the phase may only last until the minimum green timer has expired; but if the demand is high, the phase may be extended (by a series of intervals equal to the vehicle extension time for each vehicle that arrives at the intersection) until the maximum green time has been reached. Thus the green duration and the cycle length will vary as the demand varies from one cycle to the next.

The Highway Capacity Manual or HCM (Transportation Research Board, 2010) includes a method for synthesizing the green interval duration based on demand and some of the timing parameters described in this section. However, in order for you to develop a signal timing plan, it is important for you to understand these actuated controller timing parameters and the process followed by each.

Learning Objectives

When you have completed the activities in this chapter, you will be able to

- Describe the actuated controller timing processes
- Complete a traffic control process diagram describing the response of the detectors, the timing processes, and the displays to a pattern of vehicle demand
- Describe the range of information provided in the ASC/3 controller
- Describe the effect of detector calls on controller timing processes
- Describe the two primary methods of terminating a traffic phase at an isolated intersection
- Describe the actuated controller timing processes
- Infer signal timing parameter values through field observations
- Contrast the signal timing terms that are presented in this chapter with those described in the *Traffic Signal Timing Manual*

Chapter Overview

This chapter begins with a *Reading* (Activity #17). The reading describes and illustrates the three basic actuated controller timing processes, establishing your initial knowledge base of these processes. This reading covers the minimum green time, the passage time, and the maximum green time. Each of these three is a timing parameter, the value of a timing parameter, and a timing process. In Activity #18, you will assess your understanding of the material presented in the *Reading* by constructing three traffic control process diagrams. In Activity #19, you will study the ASC/3 traffic controller to learn about how basic actuated timing processes operate and are displayed in the controller front panel. In Activity #20, you will use a side-by-side movie of a VISSIM simulation to observe the two different ways in which a phase first times and then terminates. Activity #21 leads you through additional studies of the traffic controller and its timing processes using an emulator run in an Excel spreadsheet. In Activity #22, you will continue to test your understanding of various traffic conditions and timing parameters by constructing a series of traffic control process diagrams, each illustrating a basic concept of signal timing. You will collect field data in Activity #23 to infer the values of the three signal timing parameters. The chapter concludes with Activity #24, in which you will link what you have learned in the chapter with material from the *Traffic Signal Timing Manual*.

Activity List

Number and Title	Type
17 Controller Timing Process	*Reading*
18 What Do You Know About Controller Operations?	*Assessment*
19 The ASC/3 Traffic Controller	*Discovery*
20 How a Traffic Phase Times and Terminates	*Discovery*
21 Exploring a Controller Emulator	*Discovery*
22 Constructing a Traffic Control Process Diagram	*Discovery*
23 Inferring Signal Timing Parameter Values	*Field*
24 Signal Timing Parameters	*In Practice*

ACTIVITY 17 Controller Timing Process

PURPOSE

The purpose of this activity is to give you the opportunity to learn more about basic actuated controller timing processes.

LEARNING OBJECTIVE

- Describe the actuated controller timing processes

DELIVERABLES

- Define the terms and variables in the Glossary
- Prepare a document that includes answers to the Critical Thinking Questions

GLOSSARY

Provide a definition for each of the following terms. Paraphrasing a formal definition (as provided by your text, instructor, or another resource) demonstrates that you understand the meaning of the term or phrase.

gap out	
max out	
maximum green	
minimum green	
passage time	

Critical Thinking Questions

When you have completed the reading, prepare answers to the following questions.

1. What are the two types of phase termination and what are the factors that result in each of these two types?

2. What happens if the passage timer expires before the minimum green timer expires?

3. What is a *traffic control process diagram* and what processes does it illustrate?

Information

In Chapter 1, you were introduced to the traffic control process diagram. This diagram, which is represented in Figure 73, shows the four processes or components of the traffic control system and how they interrelate:

1. The user arrives at the intersection and is detected.
2. The detector sends a call to the traffic controller.
3. The controller determines the signals to display.
4. The user responds to the signal that is displayed, shown with the feedback loop on the left of Figure 73.

In this chapter, you will learn about the three most important timing processes that govern the operation of the actuated controller, and the logic that is used to determine how long a phase remains active ("is timing") and when and how the phase will terminate.

Let's first define these parameters and the process that each follows. It should be noted that we will describe both a timing parameter and a process, each with the same name. This may seem a bit confusing at first!

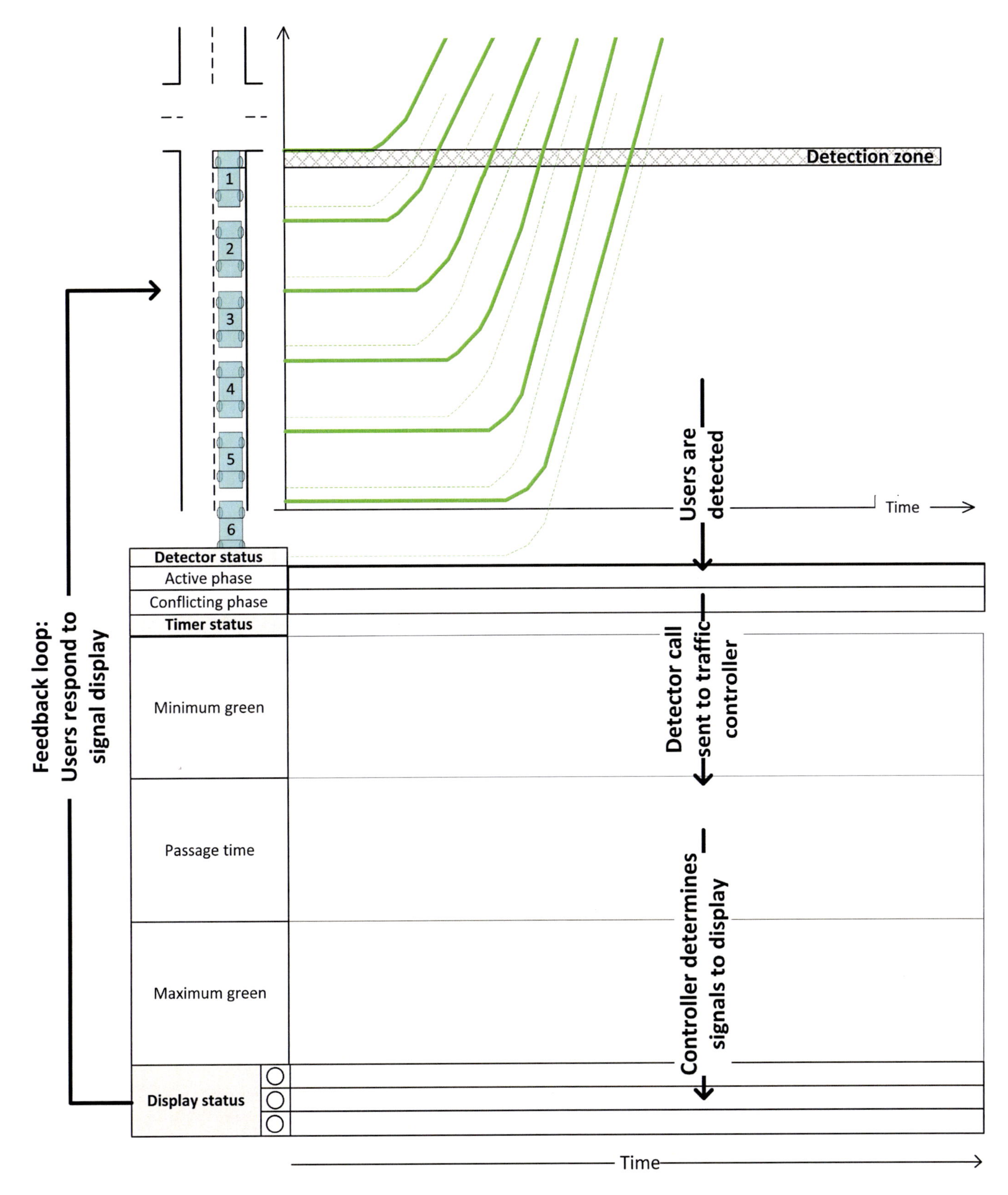

Figure 73. Traffic control process diagram

The minimum green time is the minimum time that the display will remain green for a phase no matter what else occurs. The minimum green timer is initially set to a value equal to the minimum green time. When the phase begins timing, the minimum green timer begins to time down and it expires when its value reaches zero, as shown in Figure 74. You will learn more about determining the length of the minimum green time in Chapter 6.

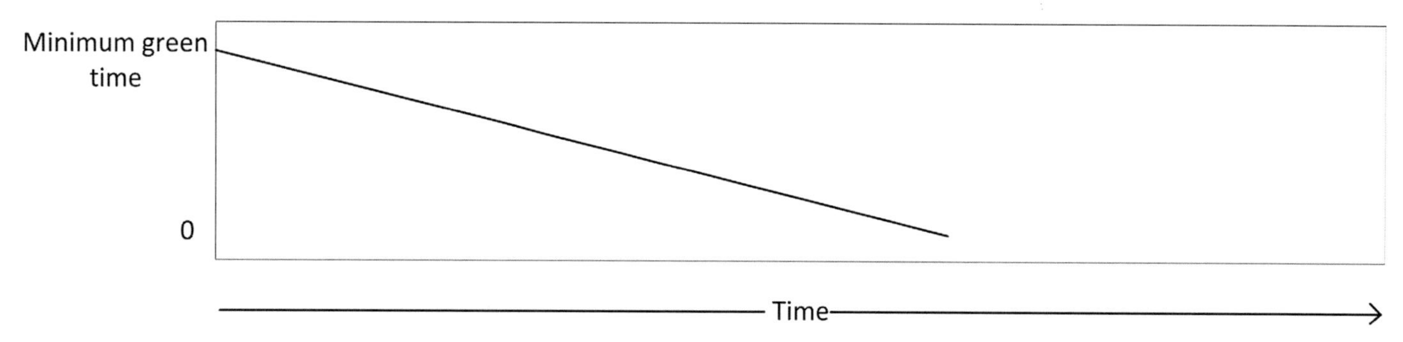

Figure 74. Minimum green timer process

The purpose of the passage timer (sometimes called the vehicle extension timer) is to extend the green until a gap of a pre-determined size is reached. The passage time is the maximum time that a detector can remain unoccupied before the passage timer expires. The significance of the relationship between the passage time and the maximum allowable headway will be described in Chapter 6. As long as a vehicle remains in the detection zone (or, "a call is active"), the timer will remain at its initial value or setting. Once a vehicle leaves the zone, the timer begins to time down. When a subsequent vehicle enters the zone, the timer is reset to its initial value. We will see in Chapter 6 the relationship of the passage time to the maximum allowable headway, the maximum headway that we will tolerate before allowing the phase to terminate. We will also see that this relationship is dependent on the length of the detection zone.

The following figures show example timing processes for the passage timer. In Figure 75, the passage timer begins to time down when a vehicle leaves the detection zone. In this example, it expires because no subsequent vehicle resets the timer.

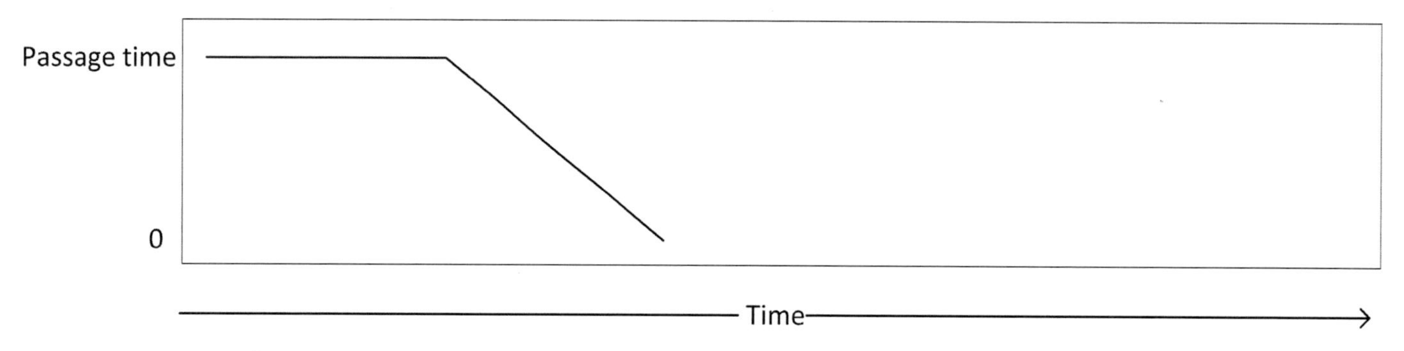

Figure 75. Passage timer process

By contrast, in Figure 76, the passage timer is reset several times, as one vehicle leaves the zone and a subsequent vehicle arrives in the zone, before the timer reaches zero.

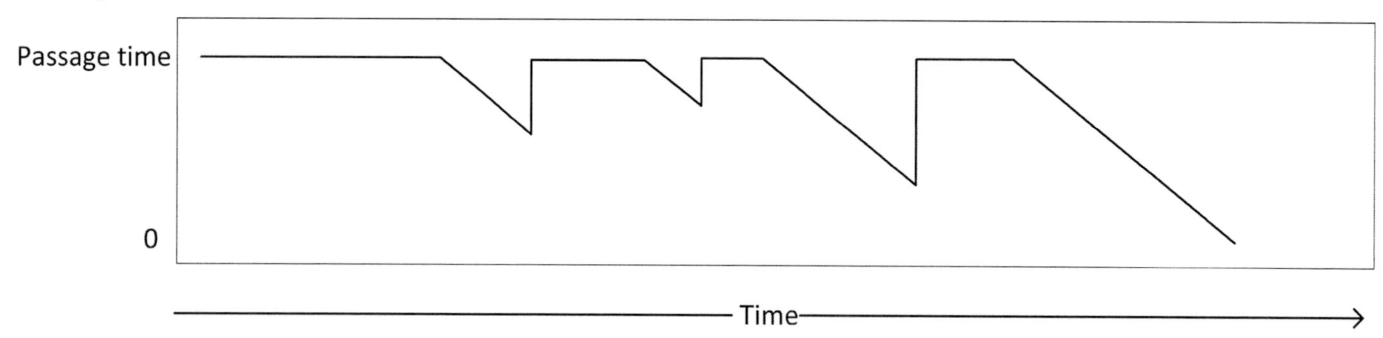

Figure 76. Passage timer process

The purpose of the maximum green time is to produce a maximum cycle length that keeps delay at a reasonable level. The maximum green time is the maximum duration that the signal display will remain green after a call has been received on a conflicting phase. When such a call is received, the timer will begin to time down and continue until it reaches zero as shown in Figure 77. You will learn more about setting the value for the maximum green time in Chapter 7.

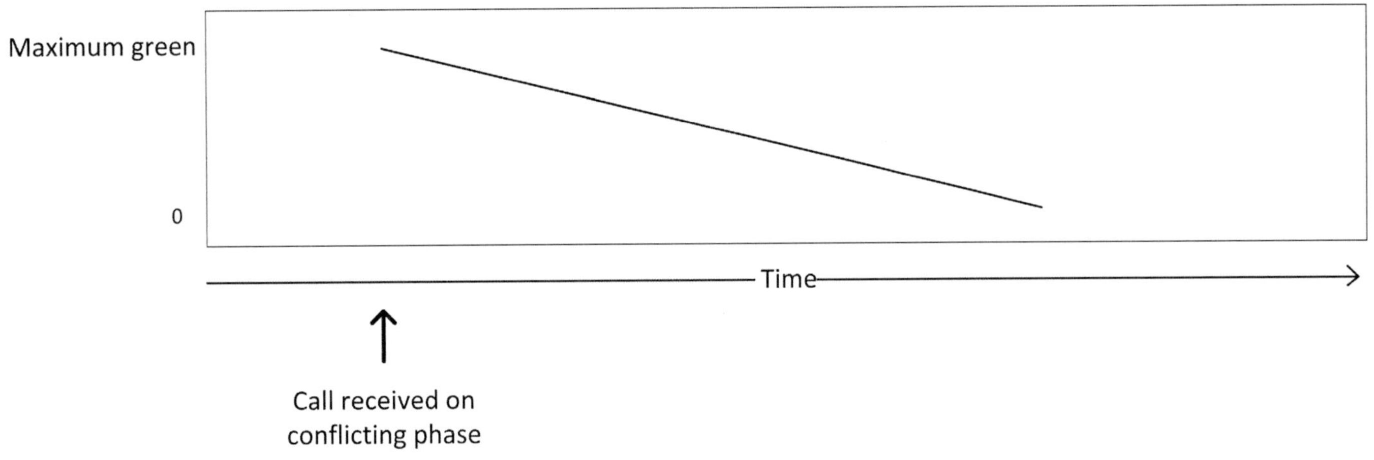

Figure 77. Maximum green timer process

The phase termination logic in a controller determines how long a phase will time and when it will terminate. The timing and termination logic is covered in more detail in Chapter 6 of this book. For standard actuated traffic control, a phase will continue to time until one of two possible events occur, either a gap out or a max out.

A gap out occurs when both the minimum green timer and the passage timer have expired. An example of a gap out is shown in Figure 78. While the maximum green timer is still active and timing down, once the passage timer expires, the phase will gap out.

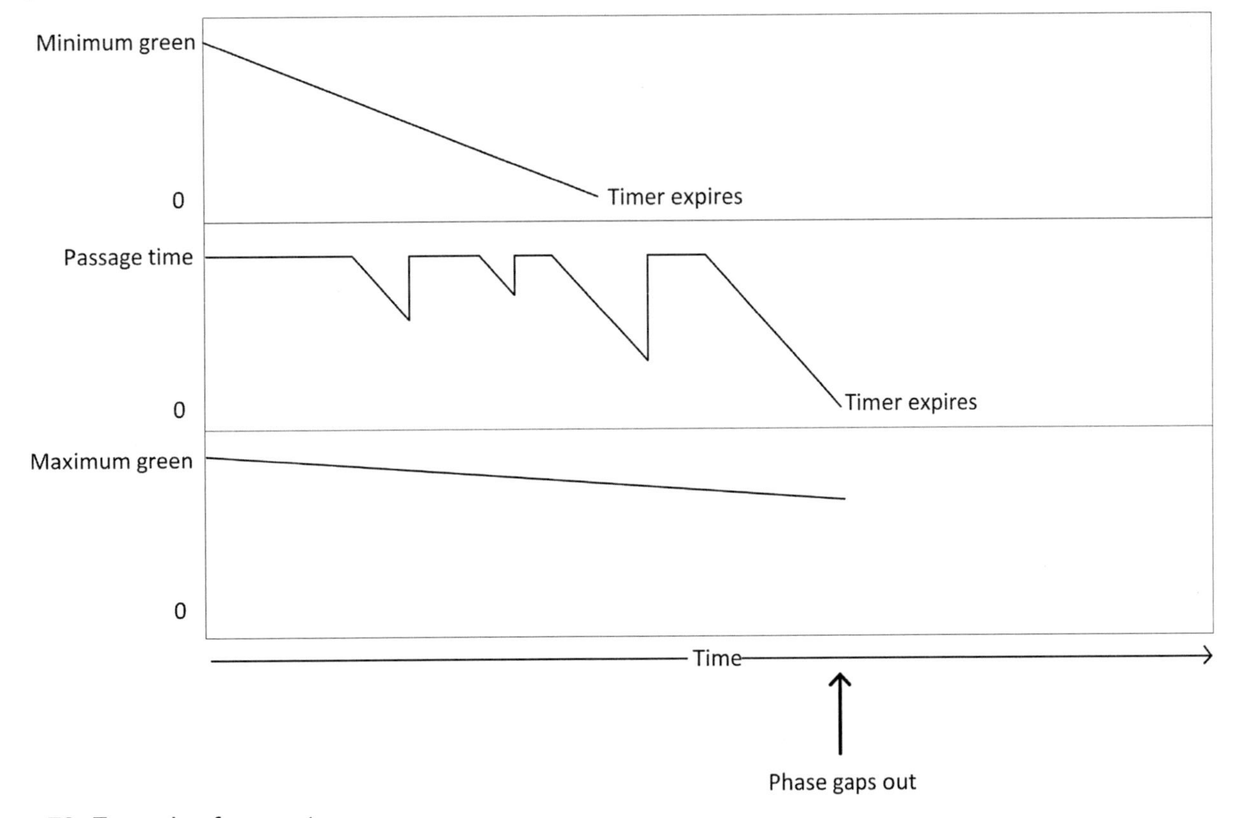

Figure 78. Example of gap out

A max out occurs when the maximum green timer expires. An example of a max out is shown in Figure 79. While the minimum green timer has expired, continuing demand extends the passage timer, as it resets each time a new vehicle is detected. However, the phase terminates when the maximum green timer expires.

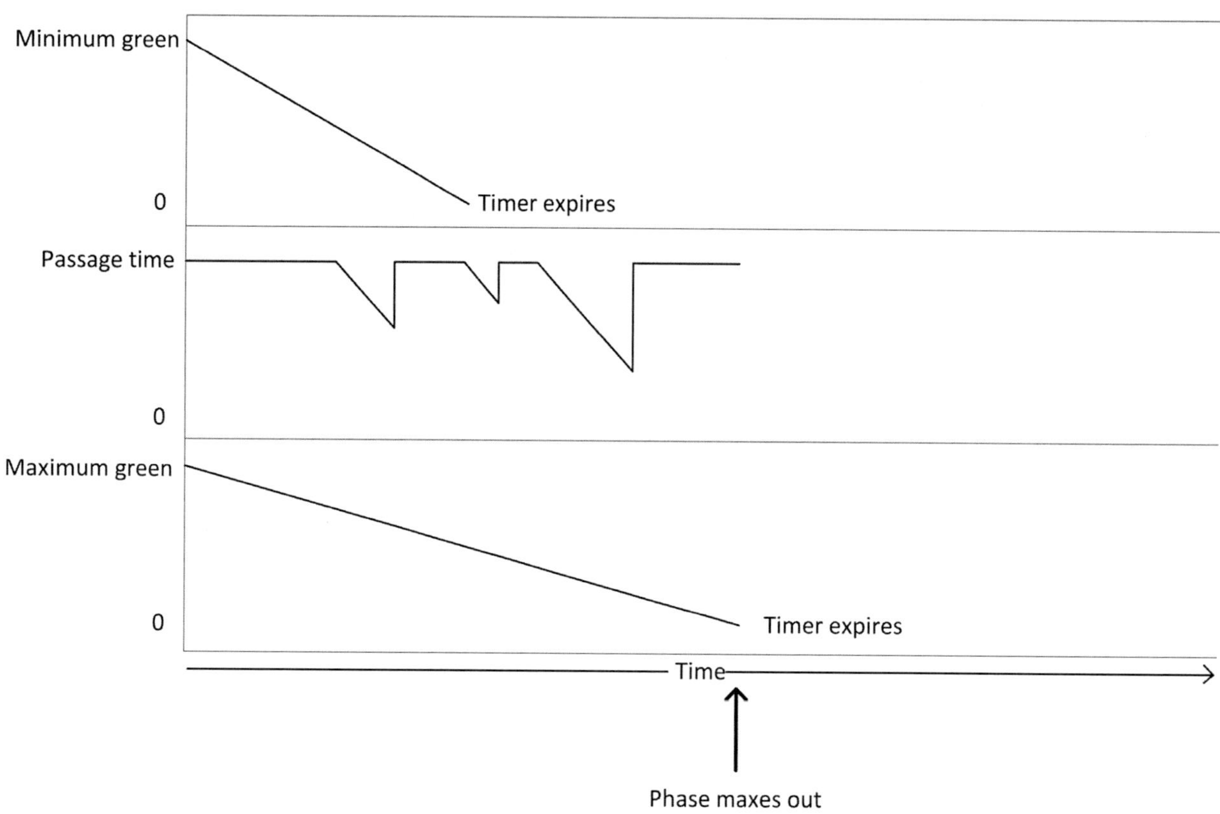

Figure 79. Example of max out

ACTIVITY

18 What Do You Know About Controller Operations?

PURPOSE

The purpose of this activity is to test your understanding of the basic timing processes for an actuated traffic controller.

LEARNING OBJECTIVE

- Complete a traffic control process diagram describing the response of the detectors, the timing processes, and the displays to a pattern of vehicle demand

DELIVERABLE

- Prepare a document with copies of the completed charts as required from Tasks 1 and 2

TASK 1

The traffic control process diagram (Figure 80, following page) shows the vehicle trajectories in a time space diagram format as well as the timing parameter values (bottom right of figure). Draw the detector status, the timer status, and the display status. Show the graphs for the values of the three timing processes in the spaces provided, noting the maximum and minimum values of the processes on the y-axis. The resulting signal display may change some of the vehicle trajectory plots. Note on the figures where you think that these changes will occur. Assume a yellow time of 3 seconds. Assume that the green time starts at $t = 0$. Also assume that the conflicting call begins at $t = 0$ and continues throughout the green duration.

TASK 2

Figure 81 and Figure 82 (see following pages) show traffic control process diagrams without the vehicle trajectories, but with the detector status data for both the active and the conflicting phases. The values of the timing parameters are given in the lower right of the two figures. Show the resulting timing processes in the form of a chart showing the value of the timing parameter as long as the green is active for that phase. Show the resulting signal display status, noting only when the display changes. State how the phase terminates in each case. Assume a yellow time of 3 seconds. Assume that the green time starts at $t = 0$.

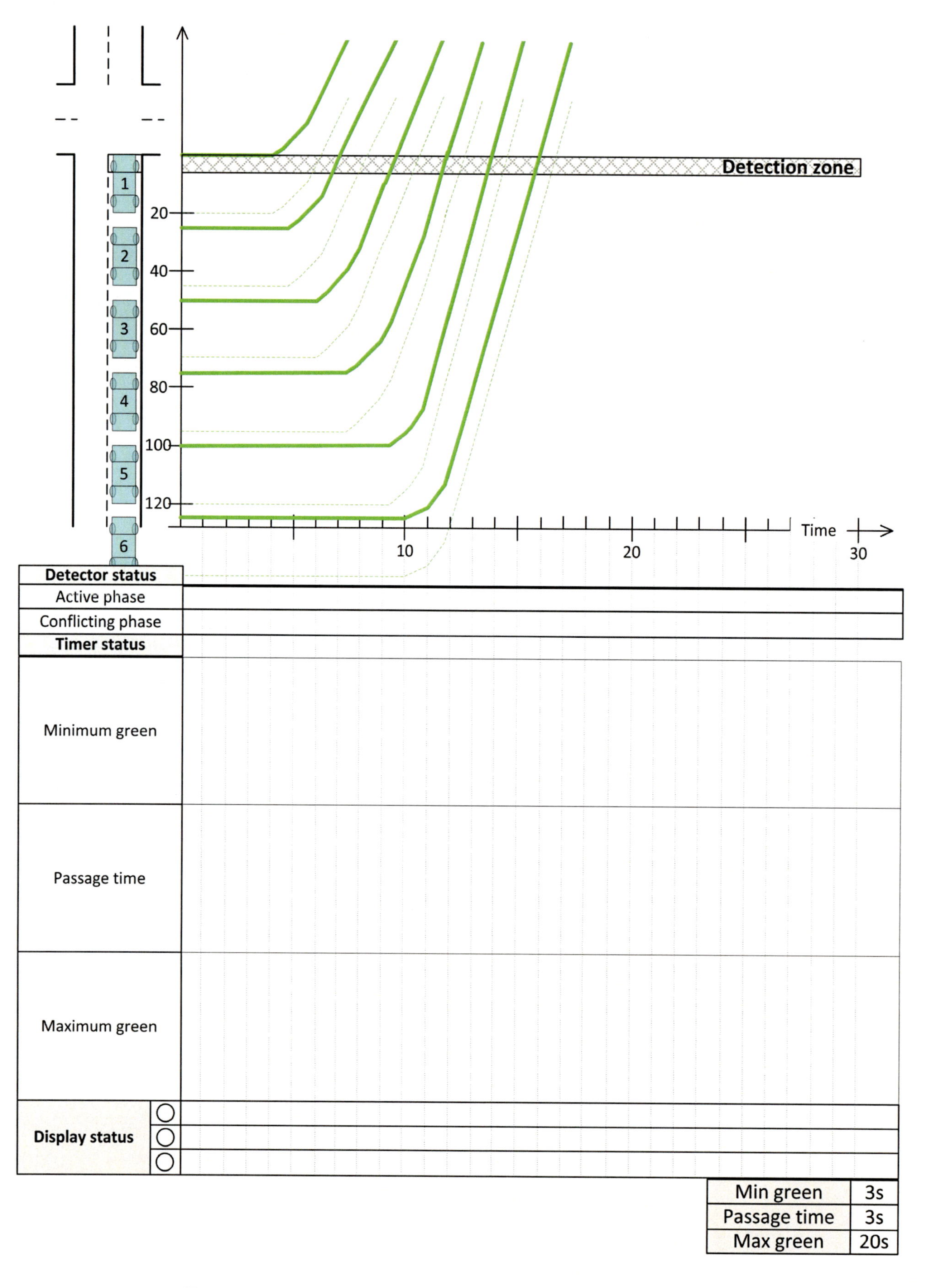

Figure 80. Traffic control process diagram with vehicle trajectories

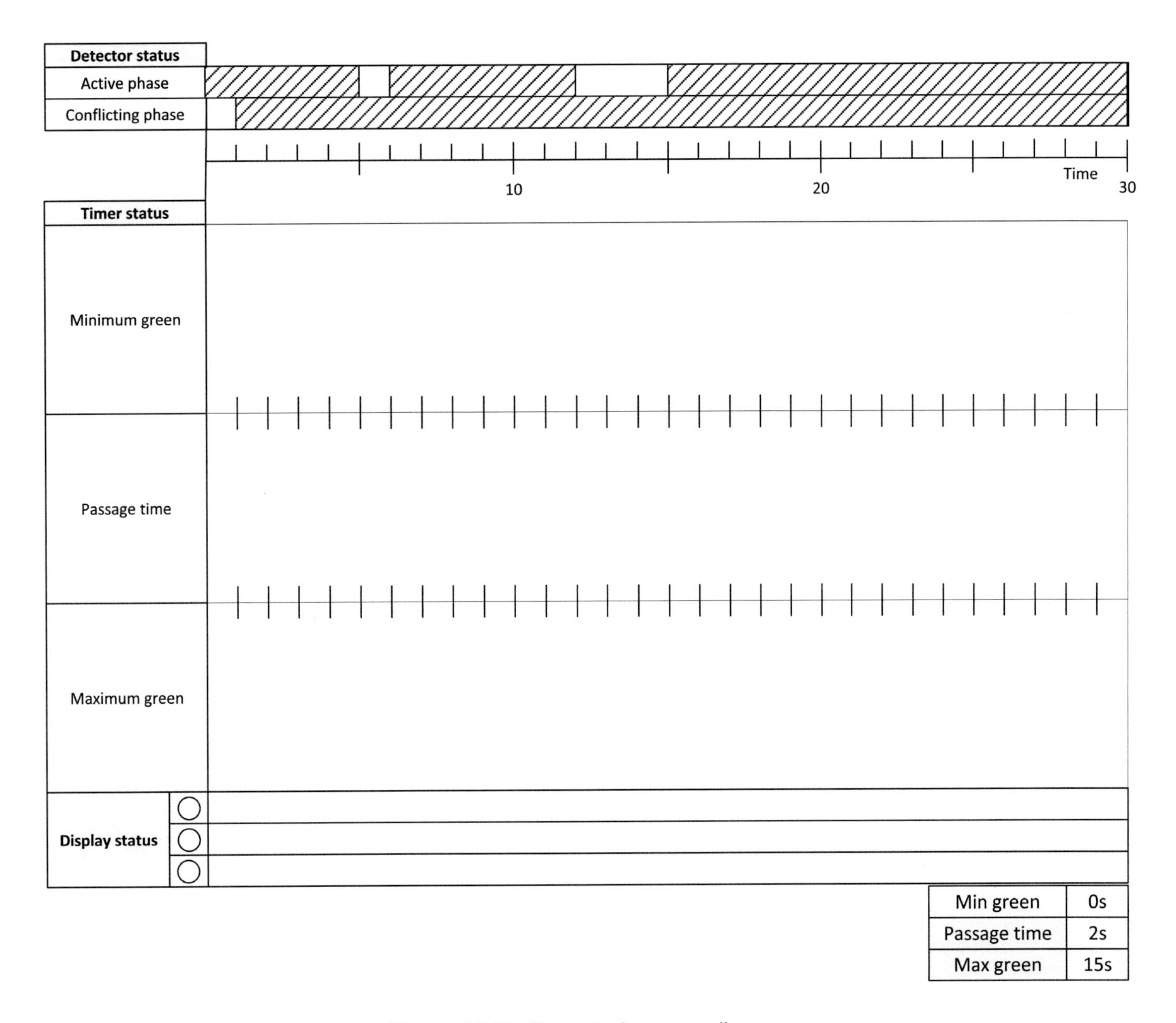

Min green	0s
Passage time	2s
Max green	15s

Figure 81. Traffic control process diagram

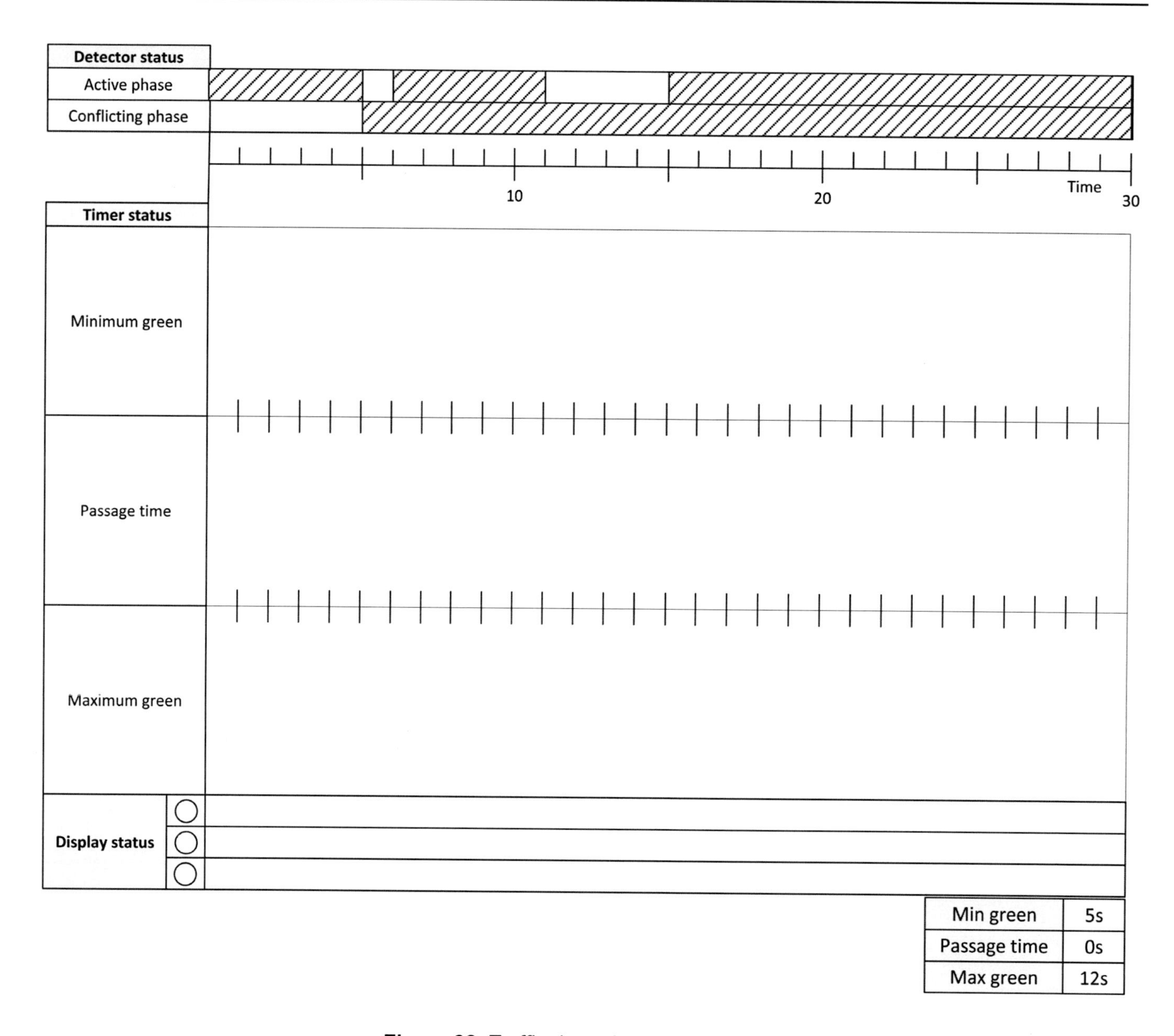

Figure 82. Traffic dontrol process diagram

ACTIVITY

19 The ASC/3 Traffic Controller

Purpose

The purpose of this activity is to give you the opportunity to learn more about the operation of an actuated traffic controller.

Learning Objectives

- Describe the range of information provided by the ASC/3 controller
- Describe the effect of detector calls on controller timing processes

Required Resource

- Movie file: A19.mp4

Deliverable

- Prepare a document that includes your answers to the Critical Thinking Questions

Critical Thinking Questions

1. What are examples of the data provided in the controller status display screen?

2. How many rings can be accommodated by the ASC/3 controller?

3. How do you know whether a gap out or a max out has occurred?

4. How can you verify that a vehicle call has been placed?

5. Describe some of the observations that you have made on the response of the controller timing processes to vehicle calls.

INFORMATION

There are a number of traffic controllers that are in common use today, including those manufactured by McCain, Siemens, Peek, Naztec, and Econolite. There are three primary types, Type 170 and 2070 controllers, NEMA controllers, and advanced traffic controllers (ATC). In this activity, you will discover (explore) a NEMA controller produced by Econolite known as the ASC/3.

The features and functions of the ASC/3 are not dramatically different from other actuated traffic controllers. Using the ASC/3 controller emulator, including its functionality and the kinds of information that it displays, will help you to understand more about traffic signal timing and the controller processes that determine how effective a timing plan will be. While other types of controllers are used in this book, the ASC/3 controller emulator provides a transparent way to learn more about traffic controllers in general.

The video that you will watch as a part of this activity shows the ASC/3 controller in operation (see Figure 83). The left portion of Figure 83 shows the display for a tool that is commonly used to test the functions of a traffic controller known as a "suitcase tester." This tester derives its name because it is often stored in a suitcase box for easy portability. The display status for each vehicle phase is shown at the top of the screen (green, yellow, red). In this example, phase 2 is currently green while the other phases display red. Detector calls can be placed by clicking on the boxes at the bottom of the left portion of the screen (where the cursor is shown in the figure).

The right portion of the figure shows the status display screen for an ASC/3 controller, as you would see it in the lab or in the field. For each phase, you can see the phase status and the presence (or not) of a vehicle call. Phase 2 is currently active ("G" for green). You can also observe the status for various timers for rings 1 and 2, and as well as the phase that is currently active. Here, phase 2 is timing in ring 1 and the vehicle extension timer (EXT1) is at 2.8 seconds. The maximum green timer (MAX1) is at zero.

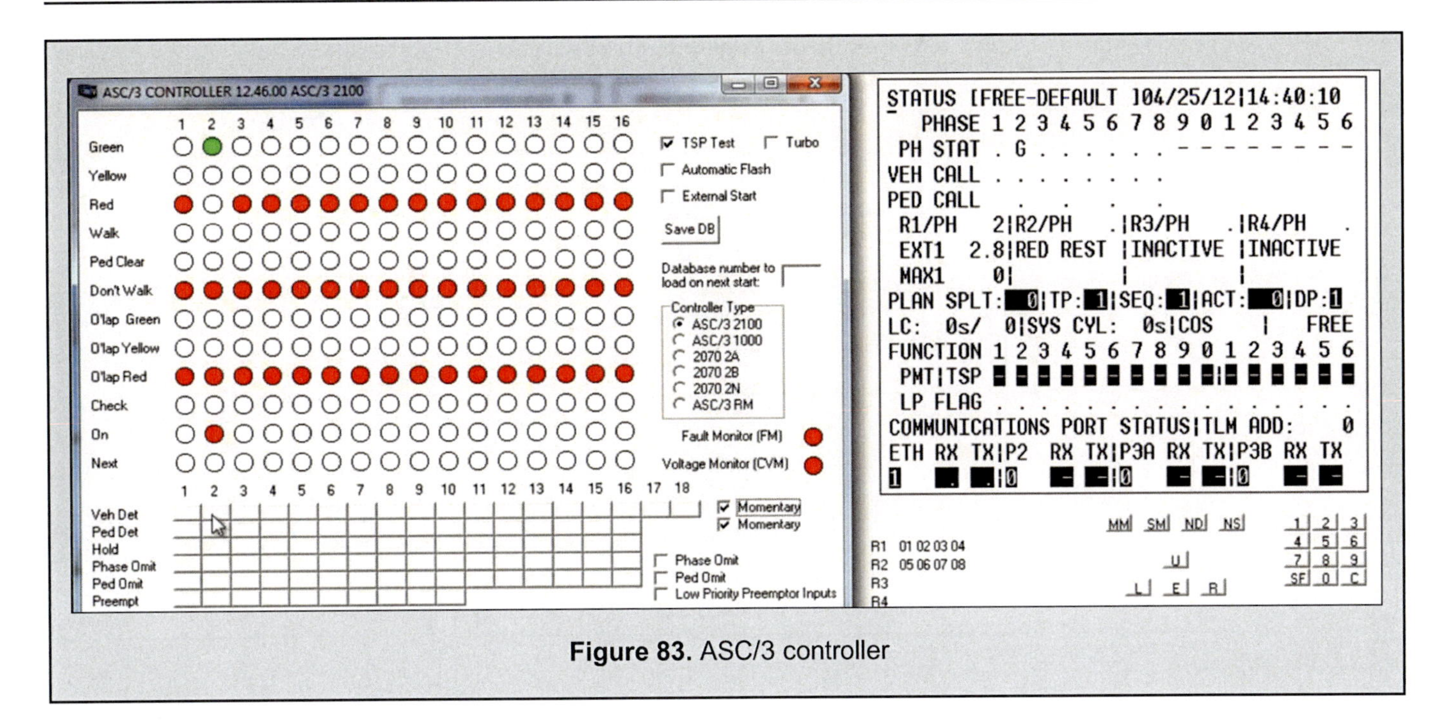

Figure 83. ASC/3 controller

Task 1

Open the movie file A19.mp4. Watch it from the beginning through t = 0:55. Familiarize yourself with both the suitcase tester display (on the left portion of the video screen) and the controller status display screen (on the right side of the video screen). Take notes on what you observe and record these notes in Table 9 (see the following page).

Task 2

The video from t = 0:58 to t = 2:30 shows a series of actions (detector calls placed) and responses (timing process changes). For the first part of this segment (0:58 – 1:12) constant calls are placed on phases 2, 4, 6, and 8. For the second part (1:13 – 1:20), there are responses to these calls. Take notes on the responses that you observe.

Task 3

From t = 2:40 to t = 5:00, there are a series of detector calls placed and controller responses that result. Closely observe the four segments in this time interval and record the controller responses that you observe in Table 9.

Task 4

Document your answers to the Critical Thinking Questions.

Table 9. ASC/3 controller observations

Video time interval	Detector calls	Controller responses/other notes
0:00 – 0:55		
0:58 – 1:12		
1:13 – 1:20		
2:40 – 3:00	Call on phase 2 at t = 2:55	
3:15 – 3:30	Calls on phases 4 and 8 at t = 3:21 – 3:23	
3:30 – 3:40	Call on phase 2 at t = 3:37	
3:40 – 4:40	Calls on phases 2, 4, and 6 between t = 3:46 and 4:35	

How a Traffic Phase Times and Terminates

PURPOSE

The purpose of this activity is to give you the opportunity to observe the timing of a traffic phase and the method by which the phase terminates.

LEARNING OBJECTIVE

- Describe the two primary methods for the termination of a traffic phase at an isolated intersection

REQUIRED RESOURCE

- Movie file: A20.wmv

DELIVERABLE

- Prepare a document that includes answers to each Critical Thinking Question based on your observations from this activity

CRITICAL THINKING QUESTIONS

1. Why does the phase terminate for each of the two cases that you observe?

2. What is the process followed by the minimum green timer from the beginning of the green indication, until the timer expires?

3. What is the process followed by the vehicle extension timer from the beginning of the green indication, until the timer expires?

4. What is the process followed by the maximum green timer from the beginning of the green indication, until the timer expires?

5. What are the two conditions that separately cause the termination of the green indication?

6. Reflect on what you have observed on how a phase terminates. Write a summary of your observations.

Information

You will observe the southbound approach (phase 4) of the intersection of State Highway 8 and Line Street. This approach (Line Street) has two lanes, a left turn lane and a through/right turn lane. State Highway 8 is the major street and serves as a primary east-west route through the city. It also serves as the major access to a university. See Figure 84. You will monitor traffic on the through/right turn lane of this approach.

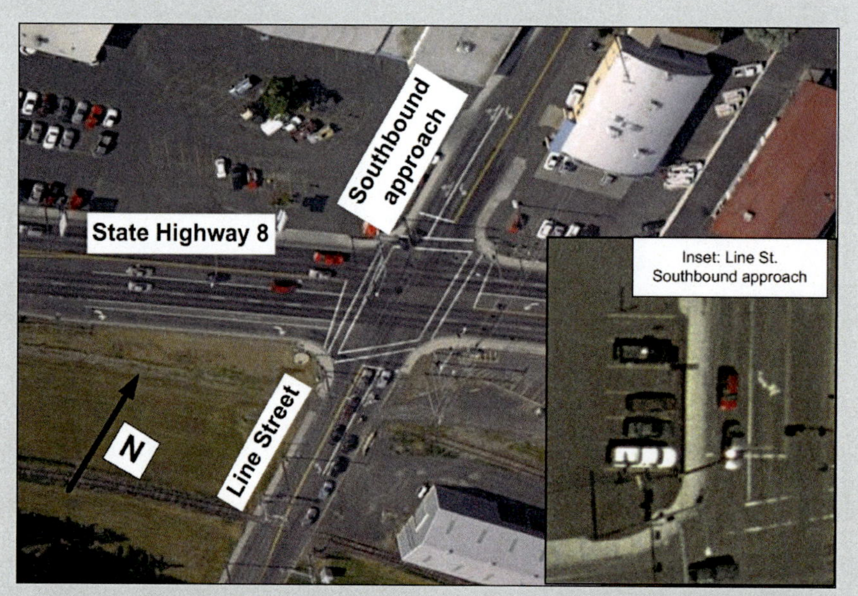

Figure 84. Aerial photograph of State Highway 8 and Line Street

In this activity, you will consider two cases, each illustrating a different method for the termination of phase 4 (which serves the SB through/right turn movements). You will observe how the phase times (the timing processes for the minimum green, vehicle extension, and maximum green timers), and how it terminates for each case. The two cases have been placed side-by-side in a movie format so that you can observe the traffic flow and timing processes at the same time. The simulation has been set to run at less than real time, slow enough so that you can observe all timing and traffic flow processes.

Task 1

Open the movie file: "A20.wmv." Pause for a moment when you've opened the movie file. Look at the screen and see what you can observe. Note that there are always four kinds of information:

- The traffic flow conditions
- The status of the detectors (active or off)
- The status of the controller and the various timing processes
- The status of the signal display (red, yellow, and green)

Make a list of the various items that you see in this screen. Why do we record the observations that you make? One of the most important skills that we want you to develop is to learn to observe and make judgments about how well (or not) traffic is flowing based on what you observe about the traffic flow, the detector status, the controller processes, and the signal display.

Task 2

Observe the status at the beginning of phase 4 green. Move the animation to t = 45.6 seconds (which is equivalent to about 00:23 on the Windows Media Player clock). See Figure 85. Observe the following conditions for the scene on the left for the southbound approach.

- Two vehicles in queue, one of which is in the detection zone
- The red indication showing for the southbound movement (which is about to end)
- The ASC/3 controller status screen showing that phase 4 is just about to begin timing ("T") and has an active call ("C") from the detector
- The ASC/3 controller status screen showing that phase 4 is active in ring 1, that the minimum green timer is at 5.0 seconds, and that the maximum green timer is not active ("0.0")

Note: There is a slight time delay between the ASC/3 controller and VISSIM, the simulation model. While the controller is now timing minimum green, the simulation will not be updated for a fraction of a second. As soon as you move forward from this instant in time, the simulation will show a green indication. These brief differences between what the controller is doing and what the simulation is displaying will only occasionally be noticeable (as when the simulation is paused).

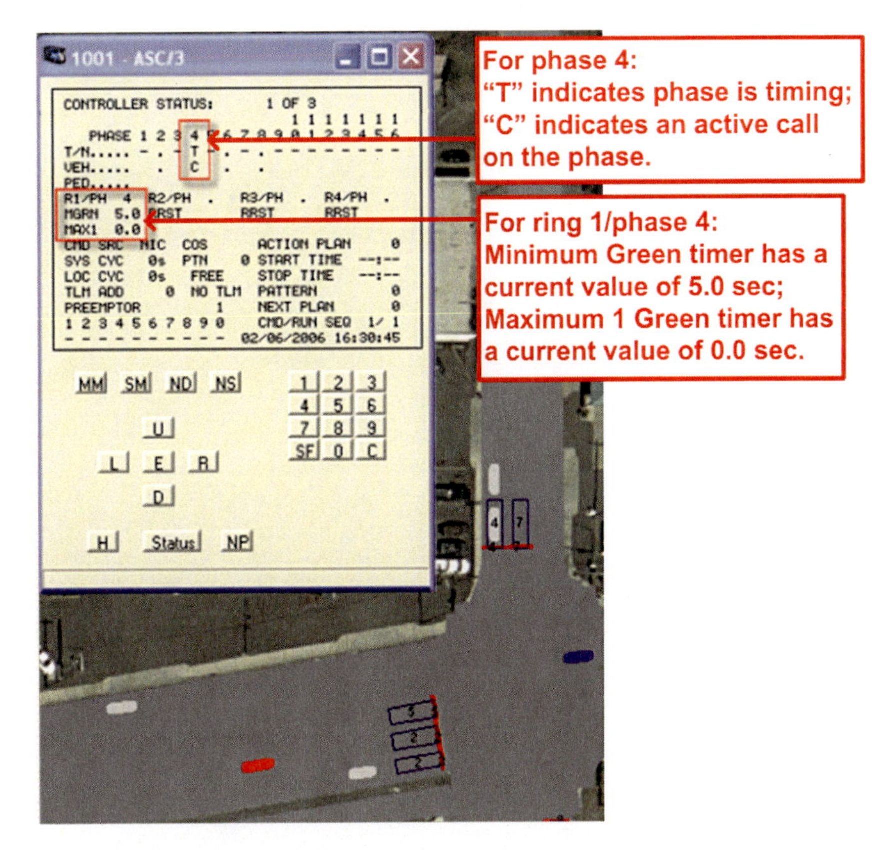

Figure 85. Traffic flow and controller status at t = 45.6

TASK 3

Observe the two cases for one green indication. [Note: Remember, the animation that you are about to observe plays at much slower than real time. This will allow you to monitor the traffic flow and the timing processes at the same time!]

- Start the Windows Media Player animation
- First, watch the traffic flow, the detector status, and the timing processes for phase 4 for the case on the left. Monitor these conditions for the entire green indication
- When this green indication is nearly complete for the case on the left, consider the reason for the phase terminating
- Now turn your attention to the case on the right. The green indication is just beginning for this case. Again, monitor these conditions for the entire green indication.
- Note when the green indication ends for each of the two cases, and why

ACTIVITY

21 Exploring a Controller Emulator

PURPOSE

The purpose of this activity is to improve your understanding of the operation of an actuated traffic controller system by studying a traffic controller emulator.

LEARNING OBJECTIVE

- Describe actuated traffic controller timing processes

REQUIRED RESOURCE

- Spreadsheet file: A21.xlsm

DELIVERABLE

- Prepare a document that includes a brief summary of what you've learned from studying the controller emulator, including answers to the Critical Thinking Questions

INFORMATION

You will use an Excel spreadsheet emulator to learn more about the operation of an actuated traffic signal controller. The spreadsheet (Figure 86) shows two intersecting one-way streets, a detector for each street, and the timers that are activated when a detector call has been initiated. The detection type is pulse, which means that the call is made and is not held after each actuation.

CRITICAL THINKING QUESTIONS

1. How and when do the phases terminate when no detector calls have been placed?

2. When calls are placed continuously only on the NB approach, how and when does the northbound phase terminate?

3. When calls are placed continuously on both the northbound and westbound approaches, how and when does the northbound phase terminate?

4. How does pulse detection differ from presence detection and how does this difference affect the timing processes that you see in this controller emulator?

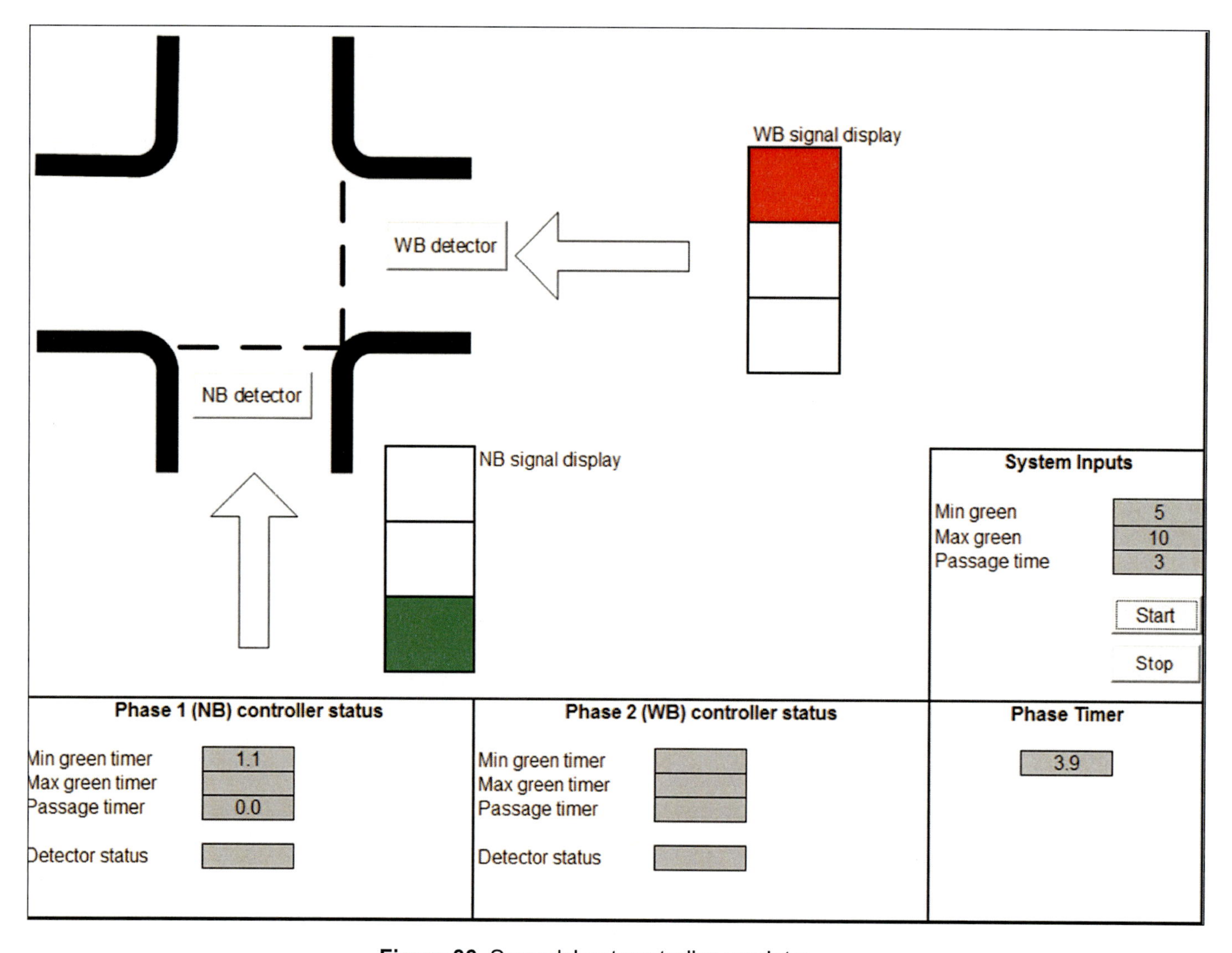

Figure 86. Spreadsheet controller emulator

Task 1

Run the Excel emulator without any detector calls. When and how do the phases terminate?

Task 2

Run the Excel emulator placing continuous calls only on the northbound approach. How long does the northbound phase run and why does it terminate?

Task 3

Run the Excel emulator placing continuous calls on both the northbound and westbound approaches. How long does the northbound phase run and how does it terminate?

Constructing a Traffic Control Process Diagram

Purpose

The purpose of this assignment is to help you improve your understanding of the operation of an actuated traffic controller system by studying eight cases of signal timing settings and preparing a traffic control process diagram for each case.

Learning Objective

- Describe actuated traffic controller timing processes

Deliverable

- Prepare a document that includes completed charts (Figure 87 through Figure 94), your answers to the Critical Thinking Questions, and a brief summary of what you have learned about the interrelationship between the detector and timing components of an actuated traffic controller.

Critical Thinking Questions

1. What questions do you still have on signal timing processes after completing this activity?

2. Can unused green time (the time after the last vehicle passes through the intersection and the onset of yellow) be effectively used? Describe some of the issues that you considered in your answer.

3. If you have to redraw any of the vehicle trajectories, how does this relate to the four interrelated steps in the traffic control process diagram that is first described in Chapter 1?

INFORMATION

traffic control process diagrams are presented for eight different cases. For each of these cases you are given the trajectories of one or more vehicles approaching and traveling through the intersection as well as the values for three timing parameters: the minimum green time, the passage time, and the maximum green time. For four of the cases, the detection zone for the active phase is six feet; for the other four cases, the length of the zone is 40 feet. One of the most important concepts in completing a traffic control process diagram is to note (as we first did in Chapter 1) the interrelationships of each of the components:

- The user is detected
- The detector sends this information to the controller
- The controller (through timing processes and control logic) determines the appropriate display
- The user responds to the display

TASKS

Notes:

(1) You may have to redraw the vehicle trajectories in response to changes in the display status.

(2) The status of the detector for a conflicting phase is given: there is an active call on a conflicting phase when the area is hatched; there is no active call when the area is blank.

TASK 1

Complete the detector responses, timer responses, and signal display responses for each of the eight cases that follow. The conditions for each case are shown in the lower right of each figure. Assume that the green time begins at $t = 3$ and that yellow time = 3 seconds and red clearance time = 1 second.

TASK 2

Record the unused green time and the percentage of vehicles that are served for each of the eight cases.

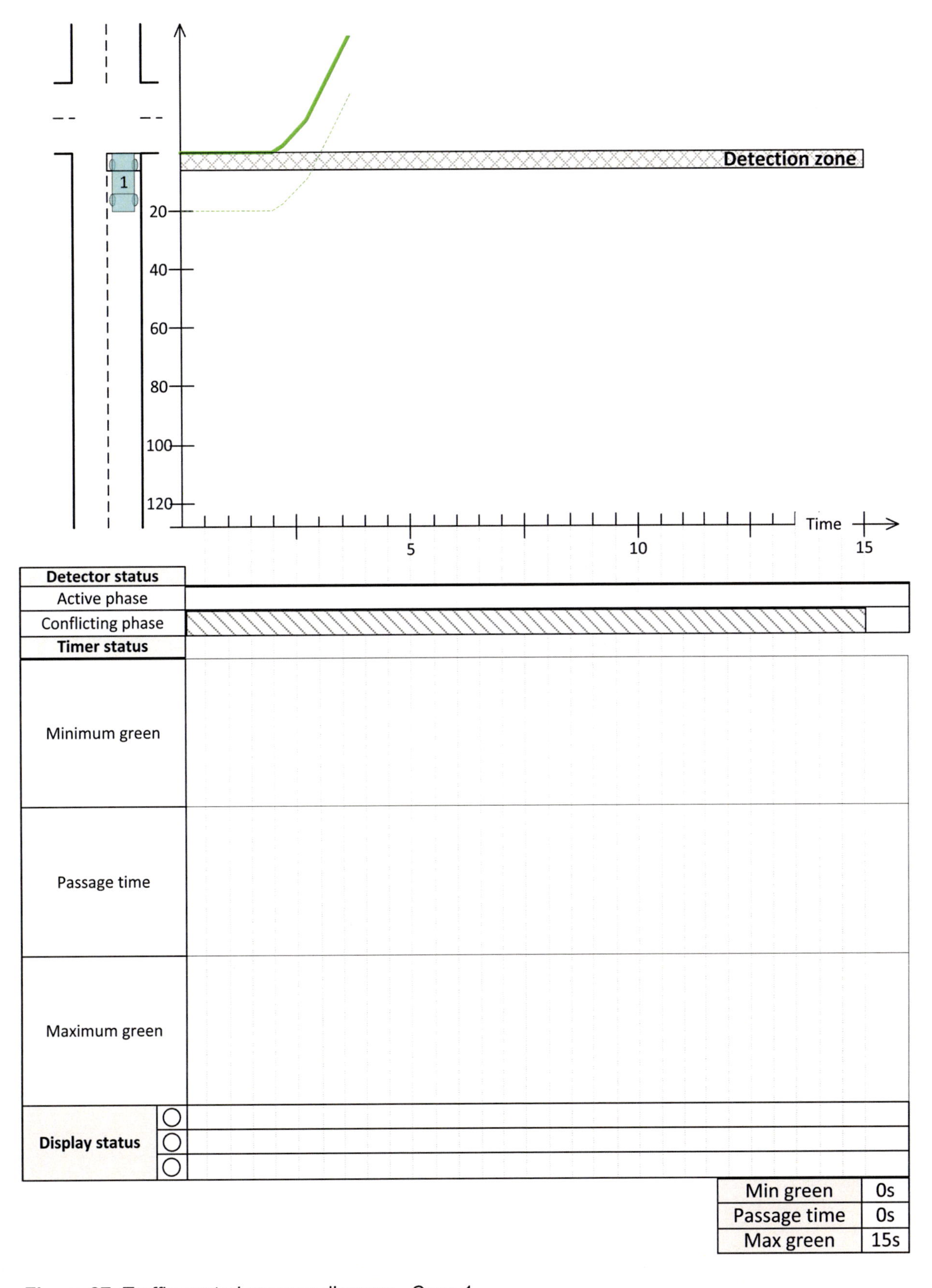

Figure 87. Traffic control process diagram - Case 1

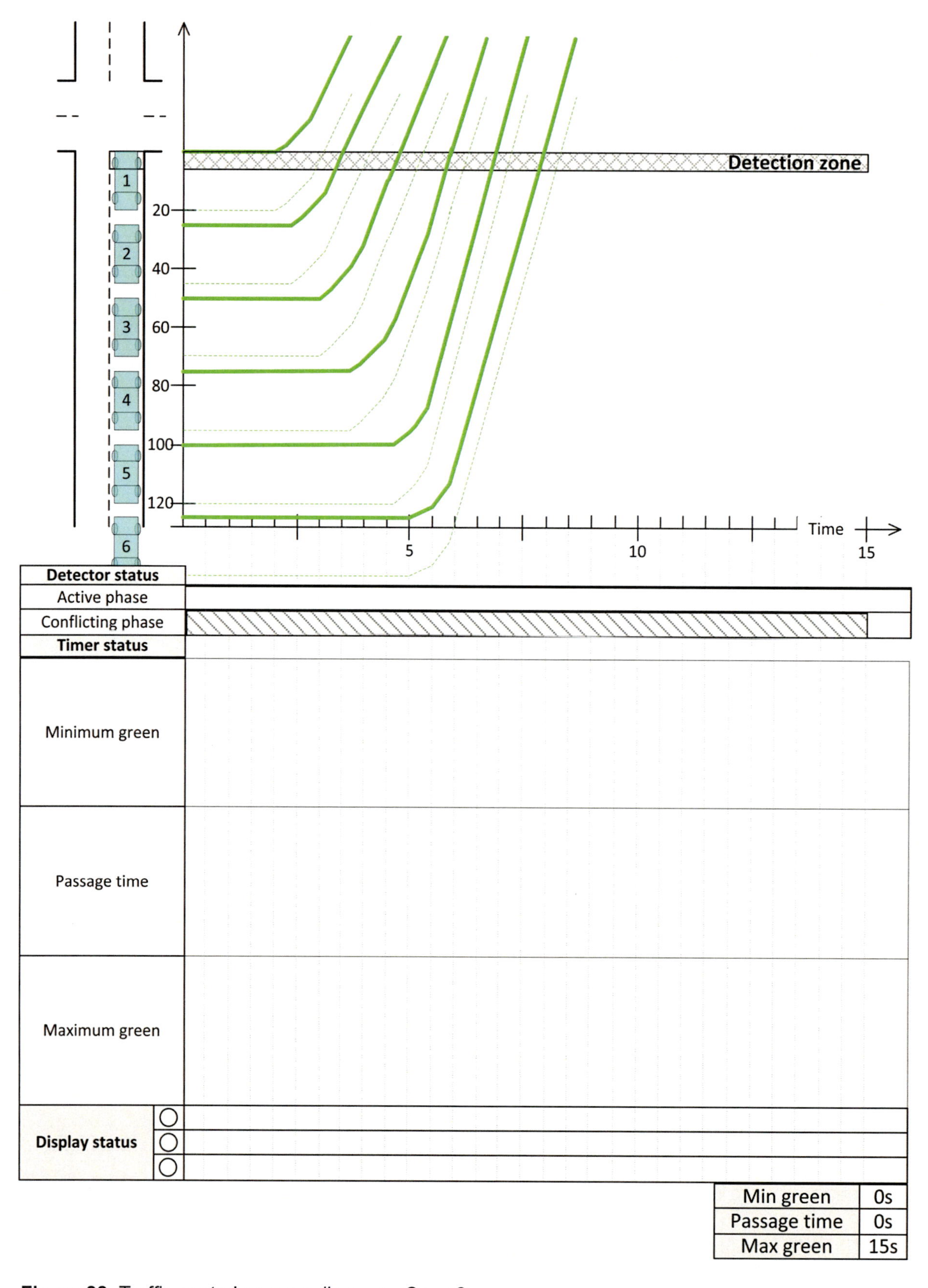

Min green	0s
Passage time	0s
Max green	15s

Figure 88. Traffic control process diagram - Case 2

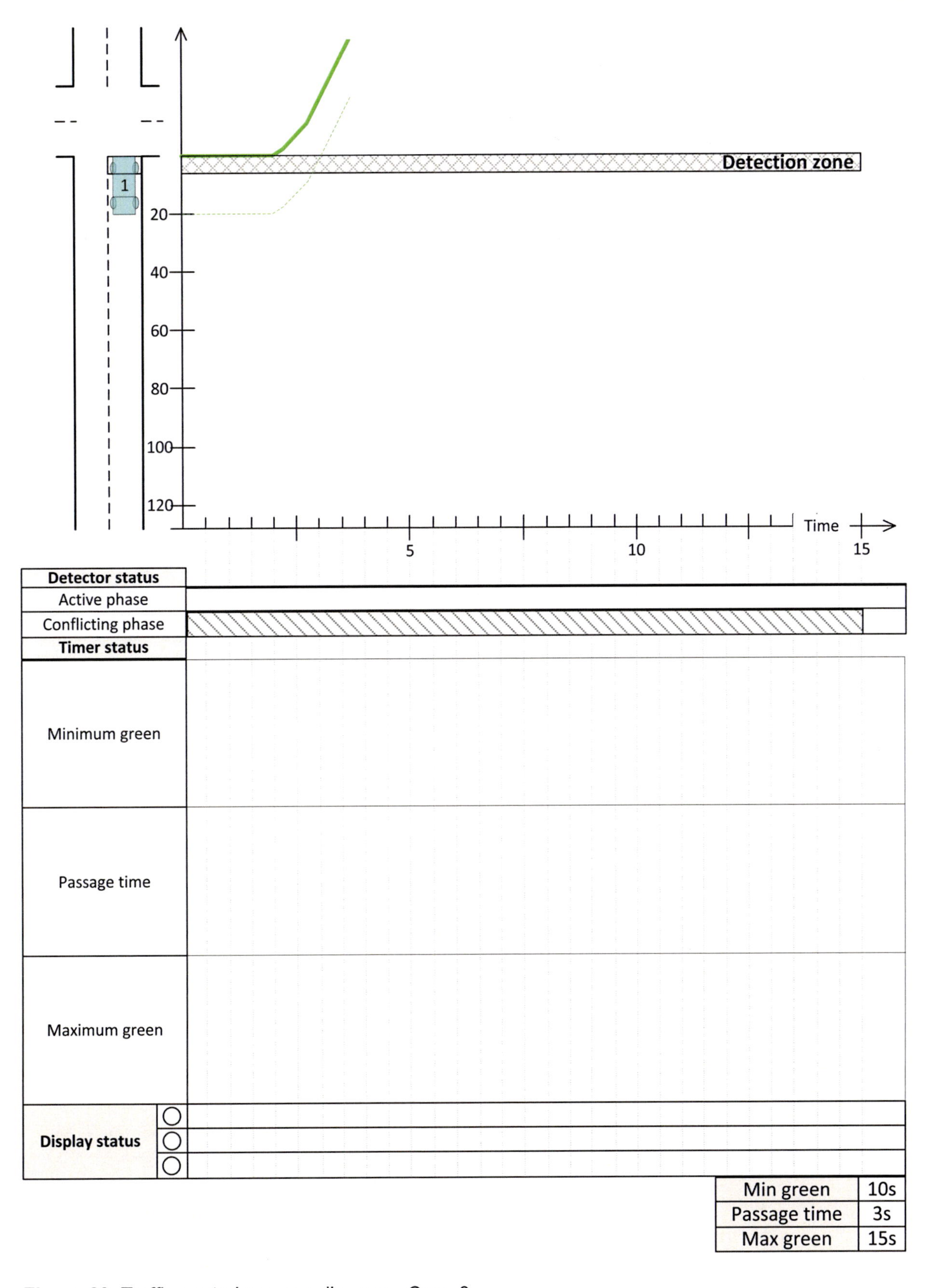

Min green	10s
Passage time	3s
Max green	15s

Figure 89. Traffic control process diagram - Case 3

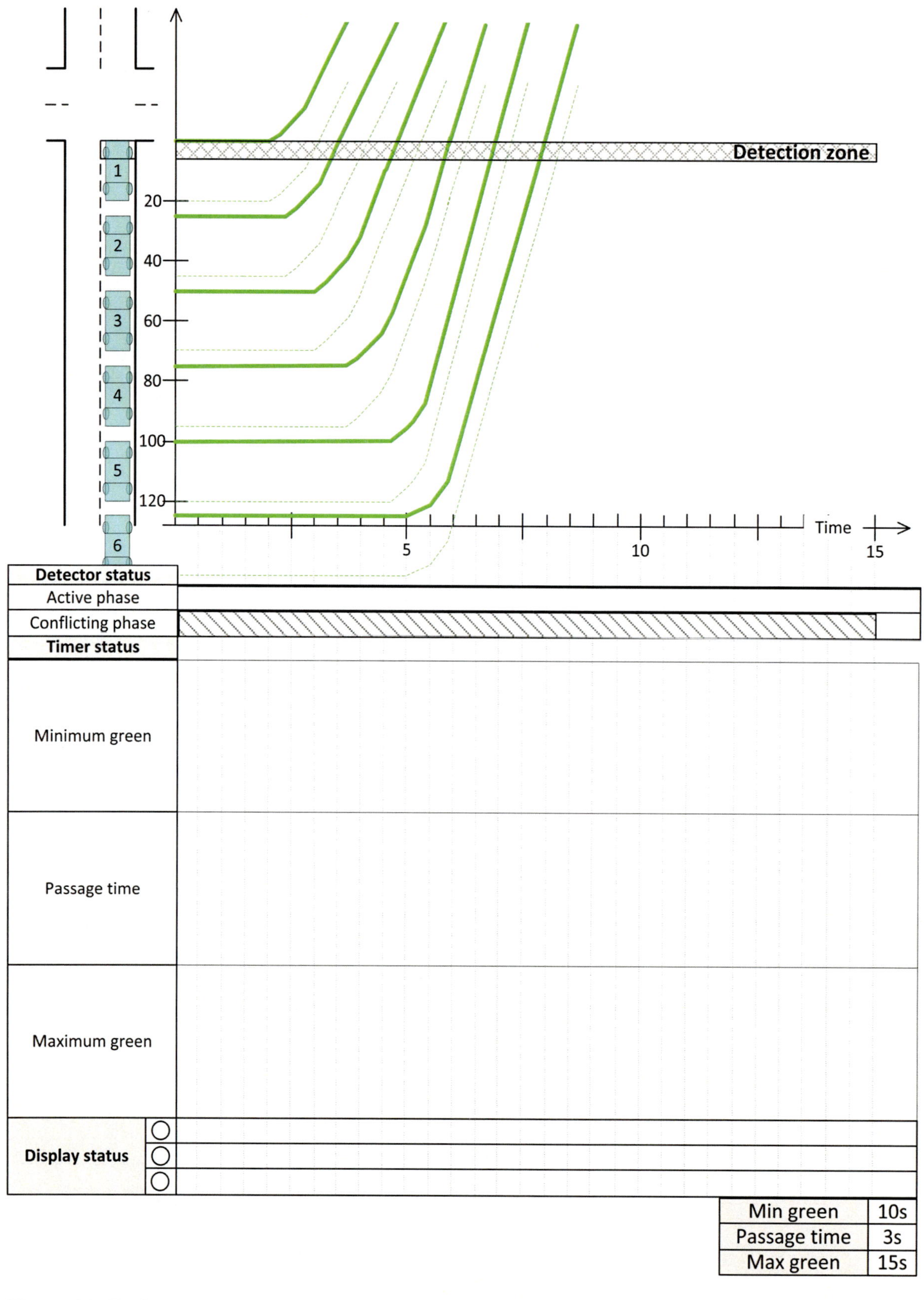

Figure 90. Traffic control process diagram - Case 4

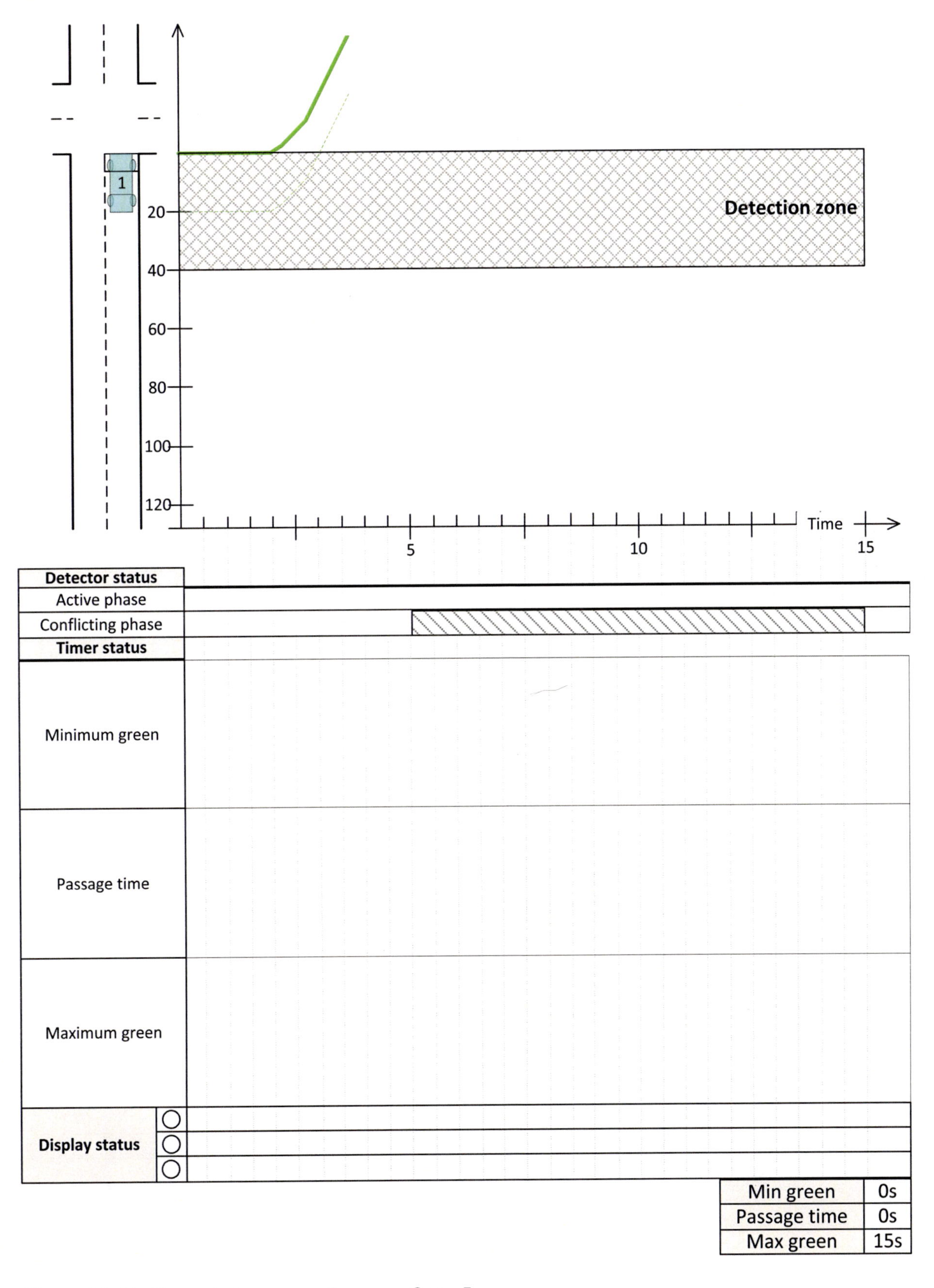

Figure 91. Traffic control process diagram - Case 5

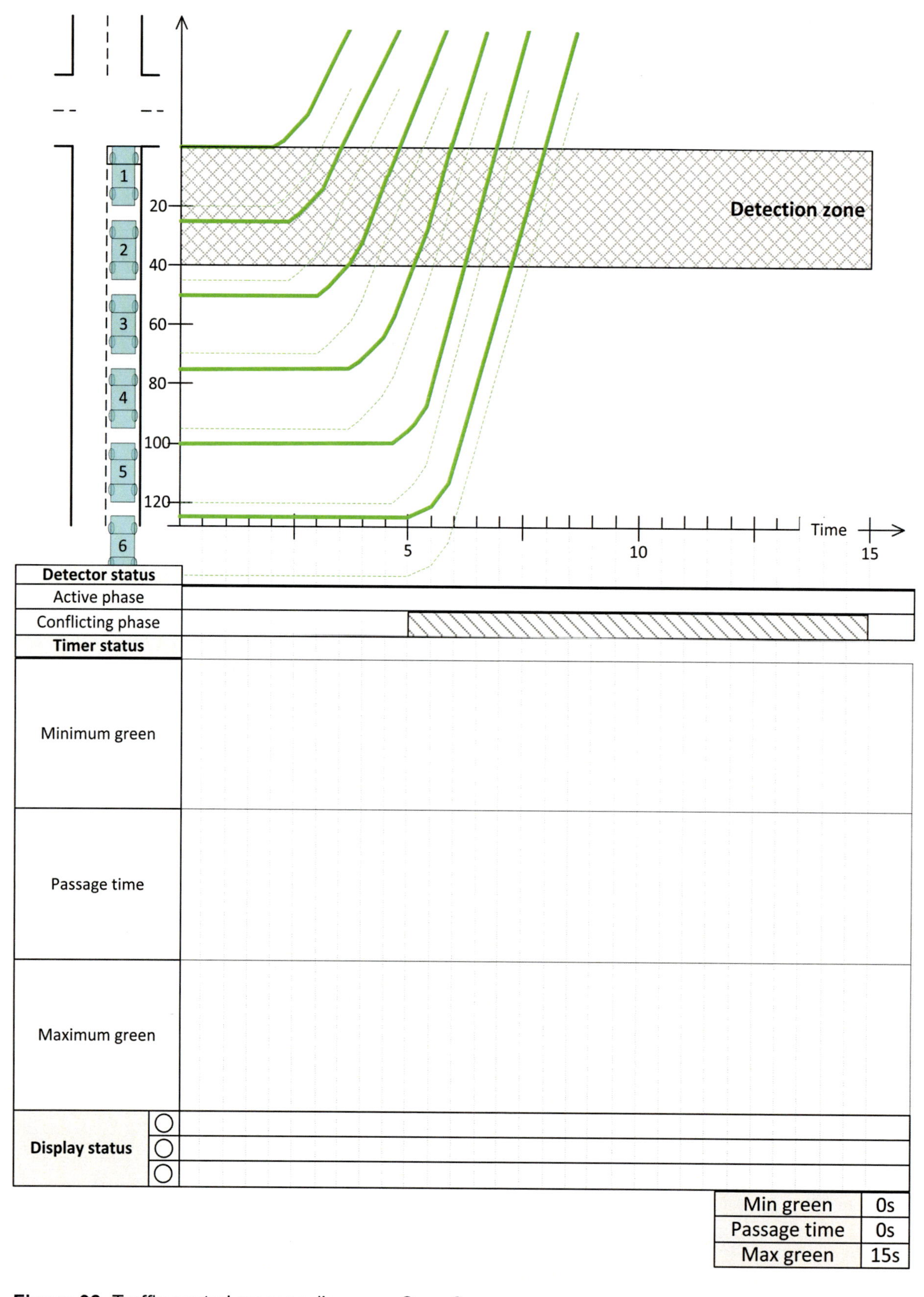

Figure 92. Traffic control process diagram - Case 6

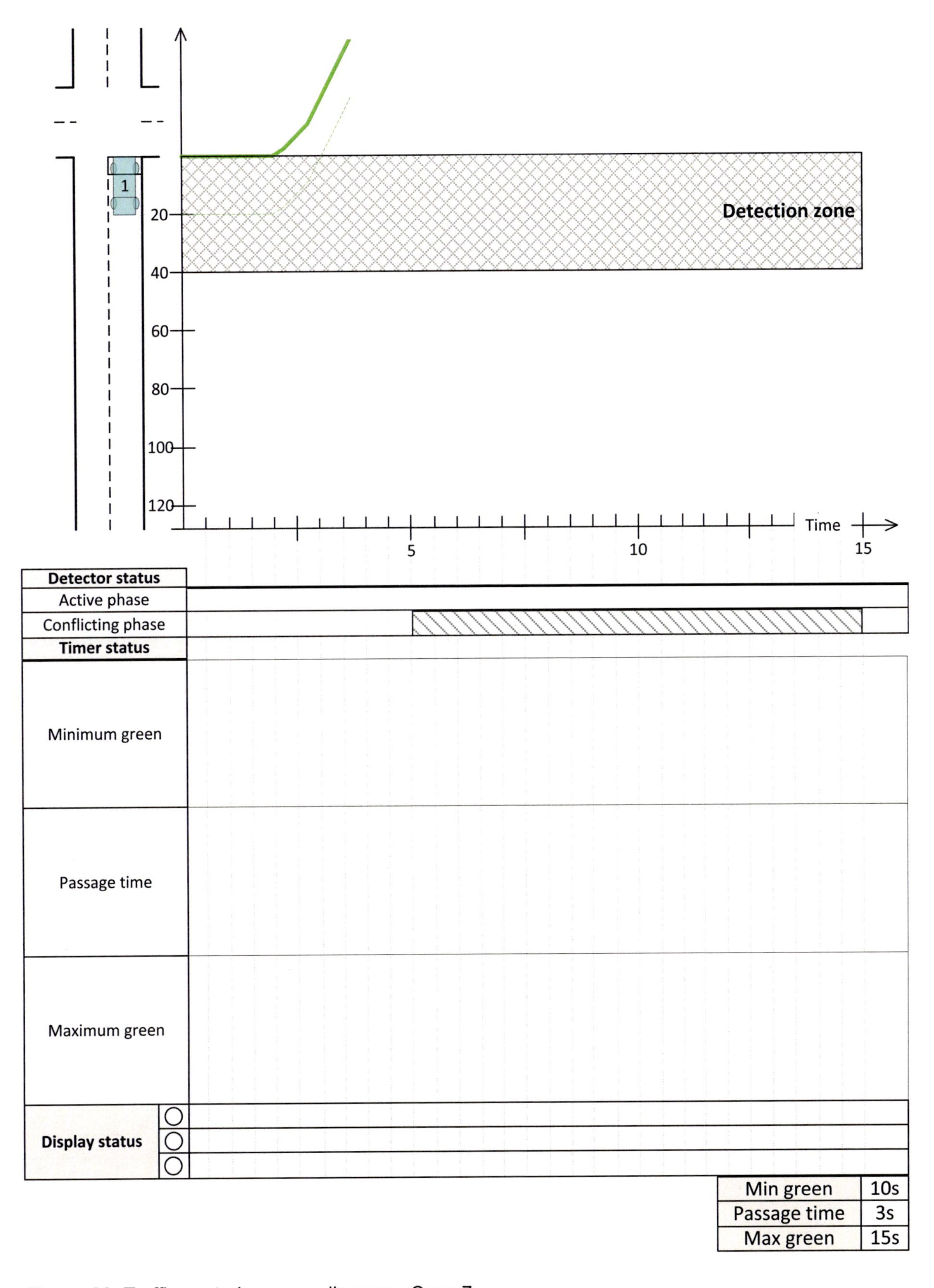

Min green	10s
Passage time	3s
Max green	15s

Figure 93. Traffic control process diagram - Case 7

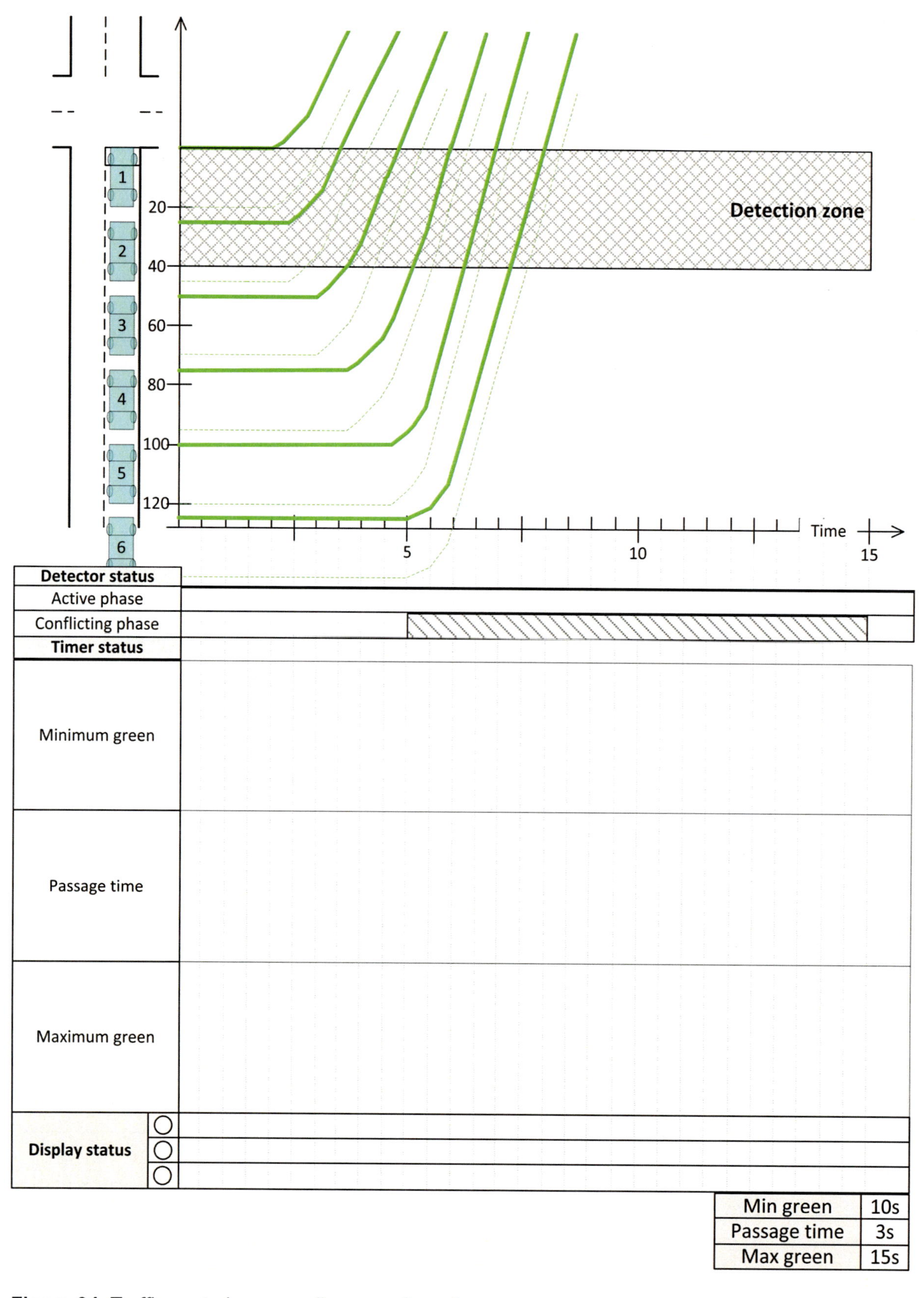

Min green	10s
Passage time	3s
Max green	15s

Figure 94. Traffic control process diagram - Case 8

ACTIVITY 23 Inferring Signal Timing Parameter Values

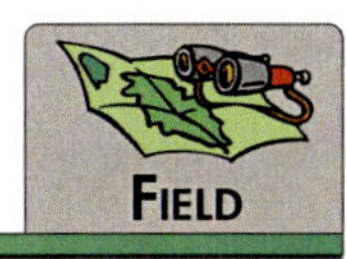

PURPOSE

The purpose of this activity is to test your understanding of traffic controller operations by inferring the value of the three standard signal timing parameters by observing traffic flow and signal displays in the field.

LEARNING OBJECTIVE

- Infer signal timing parameter values through field observations

DELIVERABLE

- Prepare completed charts with data as required for the following tasks, including inferred values of the signal timing parameters

INFORMATION

You will make field observations for one approach of the same intersection that you studied in Activity #15 (in Chapter 3) and for which you constructed a ring barrier diagram. You will record the detector status data and the signal display data for one approach of that intersection using forms shown in Table 10a and Table 10b. A major challenge of this activity is to infer the values of the timing parameters solely through observation of the vehicle arrival patterns and the signal displays on the approach that you are observing.

As you are collecting your data, look for traffic conditions that would allow you to observe the minimum green time, such as a queue of just one vehicle. Also, note that pedestrian calls may affect your timing observations, so it is best to not collect data when there is a pedestrian call.

CRITICAL THINKING QUESTIONS

1. One complicating factor in the determination of your timing values is the interaction of your approach with the opposing approach. Describe how this interaction may affect the conclusions that you make about the timing values that have been set for your approach for the vehicle extension time.

2. Why is the maximum green time difficult to determine if the volumes on your approach are low?

3. What changes would you make in the way in which you made your observations and collected your data to improve the precision of your estimates?

Task 1

Document the location of the detector or detectors on one approach of your intersection.

Task 2

For a period of 15 minutes, record the following data for this approach:

- The times that each detector changes its state (from on to off, and from off to on) based on your knowledge of where the detection zone is placed on the approach
- The times that the signal display changes its state (when green starts, yellow starts, and red starts)

Task 3

Prepare traffic control process diagram sketches (see Figure 95) for enough cycles for which you've collected data to show how you've inferred the signal timing data, noting the detector status as well as the signal display status data. Based on these sketches, estimate the values for the minimum green time, the passage time, the maximum green time, the yellow time, and the red clearance time.

Table 10a. Detector state data collection form

Time	Detector state (on/off)

Table 10b. Display state data collection form

Time	Display/color indication

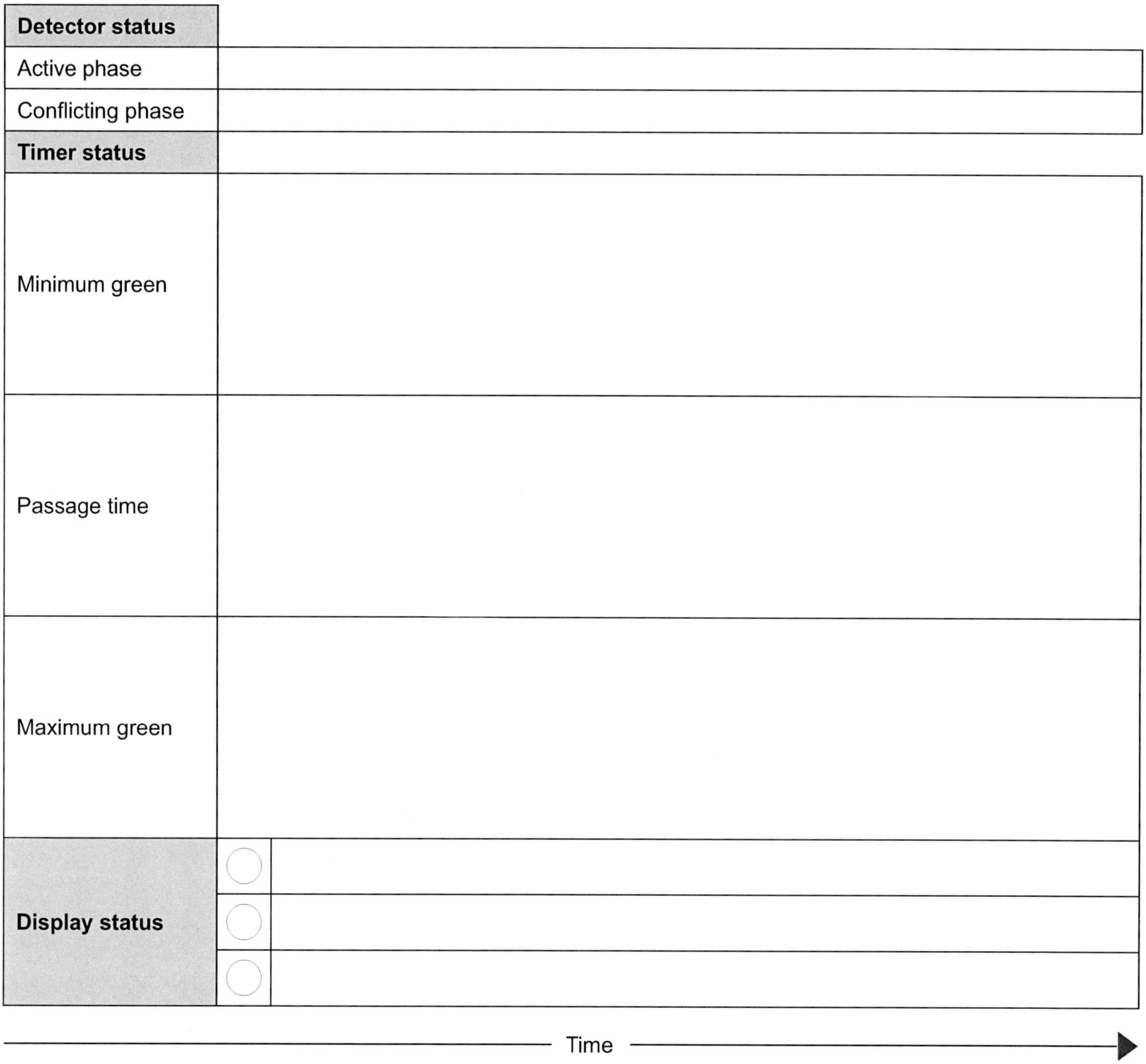

Detector status		
Active phase		
Conflicting phase		
Timer status		
Minimum green		
Passage time		
Maximum green		
Display status	◯	
	◯	
	◯	

Time →

Figure 95. Traffic control process diagram for field data observations

ACTIVITY

24 Signal Timing Parameters

PURPOSE

The purpose of this activity is to give you the opportunity to learn how the *Traffic Signal Timing Manual* describes some of the signal timing parameters that you studied in this chapter.

LEARNING OBJECTIVE

- Contrast the description of the signal timing terms that are presented in this chapter with those described in the *Traffic Signal Timing Manual*

REQUIRED RESOURCE

- *Traffic Signal Timing Manual*

DELIVERABLES

Prepare a document that includes

- Answer to the Critical Thinking Question
- Completed Concept Map

LINK TO PRACTICE

Read the sections of the *Traffic Signal Timing Manual* assigned by your instructor.

CRITICAL THINKING QUESTION

When you have completed the reading, prepare an answer to the following question:

1. Describe how the timing processes that you observed in the field (Activity #23) compare with their descriptions in the *Traffic Signal Timing Manual*.

In My Practice...

by Tom Urbanik

Stop bar detection can be very efficient in allocating green time if the detection zone is large and the passage time is small (zero or close to zero). If the passage time is zero, the approach turns yellow as the vehicle leaves the stop bar. However, the onset of yellow occurs when the vehicle is well into the intersection. To achieve an even more efficient operation (and also provide dilemma zone protection which is not discussed here), separate setback (from the stop bar) detection is used.

The design of setback detection is complex. However, to illustrate the point, consider a detector located 3 seconds from the stop bar. If the gap time is 2 seconds, the vehicle is approximately 1 second from the stop bar if it is the last vehicle to gap out. At 1 second from the stop bar, there is no question that the vehicle will enter the intersection as it will take 1 second to react to the yellow. This type of operation is safer and more efficient.

Concept Map

Terms and variables that should appear in your map are listed below.

gap out	maximum green	passage time
max out	minimum green	

Student Notes:

CHAPTER 5 The Simulation Environment: Learning to See a Traffic Signal System

PURPOSE

In Chapter 5, you will learn how the simulation environment can help you to visualize traffic control processes. Simulation, or microsimulation as it is sometimes called, is a very detailed and realistic representation of a transportation system. It is detailed enough that you can directly view individual vehicles traveling along an arterial or through an intersection. Many simulation models even provide for a three-dimensional perspective of a network. In microsimulation models, the decisions of individual drivers are modeled: how fast to travel, how far behind a leading vehicle to follow, when to change lanes, when to stop when the yellow indication is displayed, and when to accelerate when the green indication is displayed. These are the kinds of driver decisions that are modeled in a microsimulation model such as VISSIM, often every tenth of a second in simulation time.

But your work is not about learning how to use a specific simulation model, though the activities that you will complete are conducted using the VISSIM microsimulation model. Nor is it about using a specific traffic signal controller. Rather it is about observing the flow of individual vehicles and how they interact with the individual controller timing processes. This is the perspective that the traffic engineer has when he or she is standing in the field: how is the traffic flowing and what are the timing processes that we can change in the traffic controller to help make traffic flow better?

You will use this simulation environment to directly see the results of the phasing plans and timing parameters that you select. Using VISSIM's animation and movie files, you will visualize the duration of a green interval, the length of a queue, or the delay experienced by vehicles traveling through a signalized intersection with the phasing and timing plan that you design. And you will use this information to make judgments about the quality of intersection performance, and whether you need to make further adjustments to the signal timing to improve intersection operations. It is almost as good as standing out at an intersection, with one eye on the traffic and the other on what is happening in the controller cabinet.

LEARNING OBJECTIVES

When you have completed the activities in this chapter, you will be able to

- Describe the common categories of transportation models and their attributes
- Describe the characteristics of a microscopic simulation model
- Contrast the performance measures produced by a simulation model
- Describe the categories of traffic analysis tools that are commonly used by a transportation engineer
- Describe the application of a simulation model
- Describe the basic features of VISSIM
- Build and use a simulation model network
- Describe the categories of traffic models

CHAPTER OVERVIEW

This chapter begins with a Reading (Activity #25) on microsimulation models. Three activities follow, including an assessment testing you on some of the basic elements of simulation models (Activity #26), an overview of the VISSIM simulation environment (Activity #27), and the preparation of a complete VISSIM simulation network (Activity #28). This network will be used for much of the testing and evaluation that is a part of the design project that you will complete. In Practice (Activity #29) describes various kinds of traffic analysis tools and some of the basic elements of a transportation simulation model that are used by practitioners.

Activity List

Number and Title	Type
25 Microsimulation Models and the Traffic Control System	*Reading*
26 What Do You Know About Simulation Models?	*Assessment*
27 The VISSIM Simulation Model – Learning Your Way Around	*Discovery*
28 Building a Simulation Model Network	*Design*
29 Traffic Analysis Tools	*In Practice*

Microsimulation Models and the Traffic Control System

Purpose

The purpose of this activity is to give you the opportunity to learn more about microsimulation models and how they can be used to study the traffic control system.

Learning Objectives

- Describe the common categories of transportation models and their attributes
- Describe the characteristics of a microscopic simulation model
- Contrast the performance measures produced by a simulation model

Deliverables

- Define the terms and variables in the Glossary and complete all Glossary items
- Prepare a document that includes answers to the Critical Thinking Questions

Glossary

Provide a definition for each of the following terms. Paraphrasing a formal definition (as provided by your text, instructor, or another resource) demonstrates that you understand the meaning of the term or phrase.

microsimulation model	
network	
performance measure	
traffic analysis tool	
VISSIM	

Critical Thinking Questions

When you have completed the reading, prepare answers to the following questions.

1. Why do you think it is important to use a microsimulation model to evaluate the performance of your design network?

2. What performance measures do you think are important to evaluate the signal timing alternatives that you develop?

3. What can you learn from the numerical performance data and the visual observations of the simulation model to help you evaluate the performance of a signal timing alternative?

Information

What is a Model?

A model is a representation of reality that allows us to study a system, to ask questions about the system and its components, and to change the conditions or features of the system and to observe how it will then behave. A model is especially useful for studying transportation systems since we are often asked to consider a range of possible solutions to a given problem and to see how the system will behave under the conditions of each solution. We can classify a model according to a set of categories, each describing attributes or features of a model. Table 11 lists seven ways in which a transportation model can be categorized.

Table 11. Model categories (adapted from Byrne, de Laski, Courage, & Wallace, 1982)

Model categories	Attributes and contrasts
Computational or simulation	○ **Computational:** Directly computes results from equations or tables
	○ **Simulation:** Tracks events and processes
Empirical or analytical	○ **Empirical:** Based on field data
	○ **Analytical:** Based on theory
Deterministic or stochastic	○ **Deterministic:** Produces same result for given set of inputs
	○ **Stochastic:** Results can vary based on statistical distributions
Microscopic or macroscopic	○ **Microscopic:** Individual driver decisions
	○ **Macroscopic:** Aggregated flow characteristics
Event scan or time scan	○ **Event scan:** Based on status of events of interest
	○ **Time scan:** Updates made every time step
Evaluation or optimization	○ **Evaluation:** Performance data produced
	○ **Optimization:** Objective function optimized based on performance data

We can also classify a model according to the complexity of the problem that it is designed to address, which we can see from Figure 96 is directly proportional to the difficulty of a given model to use. Some common problems considered by transportation engineers include:

- How many lanes will be needed to accommodate traffic volumes projected five years in the future?
- How much will delay increase at adjacent intersections if a new hotel or shopping center is constructed on this site?
- What signal timing plan will work best with this intersection design?

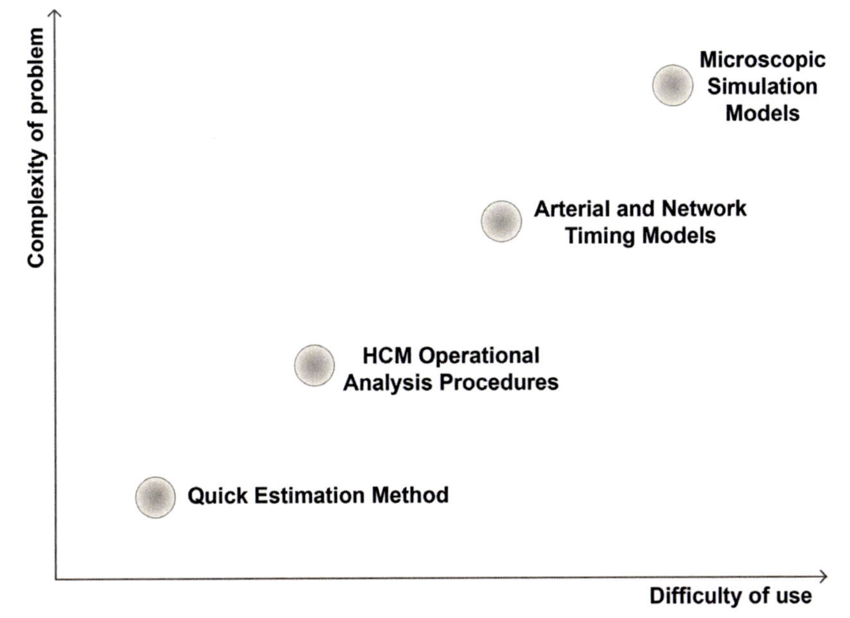

Figure 96. Model classification scheme (adapted from the *Traffic Signal Timing Manual*)

For problems that require only an assessment of whether an intersection has sufficient capacity or not, the Quick Estimation Method (sometimes called the *Critical Movement Analysis*) can be used. For single intersections or when volume is less than capacity, the Highway Capacity Manual (HCM) operational analysis procedures are appropriate. Arterial and network timing models generally produce optimized signal timing plans for a set of signalized intersections. Microscopic simulation models yield a high level of detail on system operation and performance, when details on signal timing plans are important to consider.

It should be noted that the HCM 2010 operational analysis procedure includes a realistic emulation of an actuated signal controller, in which phase sequencing and actuated timing parameters are considered. Thus there is some blurring of the distinctions that appear clear in Figure 96 as the capabilities of models and techniques are increased over time.

Why Use a Microsimulation Model

A microsimulation model *requires* a rich and detailed set of data to describe a traffic facility and its conditions. It requires calibration of the model to these local conditions. But it has the potential to *produce* a rich and complex set of data, both numbers and visuals, which represent how the facility is likely to operate and perform under the given set of conditions. The extent of the data required to run the simulation model, and the time required to develop and test a model network, means that a simulation model should be used to address only the most complex kinds of transportation problems for which very detailed information is needed on system operation and performance. Examples of these kinds of problems include:

- The demand exceeds capacity at an intersection, and the effect of this oversaturation spreads from one intersection to another
- The effect of a specific design element, such as the length of a left turn bay, needs to be tested
- The details of a timing plan for an actuated signal controller needs to be developed, tested, and evaluated

This latter problem is the primary subject of this book and the reason that you will learn to use the VISSIM simulation model.

We also can use a simulation model to visualize the operation of a facility under given conditions, something that is useful for both the traffic engineer and for the general public.

Components of a Simulation Model

A microsimulation model consists of three primary components: (1) input data that describe the facility and conditions of interest, (2) the ways in which users (drivers of passenger cars, pedestrians, or truck drivers, for example) interact with the system or facility, and (3) output data that describe how the system is expected to operate and perform.

There are three kinds of input data describing the facility and conditions, including flow characteristics of the users, the geometry of the facility, and the control system parameters. Examples of these data are given in Figure 97.

User flows	Facility geometry	Control settings
Vehicles/hour	Location of nodes	Phasing plan
Percent trucks	Number of lanes	Timing values
Pedestrians/hour	Lane assignments	Timing settings

Figure 97. Example input data for microsimulation model

A facility is represented by a network, a collection of links (representing streets or arterials) and nodes (representing intersections). Figure 98 shows an aerial view of an intersection while Figure 99 shows its link-node representation. Each link is specified by a length, free-flow travel time, capacity, and other relevant geometric and operational characteristics. A node is represented by its control characteristics such as the type of control and the elements that describe that control.

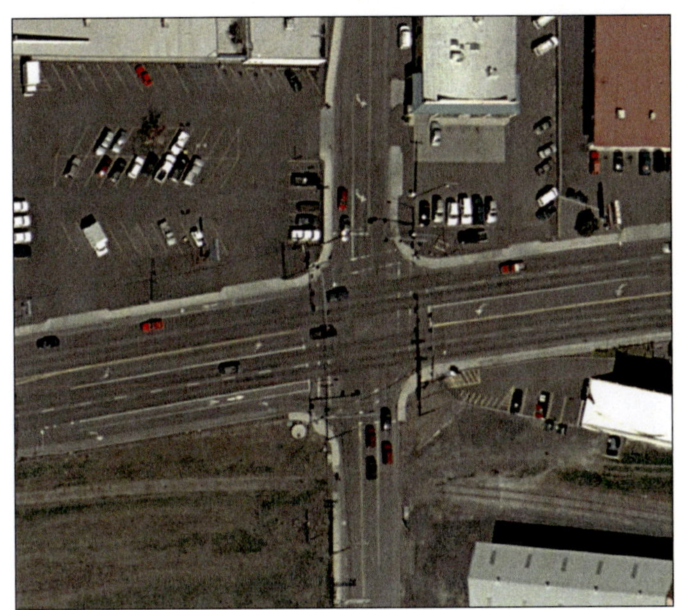

Figure 98. Aerial view of intersection

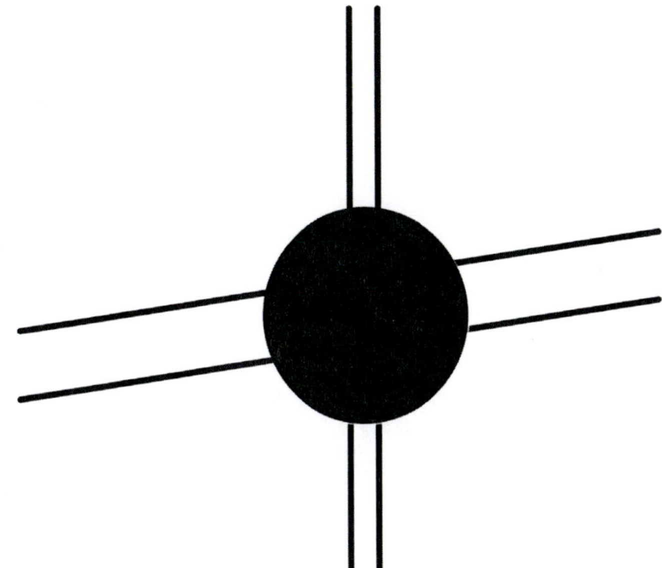

Figure 99. Link-node representation of intersection

Three of the ways in which the users interact with the facility include a range of driver behavior types, the ways in which a "following" vehicle responds to a "leading" vehicle, and the ways in which the user responds to the facility geometry and control system. For example, a user will decide to respond to a change in the control system display (a change in the display from green to yellow) depending on the driver behavior type, the speed that the vehicle is traveling, and the distance that the vehicle is upstream of the intersection. This latter factor was described earlier in this book (in Chapter 1 on the elements of the traffic control system) and is the basis for the construction of the traffic control process diagram.

Performance Analysis Using a Simulation Model

The data produced by a microsimulation model can be extremely detailed. For example, VISSIM can produce performance data for the network, for each intersection and its approaches, and for each movement on the approach. In the system view (Figure 100), the overall speed and delay are shown, as is the time that it takes to travel from one end of the system to the other. The average delay for the intersection approaches and for the individual movements (Figure 101 and Figure 102) provide additional detail that is important in evaluating how the individual components of the system are performing. These figures show a visual comparison of these three levels of data aggregation, each with an important part of the story on how the system is performing.

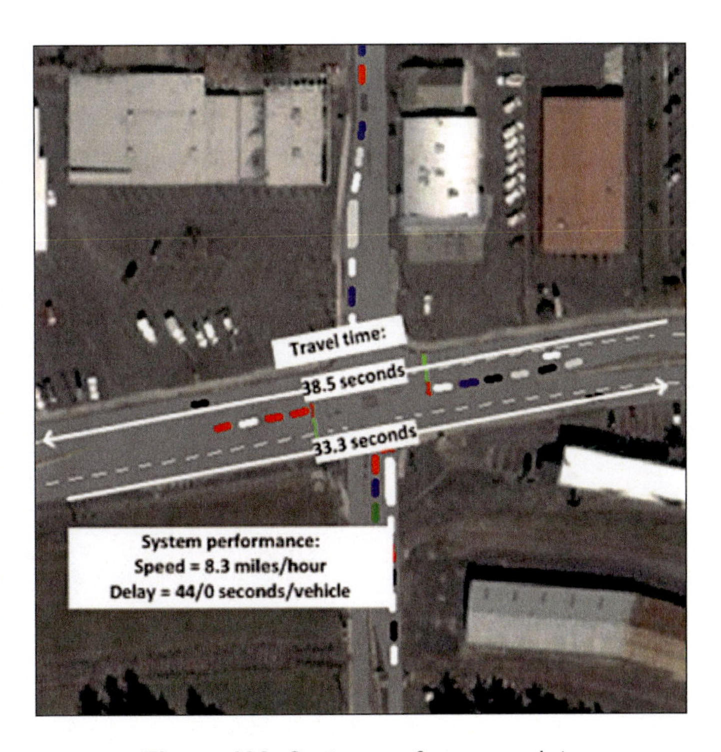

Figure 100. System performance data

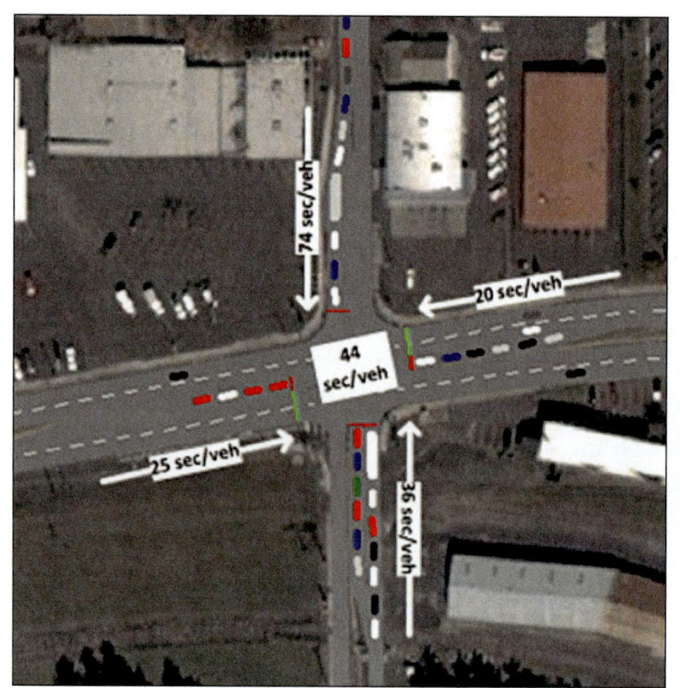

Figure 101. Intersection and approach performance data

Figure 102. Movement performance data

An even more detailed representation of the system operation is shown in Figure 103 and Figure 104, where the position and speed of a single vehicle is shown. On the left, the table shows the position and speed of the vehicle every 0.1 second over a 1.8 second time interval. The figure on the right shows the time-space representation of these data over this same time interval.

Simulation time	x-coordinates	y-coordinates	Vehicle speed
10.0	-31.5	-347.0	33
10.1	-46.3	-344.5	34
10.2	-61.5	-343.1	34
10.3	-76.7	-342.7	33
10.4	-91.5	-343.2	33
10.5	-106.0	-344.8	32
10.6	-120.3	-346.3	32
10.7	-134.7	-348.9	33
10.8	-149.3	-351.4	33
10.9	-164.2	-353.8	34
11.0	-179.2	-356.2	34
11.1	-193.8	-358.9	33
11.2	-208.2	-361.5	32
11.3	-222.3	-364.0	32
11.4	-236.6	-366.6	33
11.5	-251.0	-369.2	33
11.6	-265.8	-371.9	34
11.7	-280.7	-374.5	34
11.8	-295.5	-377.2	33

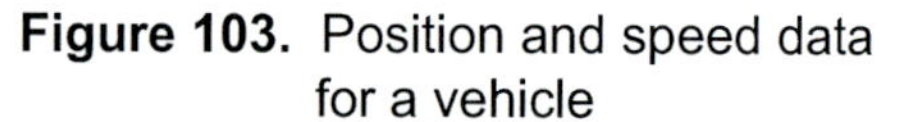

Figure 103. Position and speed data for a vehicle

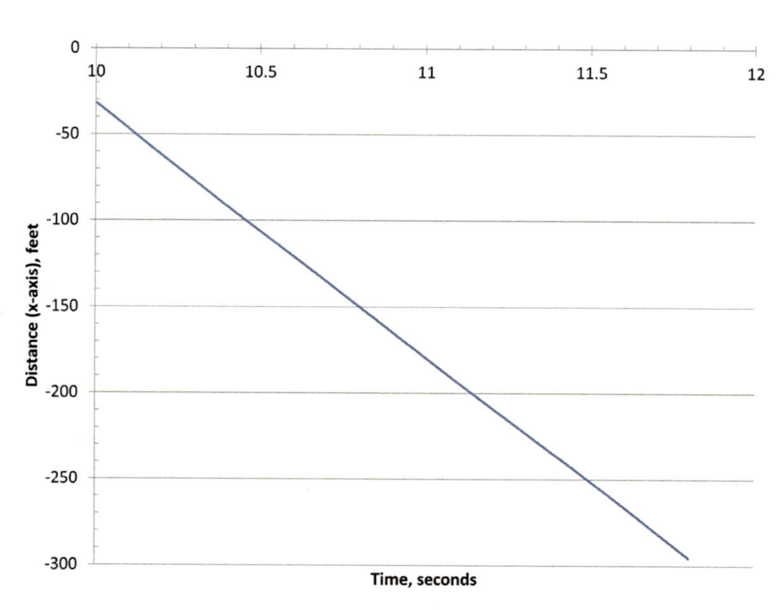

Figure 104. Time-space representation of position and speed

The fact that a microscopic simulation model produces a rich and detailed set of output data is both a strength and a weakness. The strength is clear: we can get a detailed picture of the performance of the system at a variety of levels. The downside is that the construction of this picture is difficult and often time consuming.

For example, suppose we are comparing two alternatives, a base case in which the cycle length has been set to 90 seconds, and a new option in which the cycle length is reduced to 60 seconds and the proportion of green time devoted to each movement is changed. Table 12 shows three system level performance measures in which these alternatives are compared.

Performance measure	Base case (C = 90 sec)	Option 1 (C = 60 sec; green times changed)	Percent change
Speed (mi/hr)	10	14	+37%
Travel time on arterial (sec)	29	21	–27%
Average delay/vehicle	24	16	–33%

Table 12. Example of system level performance data

It can be concluded from this comparison that the average delay per vehicle was reduced by 33 percent and that the average speed was improved by 37 percent. However, we need to look more deeply into the data to get a more comprehensive picture of the results of these changes. Evaluating the performance at the system level may mean that we miss some of the important changes and trade-offs that would occur in this change to the signal timing.

Table 13 shows the average delay per vehicle for each of the four approaches at this intersection. We can begin to see the trade-offs that would actually result from this change. Users on the main street (eastbound and westbound) would see reduction in their delay, while side street users would see an increase in theirs. Two points should be made here. First, we need to be able to define what amount of change in delay is significant and could be perceived by the user. Second, some increase in delay for the minor street may be acceptable for both policy reasons (we often emphasize the quality of flow on the main street) and because volumes are often significantly lower on the side street than on the main street.

Approach	Delay (sec/veh)	
	Base case (C = 90 sec)	Option 1 (C = 60 sec; green times changed)
Northbound	8	13
Southbound	14	21
Eastbound	22	13
Westbound	30	18
Average for system	24	16

Table 13. Example of approach level performance data

Drilling down one more level provides even more insights on the difference in performance of the base case and the signal timing change as shown in Table 14. We can now see that the main street through movements (eastbound and westbound) have improved more significantly than was evident when just looking at the approach data. Further, we may have made delay to the minor street left turn movement too high, unfairly penalizing these users. This may point to a further refinement that is needed to the timing plan that we have developed as part of Option 1.

How do we know the quality of service that is delivered to the various users of this intersection? One standard, given in the Highway Capacity Manual (HCM), provides level of service ranges that can be used to provide a relative performance standard that we can use to compare the base case with Option 1. In this case, the level of service improves for some of the main street (the east-west) movements, while it stays the same or degrades from some of the minor street (north-south) movements.

System component	Delay, sec/veh (level of service)	
	Base case (C = 90 sec)	Option 1 (C = 60 sec; green times changed)
Approach		
Northbound	7 (A)	11 (B)
Southbound	13 (B)	19 (B)
Eastbound	12 (B)	19 (B)
Westbound	13 (B)	17 (B)
Movement		
NBLT	5 (A)	10 (A)
NBTH	11 (B)	15 (B)
NBRT	5 (A)	10 (A)
SBLT	4 (A)	8 (A)
SBTH	20 (B)	30 (C)
SBRT	4 (A)	4 (A)
EBLT	10 (A)	5 (A)
EBTH	30 (C)	15 (B)
EBRT	10 (A)	10 (A)
WBLT	25 (C)	19 (B)
WBTH	35 (C)	20 (B)
WBRT	5 (A)	5 (A)

Table 14. Performance data

Visualization and Numeric Data

The final step in the description of the performance of a system, and in the comparison of a base case with a proposed change, is what you would "see" in the field and how you would describe what you see. You could use such descriptions as:

- The standing queue on the eastbound through lanes clears before the end of green
- The traffic on the southbound left turn lane backs up out of the left turn pocket, thus delaying all traffic on the southbound approach
- Much of the green time on the eastbound through lanes is not utilized indicating a possible misallocation of time for this approach and the others
- Vehicles on the westbound through lane must wait through two cycles

This gets us to the final step in the "learning to see" process, integrating the visual observations that you make (and your description of them) with the numerical data that you collect in the field or from a simulation model. You will learn in this chapter when you construct your simulation model to see both the "visual" and the "numbers" and the conclusions that you can make from one reinforcing the other.

Learning to integrate the visual and the numbers also takes us to the experience of the traffic engineer, and understanding the connection between the traffic controller and its effect on the user. What do you see in the field and thus what should you change in the traffic controller? This is learning to be in the mode of the traffic engineer standing in the field: one eye on the traffic flow and other on the controller. Making these connections is what this chapter is all about.

ACTIVITY

26 What Do You Know About Simulation Models?

PURPOSE

The purpose of this activity is to provide you with the opportunity to learn more about the variety of traffic analysis tools that are available to the transportation engineer. A traffic analysis tool is a "software-based analytical procedure and/or methodology that supports different aspects of traffic and transportation analysis" (Alexiadis, Jeannotte, & Chandra, 2004).

LEARNING OBJECTIVES

- Describe the categories of traffic analysis tools that are commonly used by the transportation engineer
- Describe the application of a simulation model

DELIVERABLE

- Prepare a document with your answers to the Critical Thinking Questions

REQUIRED RESOURCES

- *Traffic Analysis Toolbox, Volume 1: Traffic Analysis Tools Primer*
- *Traffic Analysis Toolbox, Volume III: Guidelines for Applying Traffic Microsimulation Modeling Software*

TASK 1

Read pages 1-18 from the *Traffic Analysis Toolbox, Volume 1: Traffic Analysis Tools Primer* (Alexiadis, Jeannotte, & Chandra, 2004) and pages 35 to 43 from *Traffic Analysis Toolbox, Volume III: Guidelines for Applying Traffic Microsimulation Modeling Software* (Dowling, Skabardonis, & Alexiadism 2004). Prepare answers to the Critical Thinking Questions.

CRITICAL THINKING QUESTIONS

When you have completed the reading, prepare answers to the following questions.

1. What are the categories of traffic analysis tools and what are the basic attributes of each category?

2. What is a microsimulation model?

3. Under what conditions would you use a microsimulation model?

4. What are the strengths and limitations of the Highway Capacity Manual methodologies?

5. What are the strengths and limitations of simulation?

6. What are the differences in how the Highway Capacity Manual and simulation models report performance measures?

7. What is a link/node diagram and what do links and nodes represent?

8. What kinds of driver behavior are modeled in a microsimulation model?

9. What are some of the new features available in the current version of VISSIM?

10. What kind of model would you use to determine the number of lanes needed at a signalized intersection to meet a desired level of service?

11. If a time-scan model is based on scanning time on a regular basis, what is an event scan model based on? For a signalized intersection, list some events of interest that would be the basis for an event scan model.

12. For what kinds of problems or system conditions would you consider using the following models: Critical Movement Analysis, Highway Capacity Manual, and VISSIM?

13. Which model would you use to test signal timing strategies for a congested arterial with three closely spaced signals? Why?

ACTIVITY

27 The VISSIM Simulation Model: Learning Your Way Around

PURPOSE

The purpose of this activity is to introduce you to the VISSIM simulation environment and to give you the opportunity to explore some of its features.

LEARNING OBJECTIVE

- Describe the basic features of VISSIM

REQUIRED RESOURCES

- Movie file: A27.mp4
- PTV America website: www.ptvamerica.com

DELIVERABLE

- Prepare a document that includes your answers to the Critical Thinking Questions

CRITICAL THINKING QUESTIONS

1. What are some of the components of the VISSIM model that are accessible through the toolbar?

2. What are some of the traffic signal timing parameters that are required to specify the operation of the RBC controller?

3. What kinds of evaluation data are produced by VISSIM and which ones might be most important in evaluating a design?

4. Why would you consider various simulation speeds when you run VISSIM?

5. What did you learn about VISSIM from the PTV website?

Information

VISSIM will be the primary tool with which you will test and evaluate your signal timing design options. VISSIM is one of several commonly used and very powerful simulation models on the market today. VISSIM includes components for how drivers interact with the roadway design, the control system, and with other drivers. It includes several options for emulating the traffic control system. While this book is not about learning to use any particular simulation model, including VISSIM, it is important for you to learn some of the key features of VISSIM and how you will use them as you do learn about traffic signal timing.

In this activity, you will learn about some of the basic features of the VISSIM simulation model. The main VISSIM screen (see Figure 105) shows the representation of the geometry, the flow of vehicles, and the control devices. Driver behavior is modeled based on car-following logic, lane changing logic, and response to the status of control displays. The signal controller operation is based on actuated control timing processes that are linked to detector inputs. The status of all vehicles, the controller timing processes, and the signal displays are updated every tenth of a second. Uncertainty in driver responses is based on probabilistic modeling of driver behavior. Evaluation and performance statistics are collected during the simulation and are available to the user both during and after the simulation period.

Task 1

Open the movie file and start the movie. Write down questions that you have on what you observe. As you watch the video, take notes on what you see and the important features of the model that are shown.

Task 2

Browse the VISSIM portion of the PTV America web site. Watch at least one of the VISSIM video demonstrations on the web site. Find and read a section of the web site that describes some of the new features on the latest VISSIM release.

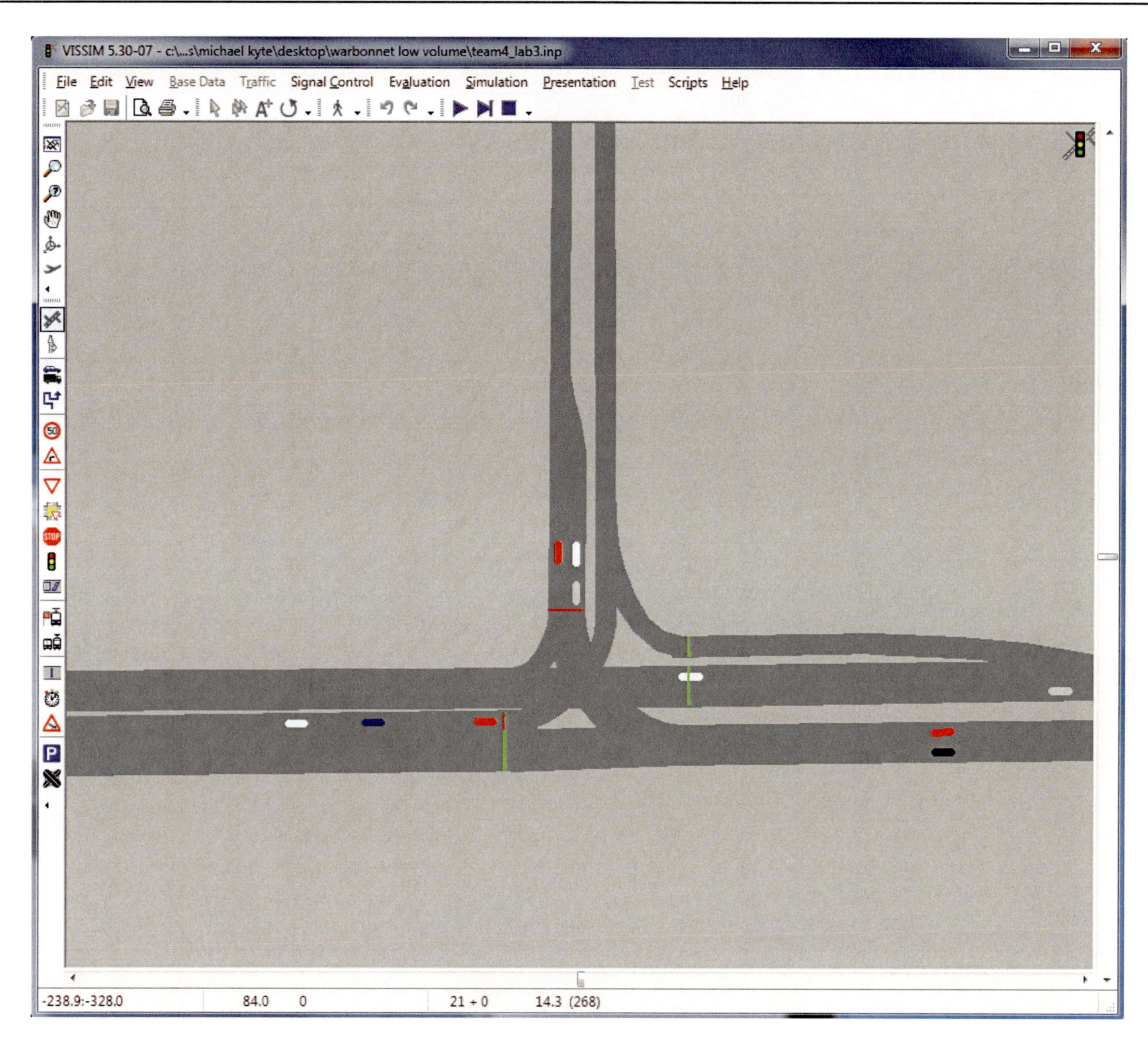

Figure 105. VISSIM display

Student Notes:

ACTIVITY

28 Building a Simulation Model Network

PURPOSE

The purpose of this activity is to give you the opportunity to learn how to build and use a simulation model network. You will use this network as you prepare a design for an isolated signalized intersection.

LEARNING OBJECTIVE

- Build and use a simulation network

REQUIRED RESOURCES

- Base VISSIM network (VISSIM .inp and other files)
- VISSIM tutorial

DELIVERABLES

- Modified VISSIM files
- Prepare an Excel spreadsheet that includes the data that you have collected, the sketches that you made, and your observations of the data and the simulation. This information represents the base data for the beginning of your design work, and against which you will compare the design plans that you develop. The spreadsheet should include tabs for each of the following:

  Tab 1: Title page with activity number and title, authors, and date completed

  Tab 2: Sketches as per Task 1

  Tab 3: Evaluation data as per Task 6

  Tab 4: A summary of the traffic flow characteristics that you observe in the animation

  Tab 5: A summary of the evaluation data that you have collected, including your comments on the evaluation data and what you learn from these data

  Tab 6: Answers to questions from Task 6

CRITICAL THINKING QUESTIONS

1. Are your observations of the animation consistent with the output data produced in Task 6?

2. What traffic problems do you observe? (Some example traffic problems might include: queues not clearing before the end of green, or queues spilling out of a left turn pocket and into a through lane).

Task 1

Prepare two electronic sketches for your subject intersection:

- Sketch #1: Prepare a sketch that shows the lane geometry, the numbered directional movements for each lane, and the phase numbers that control the movement in each lane. Add to this sketch a stop bar detector in each lane, numbering each detector with two digits, the first number equal to the phase number controlling that lane. The second number should correspond to the lane number beginning with one for the left most lane for that approach. For example, if you have two northbound lanes controlled by phase 2, your detectors would be numbered 21 for the left lane and 22 for the right lane as shown in Figure 106.
- Sketch #2: Prepare a ring barrier diagram that represents a standard eight (or five, if T-intersection) phase NEMA pattern, with protected leading left turns, for the intersection

Figure 106. Signal head and detector numbering example

Task 2

Download the VISSIM network that you have been assigned, unzip it, and copy it to your hard drive. Open the VISSIM network. This network includes the link and node structure, a background photograph, priority rules governing vehicle interactions, vehicle routing decisions, and traffic volumes. Load the .ini file from "View," then "Load Settings." Run the network (by selecting the "Simulation Continuous" button) and observe the operation. What is missing?

Task 3

You will need to add several components to the network, including a signal controller (with a set of given timing parameters), one detector for each lane, and a signal head for each lane. The process for adding the signal controller, the detectors, and the signal head is described in the "VISSIM tutorial."

- Add a Ring Barrier Controller (RBC) with the following signal timing parameters: yellow time = 3 seconds, red clearance time = 1 second, passage time = 5 seconds, minimum green = 15 seconds, and maximum green = 100 seconds
- Add stop bar detection in each lane with length of 22 feet, and numbered according to the sketch that you prepared
- Make sure that the detectors are correctly mapped to the phases
- Add signal heads numbered as per the detector numbers that you have prepared
- "Dual entry" should be set for through movements that have been designated as the "start phases"

Task 4

Verify that Node Evaluations have been added (to measure delay and queue length, and green time distribution). The process for adding evaluations to your network is described in the "VISSIM tutorial." Set the data collection period to begin at 300 seconds and continue to 3600 seconds. Set the interval to 3300 seconds.

Task 5

Run your network and debug as needed. When the network is debugged, complete a final run to collect your data. The following tests can be considered to determine if the network is debugged and operating correctly: (1) phases operate in sequence and for the expected durations, (2) vehicles respond to the signal displays in an appropriate manner, and (3) there are no conflicts between vehicles from different streams.

Task 6

Review the evaluation data produced by the simulation run. These data files will be in the same folder as your input files. Refer to the "VISSIM tutorial" and the VISSIM help file to learn more about the evaluation data. Your review should include an analysis of the delay data, the queue length data, and the green time distribution data.

- Prepare a table showing a summary of the delay and queue length data
- Prepare a histogram of the length of the green intervals for the through and left turn phases of your network, choosing the bin size so that your graph conveys a clear picture of the distribution. Note the mean green duration for each phase. Discuss the variations about the mean based on your histogram plot and why this variation occurs.
- These data will be used as the base case evaluation for comparison with your final design (that you will finish in Chapter 10)

Task 7

Demonstrate your completed (and working) network to your instructor. "Working network" means that traffic flows in response to displays, detectors respond to traffic, the controller responds to detectors, and the displays respond to the controller.

Things to Check (Common Problems)

Preparing a simulation file for the first time is a complex process requiring attention to a number of details. Following is a list of common problems that are often encountered during the preparation of a VISSIM simulation file:

1. Check carefully that the detectors have been mapped to the correct phases.
2. Make sure that you add the correct signal group (phase) number to the signal heads.
3. Don't double extend phase calls (note process in RBC controller setup).
4. Detectors shouldn't be in connectors.
5. Verify that evaluation nodes have been added.
6. Be sure that the signal heads are placed in front of a connector.

Student Notes:

ACTIVITY

29 Traffic Analysis Tools

Purpose

The purpose of this activity is for you to learn what kinds of models are used in practice.

Learning Objective

- Describe the categories of traffic models

Required Resource

- *Traffic Signal Timing Manual*

Deliverables

Prepare a document that includes

- Answers to the Critical Thinking Question
- Completed Concept Map

Link to Practice

Read the sections of the *Traffic Signal Timing Manual* assigned by your instructor.

Critical Thinking Questions

When you have completed the reading, prepare answers to the following questions:

1. How does the description of models from the Traffic Analysis Toolbox compare and contrast with the discussion in the *Traffic Signal Timing Manual*?

2. What could you learn from the four model types described in the *Traffic Signal Timing Manual* if they are used to evaluate a signal timing plan?

IN MY PRACTICE...

by Michael Kyte

Computer modeling tools are commonly used by the transportation engineer to both evaluate alternative designs as well as to visually demonstrate, through advanced animation techniques, how a transportation facility will perform under a given set of conditions. Even though many computer models are easy to set up, run, and produce "results", often the engineer is under pressure to produce results quickly. This means that such important steps as calibration of a model to the conditions of a local area are often skipped or not given sufficient time. The results from a model are only as good as the input data used and the time spent to fine-tune the model itself. Be skeptical of results. Ask questions before you believe what "the model says." It is a tool. You are the engineer.

Concept Map

Terms and variables that should appear in your map are listed below.

microsimulation model
performance measure
VISSIM
network
traffic analysis tool

Student Notes:

Timing Processes on One Approach

Purpose

The duration of the green interval for an actuated controlled intersection depends on the interaction of the traffic demand and the actuated control settings. In Chapter 6, you will learn about two of these settings that are so crucial to the efficient operation of the intersection: the minimum green time and the passage time. You will consider these two timing parameters in the context of a related component: the placement and length of the detection zone. You will learn that, in practice, you need to decide what conditions you will tolerate. Do you want to risk that the green will terminate too soon and leave some vehicles unserved? Or, do you want to risk that the green will not terminate soon enough, resulting in wasted green time? Balancing these risks is one of the keys to efficient and effective signal timing.

Learning Objectives

When you have completed the activities in this chapter, you will be able to

- Describe the interaction of the minimum green time, the passage time, and the detection zone length in producing efficient intersection operation
- Describe the timing processes for actuated traffic control
- Describe how the length of the detection zone affects the setting of the basic timing parameters
- Relate the length of the detection zone to the duration of the green indication
- Relate the length of the minimum green time to the efficient operation of a phase
- Describe the variation of vehicle headways in a departing queue
- Establish a desired maximum allowable headway
- Relate the maximum allowable headway to unoccupancy time
- Determine the vehicle extension time based on the length of the detection zone and the desired maximum allowable headway
- Select a maximum allowable headway
- Compare headway distributions for one lane and two lane data
- Select passage time values for one lane and two lane approaches
- Contrast design values with those recommended in practice

Chapter Overview

This chapter begins with a *Reading* (Activity #30) on the relationship between the minimum green time, the passage time (or vehicle extension time), and the detection zone length. The chapter then proceeds to eight activities including an assessment of your understanding of the basic concepts of passage time and detection zone length (Activity #31), four discovery activities in which you will learn about the timing processes through observation of simulation, particularly the factors that should be considered when the minimum green time and the vehicle extension time parameters are set, for a given length of the detection zone. In Activity #32, you will see that the detection zone itself can provide some extension of the green as vehicles arrive at the intersection and enter the zone. You will learn in Activity #33 how the minimum green time must be set long enough so that the queue begins to move but short enough so that the phase doesn't extend inefficiently when very short queues are present. You will learn about the variation in headways in a departing queue (Activity #34) and how the headways relate to the unoccupancy time and the vehicle extension time (Activity #35). As part of the two design activities, you will set the maximum allowable headway and the passage time (Activities #36 and #37). The chapter concludes with an *In Practice* activity (Activity #38) in which you compare your design results with the setting of actuated timing values described in the *Traffic Signal Timing Manual*.

Activity List

Number and Title	Type
30 Considering Minimum Green Time, Passage Time, and Detection Zone Length	*Reading*
31 What Do You Know About Detection Zone Length and Passage Time?	*Assessment*
32 Relating the Length of the Detection Zone to the Duration of the Green Indication	*Discovery*
33 Determining the Length of the Minimum Green Time	*Discovery*
34 Understanding the Variation of Vehicle Headways in a Departing Queue	*Discovery*
35 Relating Headway to Unoccupancy Time and Vehicle Extension Time	*Discovery*
36 Determining the Maximum Allowable Headway	*Design*
37 Determining the Passage Time	*Design*
38 Actuated Traffic Control Processes	*In Practice*

ACTIVITY

30 Considering Minimum Green Time, Passage Time, and Detection Zone Length

Purpose

The purpose of this activity is to provide you with the opportunity to learn more about how the selection of the basic actuated traffic control timing parameters (minimum green time and passage time) are related to the length of the detection zone.

Learning Objective

- Describe the interaction of the minimum green time, the passage time, and the detection zone length in producing efficient intersection operation

Deliverables

- Define the terms and variables in the Glossary
- Prepare a document that includes answers to the Critical Thinking Questions

Glossary

Provide a definition for each of the following terms. Paraphrasing a formal definition (as provided by your text, instructor, or another resource) demonstrates that you understand the meaning of the term or phrase.

Term	Definition
call	
detection zone	
interval	
maximum allowable headway	
occupancy time	

recall	
unoccupancy time	
h	
L_d	
L_v	
t_o	
t_u	
v	

Critical Thinking Questions

When you have completed the reading, prepare answers to the following questions.

1. Describe how passage time and the length of the detection zone are related.

2. What is one criterion for terminating a phase?

3. When using a standard loop detector with stop bar presence detection, why is it difficult to determine when a queue has cleared?

4. Explain why the passage time should be decreased when the detection zone length is increased.

5. Explain how variability in the vehicle lengths and speeds affect the determination of the passage time.

6. Describe in your own words the implications of the data presented in Figure 111.

7. Since vehicle headways vary widely and are not constant, even during periods of saturation flow, explain the risks involved in setting the passage time.

8. Summarize your understanding of the headway variability for the four time segments of vehicles departing after the start of green.

9. Describe how the problem of determining passage time changes when considering a two-lane approach.

Information

Introduction

Consider an actuated traffic control system with stop bar presence detection on a single-lane approach. A vehicle requests service by passing into the detection zone. The request is processed when the associated phase is next in the controller sequence. The timing of the phase is based on the interaction of the vehicle request (or call), the length of the detection zone, and the value of three timing parameters (the minimum green time, the passage time, and the maximum green time) as shown in the traffic control process diagram in Figure 107. In this example, six vehicles are stopped at the intersection at the beginning of green. They travel through the intersection when the green indication begins, as shown by the time-space trajectories, and the detection system responds to these vehicles (box at top left). The detector calls activate the controller timing processes (box at top right), and the signal displays respond to the controller timing processes and logic (box at bottom right). Finally, to "close the loop", the vehicles respond to the display status (box at lower left).

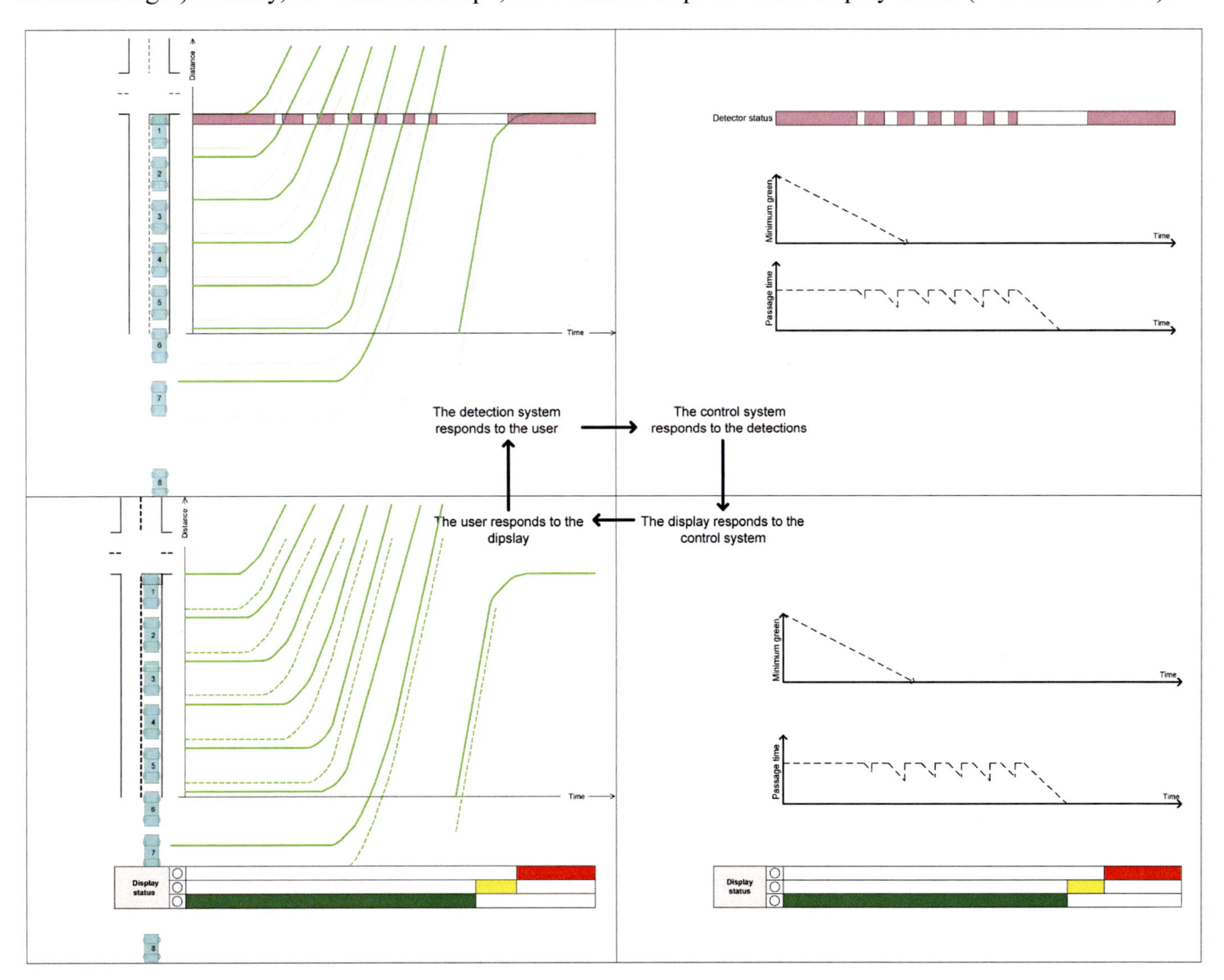

Figure 107. Traffic control process diagram

The purpose of the minimum green timing interval in the case of stop bar presence detection is to make sure the green is displayed for at least as long as driver expectation. Driver expectation can be thought of as a time so short that if the minimum green time is less than this time the public may complain or identify what they perceive to be a problem. The minimum green time does not have to account for a slow moving vehicle because as long as the vehicle is in the detection zone the call will be extended until it clears the stop bar. The minimum green time is generally lowest for left turns, more for side streets, and longest for through movements on the main street.

The purpose of the passage timer is to extend the phase as long as the headway between vehicles is less than a specified value called the maximum allowable headway (MAH). The goal is to make sure that green is displayed as long as a queue is present but then to terminate the phase when the queue has cleared.

In this chapter you will learn more about two of these timing parameters, the minimum green time and the passage time. Activities related to the maximum green time are included in Chapter 7.

A Theoretical Foundation: Traffic Flow Theory for Queue Clearance

The traffic flow process of queue clearance can be represented as a flow profile diagram, as shown in Figure 108a. In the first segment of the green interval (noted as "1" in the figure), the flow rate increases as the queue begins to move from the stop bar into the intersection. This time period is characterized by the start-up lost time parameter described in the Highway Capacity Manual (HCM), and includes the first four vehicles in queue. In the second segment of the green interval, beginning with the fifth vehicle in queue, vehicles depart at the saturation flow rate. The HCM suggests an ideal value of 1900 vehicles per hour of green for the saturation flow rate.

When the queue has cleared, vehicles arrive at and depart from the intersection at a constant rate, with no delay. This is shown as the fourth segment of the green interval where the flow is represented as uniform. The segment between the queue clearance period (the second segment) and the period after the queue has cleared (the fourth segment) is the transition, or third, segment. It is during this third segment, after the queue has been served, that the phase should be terminated and service transferred to the next phase in the controller sequence. This process can also be represented in terms of headways, as shown in Figure 108b. The headway during the second segment is the saturation headway.

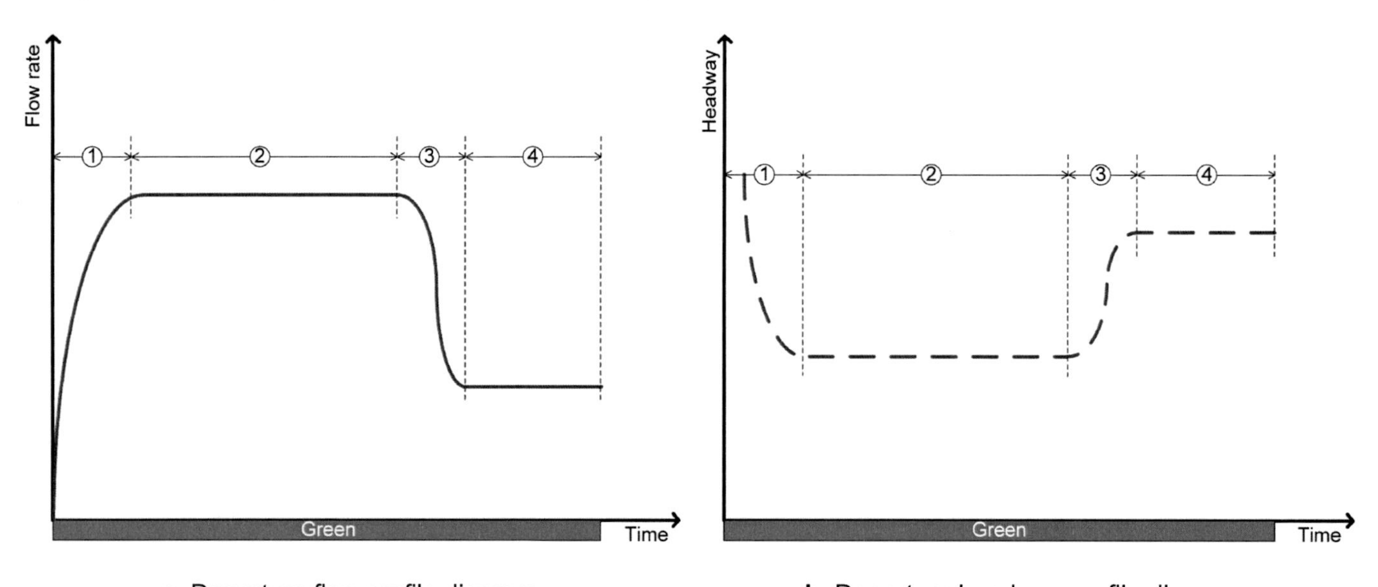

a. Departure flow profile diagram

b. Departure headway profile diagram

Figure 108. Departure flow and headway profiles

Headway and Unoccupancy Time

In the field, traffic control systems don't typically measure flow rates or headways but rather whether the detection zone is occupied or not. Unoccupancy time is defined as the time that the detection zone is not occupied by a vehicle. Figure 109 shows a time space diagram representation of vehicles departing in a queue at the beginning of green and then arriving and departing without delay after the queue has cleared for two different detection zone lengths (shown in gray shade). When the unoccupancy time reaches the passage time set in the controller, the phase is terminated. The unoccupancy time depends directly on the length of the detection zone, as well as the vehicle speed (which may vary over time) and vehicle length. In Figure 109a, with a shorter detection zone, the horizontal distance between points A and B represents the unoccupancy time, the time between the fourth vehicle leaving the detection zone and the fifth vehicle arriving in the detection zone. In Figure 109b, with a longer detection zone, the event represented by point B occurs before that represented by point A (vehicle 5 arrives in the detection zone before vehicle 4 leaves the zone), so the unoccupancy time is zero.

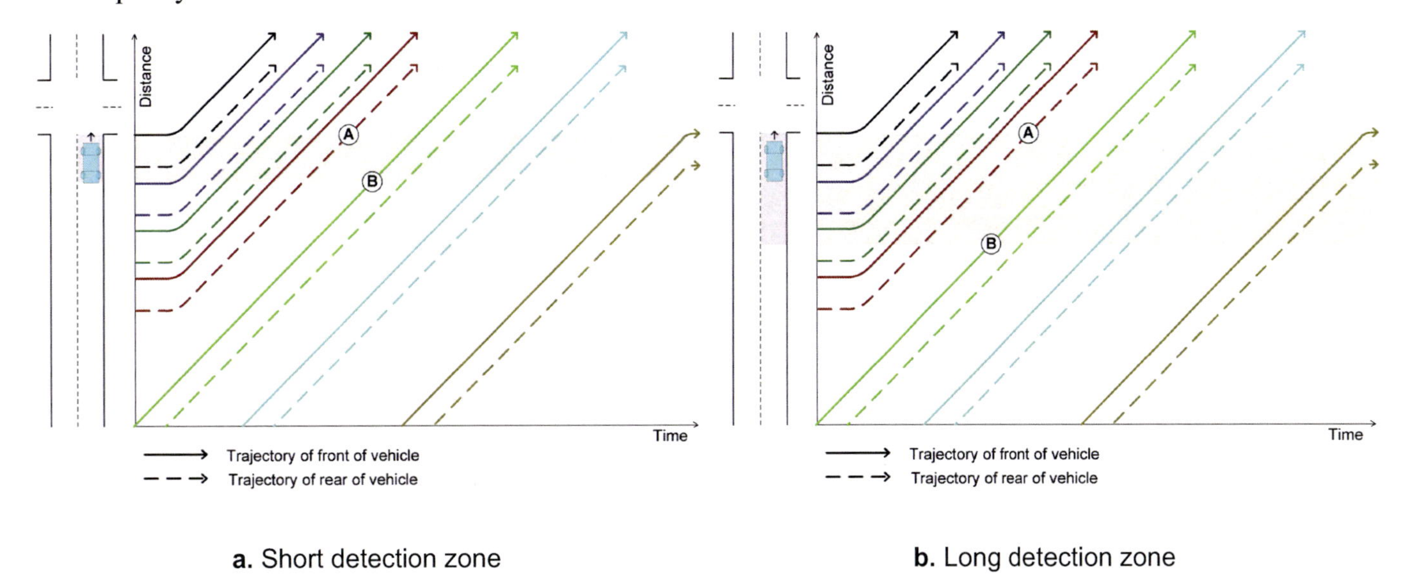

a. Short detection zone

b. Long detection zone

Figure 109. Time-space diagram for long and short detection zones

Maximum Allowable Headway (MAH)

Bonneson and McCoy (2005) established the concept of the MAH as the maximum headway that will be tolerated in the traffic stream before the phase is terminated. The analytical relationship between the unoccupancy time and the MAH can be developed as follows.

The headway (h) between two vehicles traveling on an intersection approach and through a detection zone consists of two parts, the time that the detection zone is occupied by the first vehicle and the time that the zone is unoccupied after the first vehicle leaves the zone and before the second vehicle arrives into the zone.

$$h = t_o + t_u$$

where t_o is the occupancy time and t_u is the unoccupancy time. The time that the detector is occupied (t_o) is equal to the length of the detection zone (L_d) plus the length of the vehicle (L_v), divided by the speed at which the vehicle is traveling (v).

$$t_o = \frac{L_d + L_v}{v}$$

Thus, we can write the unoccupancy time (t_u) as follows:

$$t_u = h - \frac{L_d + L_v}{v}$$

This relationship is shown graphically in Figure 110. The occupancy time (t_o) is the time that it takes vehicle 1 to travel its own length plus the length of the detection zone. The unoccupancy time (t_u) is the time from when vehicle 1 leaves the detection zone until vehicle 2 arrives in the detection zone. The headway (h) is the sum of the occupancy time and the unoccupancy time.

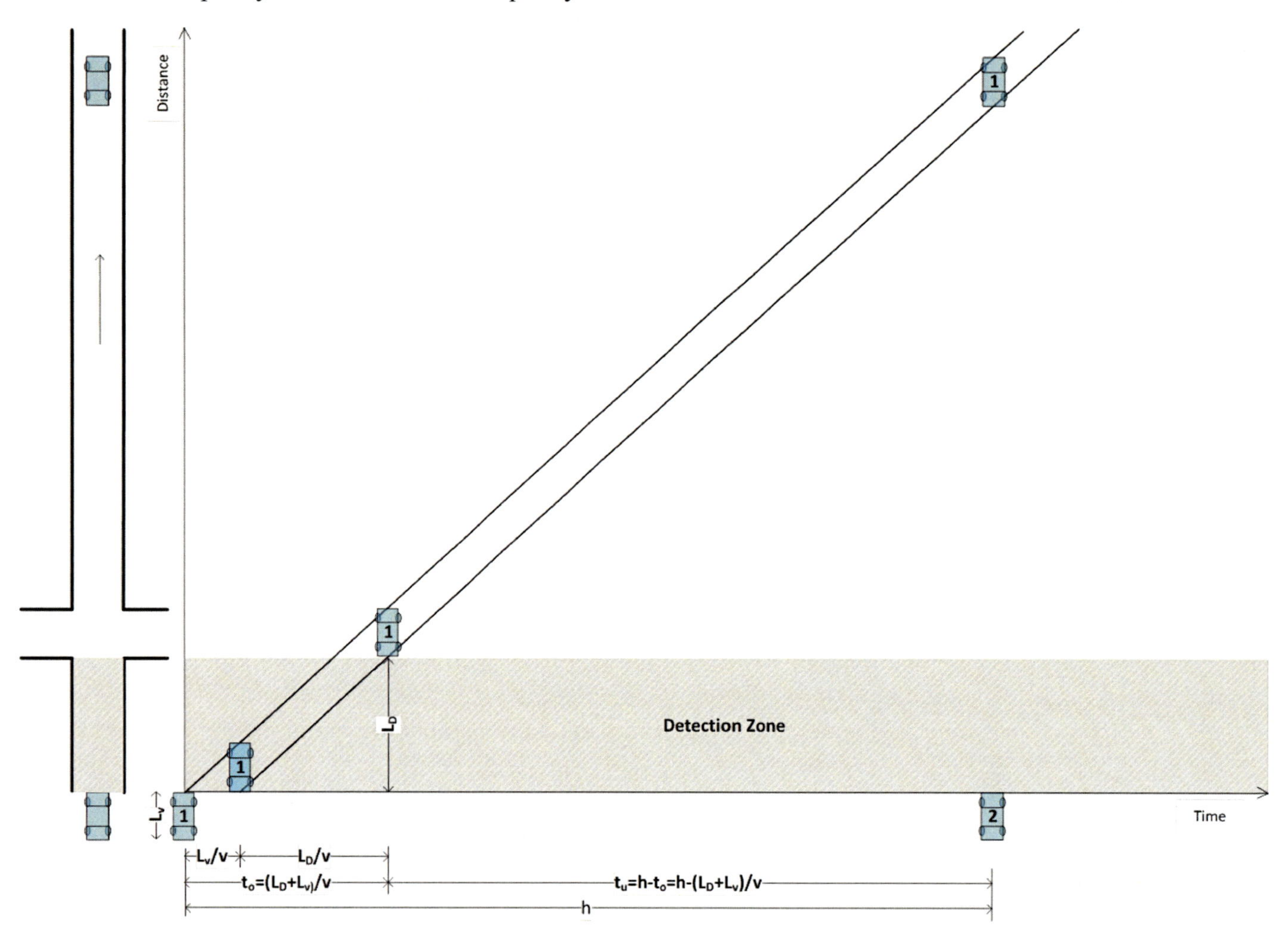

Figure 110. Headway, occupancy time, and unoccupancy time

A Stochastic Perspective Using Simulation Data

The reality is that queue clearance is a messy process with field measured values of headway and flow rate varying significantly about the theoretical values shown in Figure 108. Drivers respond to the change to green at different rates, and once they begin to move, they establish varying following distances behind the preceding vehicle. Since our desire is to extend the green only as long as a queue is present, and to terminate the green when the queue has cleared, we need to better understand the stochastic nature of this process. This involves understanding the headway, flow rate, and unoccupancy time distributions during queue clearance (time segments 1 and 2), the transition period (time segment 3) and the post-queue period (time segment 4).

To gain a perspective on this problem, the results from a set of simulation runs using the VISSIM microscopic simulation model are presented here (Kyte, Urbanik, & Amin, 2007). For these simulations, the queue at the beginning of green ranged from eight to ten vehicles. The traffic control was set to fixed time so that additional vehicles would be served after this initial queue had cleared and a comparison between headways of vehicles in the departing queue and during the post queue period could be made.

Figure 111a shows the mean headways measured for the first 25 vehicles passing the stop bar at the beginning of green. The dashed line represents the theoretical departure headways shown previously in Figure 109b. The simulation data are shown varying about this theoretical line, the kind of stochastic variation that we expect to see in the field. The headways measured during queue clearance vary in a narrow range about the theoretical line. However, the headways during the post queue period have a much wider variation with some almost as low as values that were measured during queue clearance. Figure 111b shows the mean flow rates for these same positions in queue, based on the headways shown in Figure 111a.

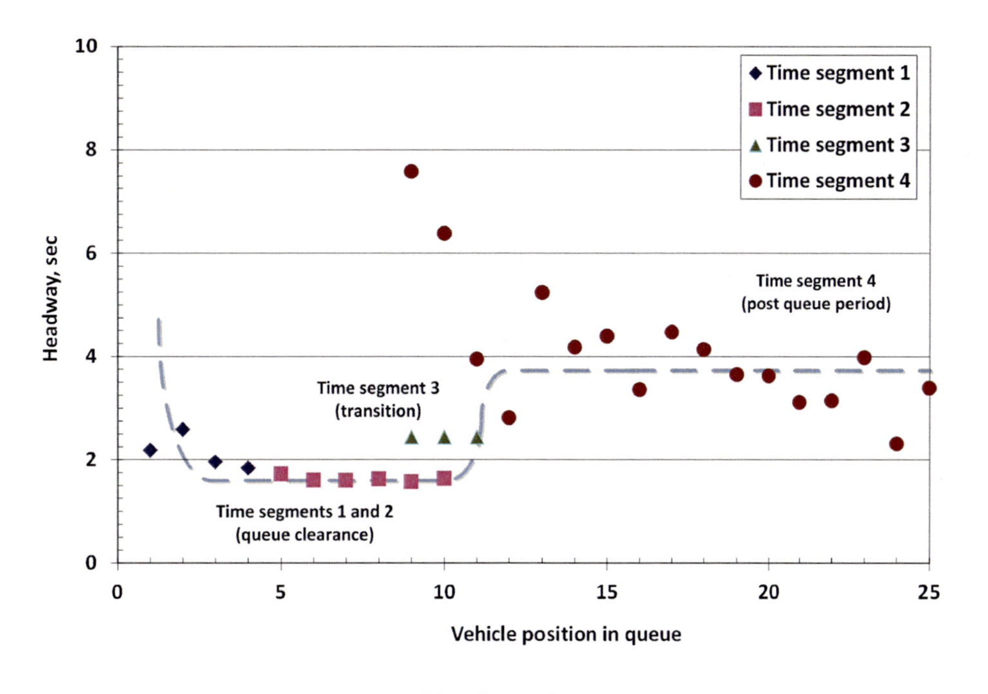

a. Headway data

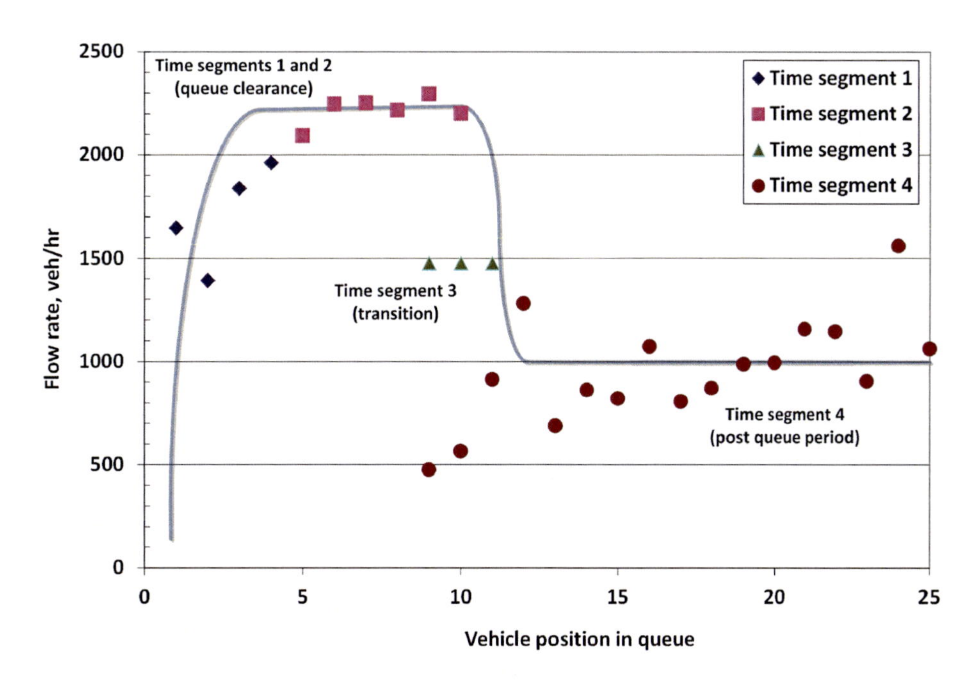

b. Flow rate data

Figure 111. Simulated data (and theoretical line) by vehicle position 111

This stochastic variation has two implications in the selection of the MAH. First, even headways in the departing queue have some variation reflecting differences in driver characteristics. Second, headways in the post queue period may be similar to those during the queue clearance period, thus making it difficult to determine, just based on headway values, which period you are observing.

Multiple lane approaches

How does the problem of setting the passage time change if there is more than one lane on the intersection approach? Often the detection scheme is such that the control system only knows that a call has been received on an approach, not which lane the call comes from. This means that the headway distribution that the traffic control system "sees" is the combined distribution of both lanes. This situation is illustrated in Figure 112 which shows an example of the departure of vehicles over a 30 second period for two lanes individually (Lane 1 and Lane 2) and then taken together (both lanes). In the figure, the headway between any two vehicles is represented by the horizontal distance between the points representing the two vehicles. Two example headways are noted in lanes 1 and 2. For lane 1, the headway is 8.0 seconds, while for lane 2 it is 3.0 seconds. However, if we measure the headways using vehicles from both lanes together (as shown for the three vehicles from lanes 1 and 2 "boxed" together), the consecutive headways would be 0.3 seconds and 2.7 seconds. The picture given with data combined from both lanes is a different one than the headways shown for lanes 1 and 2 separately. And, the conclusion would be different as well. The detection system would "see" three closely spaced vehicles (0.3 and 2.7 seconds) and conclude something different than if the headways are measured from each lane separately (8.0 and 3.0 seconds).

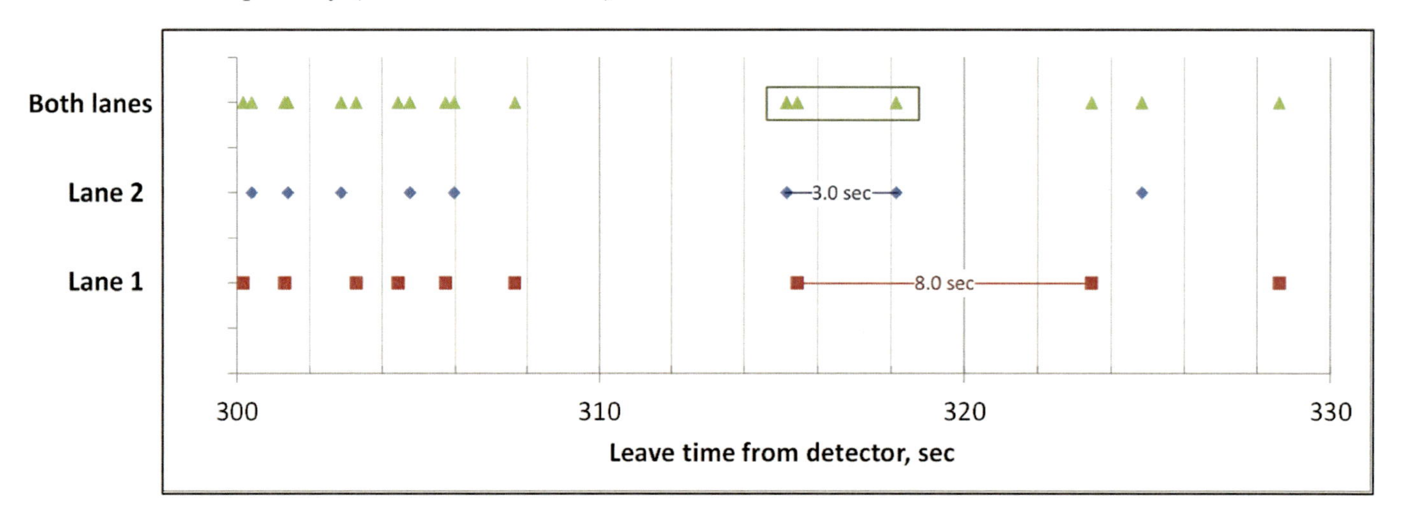

Figure 112. "Leave time" for lane 1, lane 2, and both lanes together

Let's now consider the headway density and cumulative density functions in a departing queue measured for one lane separately and for two lanes combined for an intersection approach, as shown in Figure 113 and Figure 114. While the density functions look similar, the two-lane data are shifted to the left, compared to the one-lane data. The mean value for the one lane data is 1.73 seconds and 0.84 seconds for the two lane (combined) data.

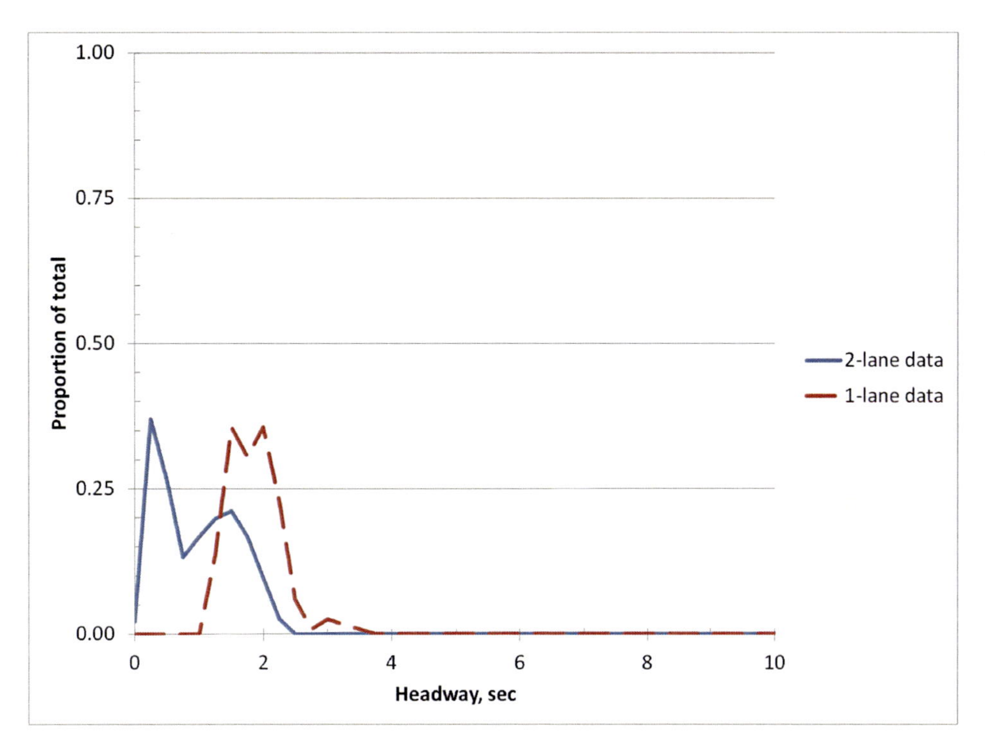

Figure 113. Headway density function, one lane and two lane data

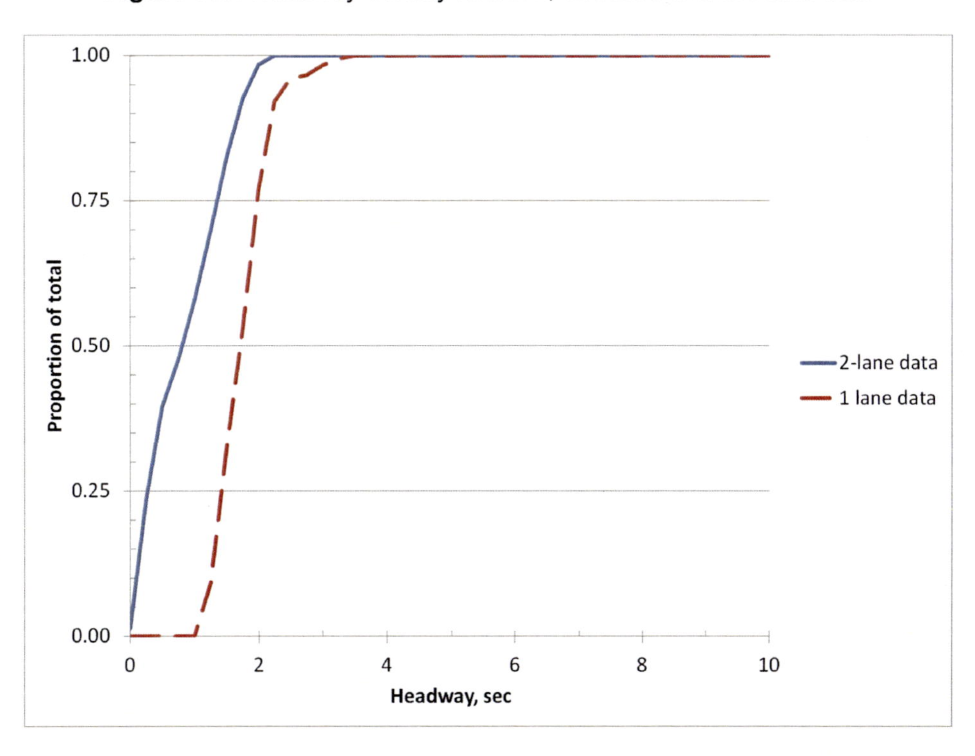

Figure 114. Cumulative headway density function, one lane and two lane data

What are the implications of these headway distribution differences when it comes to setting the passage time? If we assumed a vehicle speed of 25 miles per hour (36.75 feet per second), an average vehicle length of 20 feet and a detection zone length of 22 feet, we could calculate the resulting unoccupancy time for both the one lane and two lane conditions for a given value of headway. Let's consider three cases, in which 99, 95, and 90 percent of the vehicles in the queue would be served by a given MAH. The resulting calculations for the unoccupancy time (and thus the passage time) are shown in Table 15, based on the equation that we considered earlier for the calculation of the unoccupancy time.

Percentile	1-Lane		2-Lane	
	Headway (sec)	Unoccupancy time (sec)	Headway (sec)	Unoccupancy time (sec)
99	3.1	2.0	2.1	1.0
95	2.4	1.3	1.8	0.7
90	2.2	1.1	1.7	0.6

Table 15. Maximum allowable headways and unoccupancy times for 1-Lane and 2-Lane conditions

This example shows that the unoccupancy times (and thus the passage times) are from 0.5 to 1.0 second lower for the 2-lane case than for the 1-lane case. The clear implication is that the passage time for a 2-lane approach should be lower than for a 1-lane approach if we are to achieve the same efficiency in signal timing and meet our objective of providing sufficient green time to serve a clearing queue but not vehicles that arrive after the queue has cleared.

Conclusion

The minimum green time establishes the minimum time that the green will be displayed for a phase. The passage time parameter determines how long the green will be extended after the minimum green timer has expired and is directly related to the MAH and the length of the detection zone. The stochastic variation of headways in the vehicle stream means that the challenge in selecting the MAH (and thus the passage time) is to balance two risks. The first risk (if the MAH is too short) is that the phase will be terminated too early if the queue is still clearing. The second risk (if the MAH is too long) is that the phase will be extended past the time that the queue has cleared. Selecting the MAH, and then the passage time, is the balance in risks that the transportation engineer must determine. Finally, we need to consider shorter passage times for a 2-lane approach than the value we would consider for a one-lane approach. The activities to follow will give you specific experiences in dealing with each of these issues.

ACTIVITY

31 What Do You Know About Detection Zone Length and Passage Time?

PURPOSE

The purpose of this activity is to test your understanding of the relationship between detection zone length and the basic actuated timing parameters.

LEARNING OBJECTIVES

- Describe the timing processes for actuated traffic control
- Describe how the length of the detection zone affects the setting of the basic timing parameters

DELIVERABLE

- Prepare a completed spreadsheet with the results of your analysis from the following tasks

  Tab 1: Title page with activity number and title, authors, and date completed

  Tab 2: Answers to the Critical Thinking Questions

  Tab 3: Tool prepared in Task 1 and results from Tasks 2, 3, and 4

CRITICAL THINKING QUESTIONS

When you have completed the reading, prepare answers to the following questions.

1. What is the relationship (in equation form) between unoccupancy time and maximum allowable headway (MAH)? What are some of the issues involving the computation of the unoccupancy time for a given intersection approach? Provide your answer in a complete paragraph.

2. What is the process for setting the passage time, given the MAH? Describe in complete sentences.

3. What are the pros and cons of a detection zone that is 100 feet in length? Provide your answer in one or more complete paragraphs.

4. How would the determination of the MAH change if you considered lane by lane detection for a two-lane approach (that is, detectors in each lane, operating independently)?

TASK 1

Prepare a spreadsheet tool that implements the relationship between the MAH, detection zone length, and unoccupancy time as shown in Figure 110. The tool should accept the following parameters as input: vehicle speed (mph), detector length (ft), vehicle length (ft), and headway (s). The spreadsheet should produce the unoccupancy time(s) as an output. The spreadsheet should also show a graph that shows the relationship for two vehicles traveling at a specified headway (as shown Figure 110).

TASK 2

Using your spreadsheet tool with a MAH of three seconds, determine the unoccupancy times that would result from detection zone lengths varying from 6 feet to 90 feet. Assume a vehicle length of 20 feet and a speed of 30 miles per hour. If the length of your detection zone was 60 feet, what value of passage time would you select and why?

TASK 3

Using your spreadsheet tool, what would you set the passage time to be, given the following conditions? Describe the assumptions that you made and the method that you used to answer this question.

- L_D = 22’
- L_v = 19’ (80 percent of the vehicles), 30’ (15 percent of the vehicles), or 55’ (5 percent of the vehicles)
- v = 29 mi/hr (mean)
- h = 1.5 – 2.9 sec, mean = 2.2 sec, 85th percentile = 2.5 sec

TASK 4

Using your spreadsheet tool, what would you set the passage time to be, given the following conditions? Describe the assumptions that you made and the method that you used to answer this question.

- L_D = 60’
- L_v = 19’ (80 percent of the vehicles), 30’ (15 percent of the vehicles), or 55’ (5 percent of the vehicles)
- v = 29 mi/hr (mean)
- h = 1.5 – 2.9 sec, mean = 2.2 sec, 85th percentile = 2.5 sec

Student Notes:

ACTIVITY

32 Relating the Length of the Detection Zone to the Duration of the Green Indication

Purpose

The purpose of this activity is to give you the opportunity to learn how the detection zone length affects the operation of a phase.

Learning Objective

- Relate the length of the detection zone to the duration of the green indication

Required Resource

- Movie file: A32.wmv

Deliverable

- Prepare a document that includes your answers to the Critical Thinking Questions

Critical Thinking Questions

As you begin this activity, consider the following questions. You will come back to these questions once you have completed the activity.

1. How do you know when the detector is active and when it is inactive?

2. When does the phase terminate for the southbound direction for each of the two cases?

3. Why does the phase terminate for each of the two cases?

4. Do you think that the phase is operating efficiently or not for the two cases? Why or why not?

5. Do you think that the quality of service provided to the motorist is good or not? Why or why not?

6. If the phase terminates too early or extends too long, what solutions should be considered?

INFORMATION

In this activity, you will consider two cases, one in which the southbound approach has a 22 foot detection zone and another in which the approach has a 66 foot detection zone, each representing a zone length sometimes used in practice. In both cases, the detection zones end at the stop bar. See Figure 115. The vehicle extension time and minimum green time are set to zero. The detectors are both operating in presence mode. You will observe how and when the phase terminates for both cases, and the status of the controller at several points in the simulation.

It is important to understand that both detectors are operating in the presence mode. This means that as long as the detection zone is occupied, a call is sent to the controller for the assigned phase. If a constant call is sent to a phase that is green, it will continue to remain green as long as the phase has not "maxed out."

The vehicle extension time and minimum green time parameters have both been set to zero for phase 4, the phase serving the southbound through approach movement. This means that you need to focus only on the detection zone length and how it affects the duration of the green indication.

Vehicles are present on the southbound approach on Line Street (phase 4), and the eastbound and westbound approaches on State Highway 8 (phases 2 and 6).

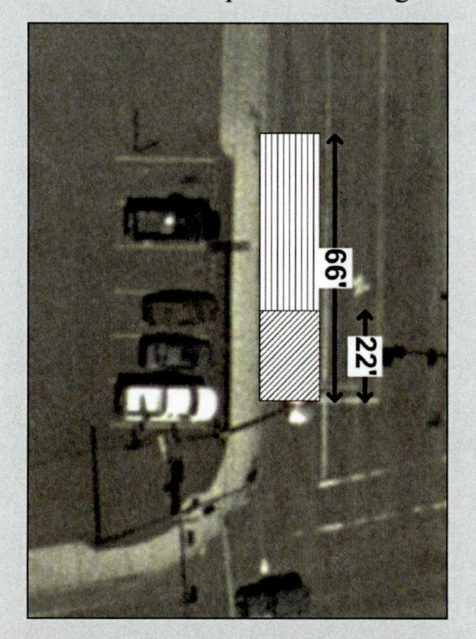

Figure 115. Two detection zone alternatives on Line Street SB approach

TASK 1

Open the movie file, A32.wmv. Pause the playback. The simulation window shows the animation of vehicles traveling through the intersection as well as other data (see Figure 116):

- The current simulation time is noted in the lower left
- The detection zones (and their numbers) are shown in each approach lane as boxes

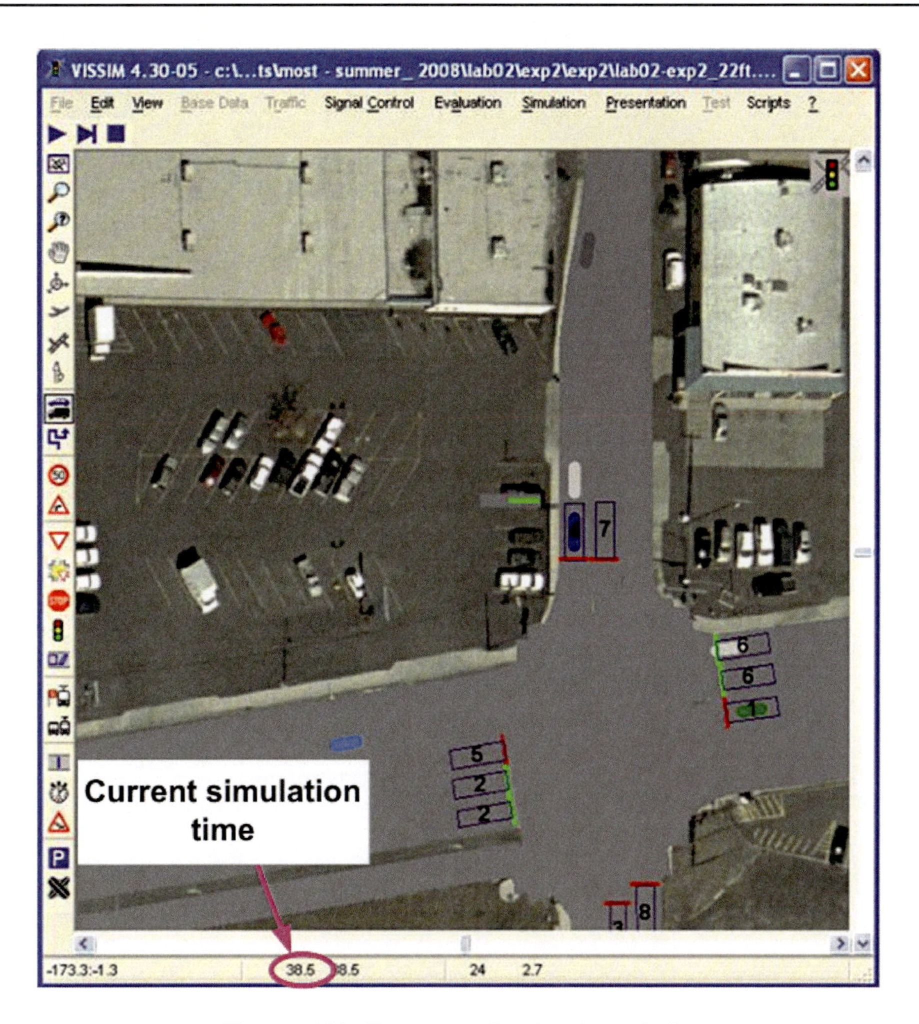

Figure 116. Features of animation window

Task 2

Start the movie file. Pause the movie when the simulation time is 49.0. Observe the conditions at the intersection and in the controller for both cases. See Figure 117.

- For the southbound approach, four vehicles are in queue for both cases
- Point B on the controller status screen shows the status for phases 2 and 6, the current active phases in rings 1 and 2. The red clearance timer is active, with a current value of 0.7 seconds. And, phases 2 and 6 have gapped out, as noted in the status screen.
- The controller status screen shows that phase 4 has an active call (noted by the "C" at point A) and is in "phase next" status (noted by the "N"). This means that when phase 2 has terminated, phase 4 will be the next phase to be served.

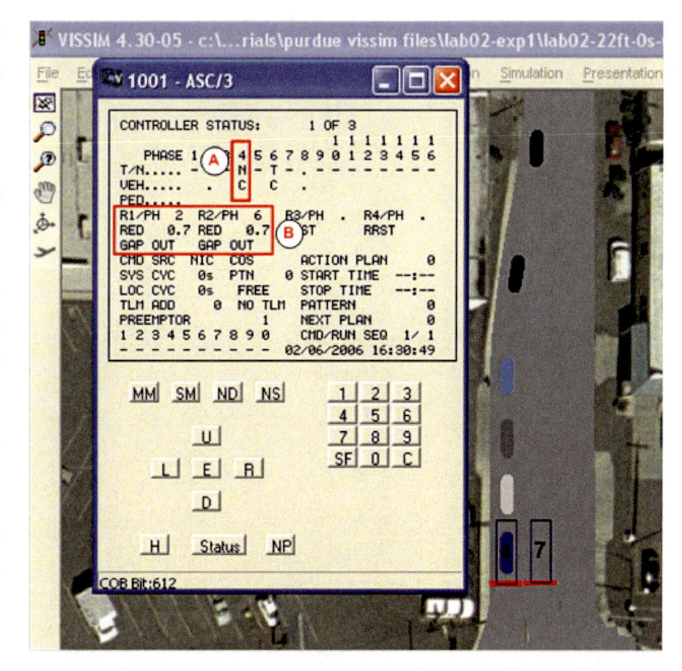

Figure 117. ASC/3 status screen at t = 49.1 for both cases

Task 3

Observe the simulation at $t = 49.9$. Record your observations on the status of phase 4. What is the color of the active indication?

Task 4

Observe the simulation from $t = 54.1$ to 54.3.

- Observe the simulation on the left of the screen (the 22 foot detector case)
- Record your observations of the controller status window, noting in particular the status of any calls, the timing status of phase 4, and the timing processes and timing parameter values for phase 4
- Also, record the status of the queue being served

Task 5

Observe the simulation from $t = 60.1$ to 61.4.

- Observe the simulation on the right of the screen (the 66 foot detector case)
- Record your observations of the controller status window, noting in particular the status of any calls, the timing status of phase 4, and the timing processes and timing parameter values for phase 4
- Also, record the status of the queue being served

ACTIVITY

33 Determining the Length of the Minimum Green Time

PURPOSE

The purpose of this activity is to help you learn to visualize the role of the minimum green time during the early portion of the green and to see how the setting of this parameter can result in efficient or inefficient timing.

LEARNING OBJECTIVE

- Relate the length of the minimum green time to the efficient operation of a phase

REQUIRED RESOURCE

- Movie file: A33.wmv

DELIVERABLE

- Prepare a spreadsheet that includes:

  Tab 1: Title page with activity number and title, authors, and date completed

  Tab 2: Your answers to the Critical Thinking Questions

  Tab 3: The data that you collected in Table 16 and Table 17

CRITICAL THINKING QUESTIONS

As you begin this activity, consider the following questions. You will come back to these questions once you have completed the activity.

1. When is the minimum green time too long?

2. How long should the minimum green time be in order to get vehicles moving during the early portion of green?

3. What are the respective roles of minimum green time and vehicle extension time in producing efficient operations?

Information

In this activity, you will see the importance of the minimum green time during the early portion of the green indication, and how you can define the roles of the minimum green time and the vehicle extension time to ensure efficient intersection operations. You will again observe the operation of the southbound approach of Line Street, at State Highway 8, and make observations about the operation. You will again consider stop bar detection, with a detection zone length of 22 feet.

Task 1

Open the movie file, A33.wmv.

Task 2

Figure 118 shows the controller screen for the ASC/3 database editor. Observe that the minimum green time is set to 5 seconds for phase 4 for the first case and 10 seconds for phase 4 for the second case. The vehicle extension time is set to 2.2 seconds for phase 4.

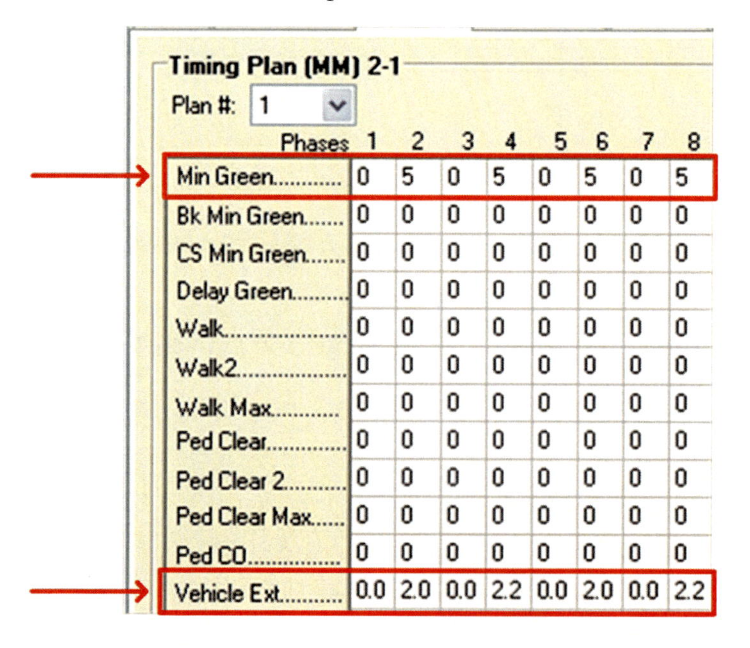

Timing Plan (MM) 2-1

Plan #: 1

Phases	1	2	3	4	5	6	7	8
Min Green...........	0	5	0	5	0	5	0	5
Bk Min Green.......	0	0	0	0	0	0	0	0
CS Min Green.......	0	0	0	0	0	0	0	0
Delay Green..........	0	0	0	0	0	0	0	0
Walk......................	0	0	0	0	0	0	0	0
Walk2....................	0	0	0	0	0	0	0	0
Walk Max............	0	0	0	0	0	0	0	0
Ped Clear..............	0	0	0	0	0	0	0	0
Ped Clear 2...........	0	0	0	0	0	0	0	0
Ped Clear Max......	0	0	0	0	0	0	0	0
Ped CO.................	0	0	0	0	0	0	0	0
Vehicle Ext..........	0.0	2.0	0.0	2.2	0.0	2.0	0.0	2.2

Figure 118. ASC/3 Database Editor

Task 3

Observe the timing and the termination of the southbound phase for the two cases.

- Advance the simulation until the start of green for phase 4 (the phase serving the southbound movement). Record the simulation time in Table 16 that corresponds to the start of green for both cases.
- Record the simulation time that the vehicle in queue leaves the detection zone
- Record the time that green ends (yellow begins) for both cases
- Reflect on the data that you recorded and the implications of these data
- Record your observations
- Remember, you will often observe a 0.1 second difference between the information shown in the controller status window and the indication status in the VISSIM window. This difference, resulting from communications latencies between the ASC/3 controller software and the VISSIM software, will not substantially affect your results.

Data to record	Case 1	Case 2
Start of green		
Back of vehicle leaves zone		
Start of yellow/end of green		
Difference between "start of yellow/end of green" and "back of vehicle leaves zone"		

Table 16. Data collection table

Task 4

Observe vehicle start-ups for case 1 only, the left side.

- Move the simulation to $t = 89.2$
- At this point ($t = 89.2$), observe the status of the traffic for the SB approach and the status of the timing process for both phase 2 and phase 4 for case 1 (on the left) only
- Record your observations
- Advance the simulation until case 1 (minimum green = 5.0 seconds) reaches the start of green for phase 4. This should occur at $t = 89.5$ seconds.
- For case 1, record the times that each of the four vehicles in the queue on the southbound approach first begin to move and when they enter the detection zone. Use Table 17 to record these data. Watch the simulation carefully to note the time step that each vehicle begins to move.

Vehicle #	Start of green	Vehicle begins to move	Vehicle enters detection zone
	89.5		
1			
2			
3			
4			

Table 17. Data collection table

ACTIVITY

34 Understanding the Variation of Vehicle Headways in a Departing Queue

PURPOSE

The purpose of this activity is to give you the opportunity to learn the degree to which vehicle headways vary in the departing queue.

LEARNING OBJECTIVES

- Describe the variation of vehicle headways in a departing queue
- Establish a desired maximum allowable headway (MAH)

REQUIRED RESOURCE

- Movie file: A34.wmv

DELIVERABLE

- Prepare a spreadsheet that includes:

  Tab 1: Title page with activity number and title, authors, and date completed

  Tab 2: Your answers to the Critical Thinking Questions

  Tab 3: The data that you collected in Table 18

CRITICAL THINKING QUESTIONS

As you begin this activity, consider the following questions. You will come back to these questions once you have completed the activity.

1. How much variation is there in the headways between vehicles in the departing queue?

2. Based on the headways that you observed in the departing queue, what is your recommendation for the desired MAH?

Information

In previous activities, you learned that when the minimum green and vehicle extension times are set to zero, the phase will terminate immediately when the detection zone becomes unoccupied. You also learned that the detection zone length alone will not guarantee good quality of service to the motorist, since the phase may terminate before the entire queue has been served. This is especially true for the shorter 22 foot detection zone, in which it is difficult to have more than one vehicle in the zone at the same time. Finally, you learned that the minimum green time should be long enough to make sure that the queue immediately upstream of the detection zone begins to move, and enters the detection zone before the green indication prematurely ends.

In this activity, the detection zone length is 22 feet. The minimum green time is set to 7 seconds. The vehicle extension time is set to 5 seconds, a very conservative (high) value. You will observe the operation of the southbound movement, developing an understanding of the normal variation of headways in a departing queue. You will also identify the desired maximum allowable headway, the longest headway in a departing queue that you are willing to tolerate without terminating the green indication. This will help you to understand how to establish the vehicle extension time, which will be covered in Activities #36 and #37.

Task 1

Open the movie file: "A34.wmv."

Task 2

Observe and record headways for one cycle.

- When the simulation time reaches $t = 66.1$ seconds, pause and observe the status of the traffic flow and the timing processes; record your observations
- Begin the simulation again
- Record the following values in Table 18 for phase 4 serving the southbound through movement
  - Record the simulation clock time that the indication changes to green ("Start of green" in the table)
  - Record the clock time that the front of each vehicle reaches the stop bar
  - Compute the headway for each vehicle (the time difference between when this vehicle enters the intersection and when the previous vehicle entered the intersection) and record your results in the table

Vehicle number	Start of green	Time front of vehicle reaches stop bar	Headway
1			
2			
3			
4			
5			
6			
7			
8			
9			
10			

Table 18. Data collection table

Student Notes:

ACTIVITY
35 Relating Headway to Unoccupancy Time and Vehicle Extension Time

PURPOSE

The purpose of this activity is to give you the opportunity to learn how to determine the vehicle extension time based on the detection zone length and the desired maximum allowable headway (MAH).

LEARNING OBJECTIVES

- Relate the MAH to unoccupancy time
- Determine the vehicle extension time based on the length of the detection zone and the desired MAH

REQUIRED RESOURCE

- Movie file: A35.wmv

DELIVERABLE

- Prepare a spreadsheet that includes the following information:

  Tab 1: Title page with activity number and title, authors, and date completed

  Tab 2: Answers to the Critical Thinking Questions

  Tab 3: A brief summary of your observations from the tasks which follow. Note any patterns that you see between the headway data and the unoccupancy time data. What would be the basis of a relationship between these two parameters?

  Tab 4: The data that you collected in Table 19

CRITICAL THINKING QUESTIONS

As you begin this activity, consider the following questions. You will come back to these questions once you have completed the activity.

1. What is your recommendation for the vehicle extension time, based on your recommended desired maximum headway? Explain your answer.

2. If the detection zone length was longer than 22 feet, would your recommended vehicle extension time value be higher or lower? Explain your answer.

INFORMATION

In Activity #34, you observed the normal variation in headways in a departing queue, and based on these observations, you selected a MAH that represents the longest headway in a departing queue that you are willing to tolerate without terminating the green indication. In this activity, you will relate this headway to its equivalent unoccupancy time. You will then select a vehicle extension time based on this unoccupancy time that, in combination with the detection zone length, ensures both efficient operation and good service quality. In this activity, the detection zone length is 22 feet and the minimum green time is set to 7 seconds. The vehicle extension time is set to 5 seconds.

TASK 1

Open the movie file: "A35.wmv."

TASK 2

Collect data.

- The minimum green time is set to 7 seconds and the vehicle extension time is set to 5 seconds
- Move the simulation time to 66.0
- Run the simulation and observe the operation of the southbound approach
- At t = 66.1 seconds, advance the simulation. Record the following values in Table 19 for phase 4 (serving the SB through movement)
    - Record the simulation clock time that the display changes to green ("Start of green" in the table)
    - Record the clock time that the front of each vehicle enters the zone and the rear of each vehicle exits the zone. The entry time for the first vehicle is noted in the table ("14.0").
    - Record the clock time that the display changes to yellow ("Start of yellow" in the table)
    - Compute the unoccupancy time for each vehicle pair and record the value in the "Unoccupancy time" column. The unoccupancy time is the difference in the clock time that the front of the vehicle enters the zone and the clock time that the rear of the previous vehicle exits the zone. If the value is negative, a zero should be entered.

Vehicle Number	Start of green	Start of yellow	Front of vehicle enters zone	Rear of vehicle exits zone	Headway	Unoccupancy time
1			14.0		1.2	
2					2.4	
3					1.7	
4					1.9	
5					1.8	
6					1.8	
7					1.7	
8					1.9	
9					1.4	
10					1.7	

Table 19. Data collection table

Student Notes:

ACTIVITY

36 Determining the Maximum Allowable Headway

Purpose

The purpose of this activity is to give you the opportunity to determine the maximum allowable headway (MAH) for your intersection.

Learning Objective

- Select an MAH

Required Resources

- VISSIM input file created in Activity #28
- Phase Termination Analysis Excel template

Deliverable

- Prepare an Excel spreadsheet that includes the phase termination analysis results and data analysis outcomes, as well as the results from Task 5. The Excel worksheet should follow the formatting outlined below:

  Tab 1: Title page with activity number and title, authors, and date completed

  Tab 2: Raw data from the MER file

  Tab 3: Headway data for both lanes

  Tab 4: Headway distribution analysis showing frequency data, frequency plot, and cumulative frequency plot for one lane

  Tab 5: Data for the phase termination analysis

  Tab 6: Phase termination analysis using template including a summary table containing the number of occurrences of each termination type

  Tab 7: Summary of phase termination analysis including selection of and justification for the MAH

Critical Thinking Questions

1. How does increasing the MAH affect the likelihood of a Type 1, Type 2, or Type 3 termination?

2. Describe the differences that you observe in the frequency distributions for queued and non-queued vehicles from your data analysis.

3. What is the result of a phase termination analysis?

4. What is the basis for your selection of the MAH? Use the results from your phase termination analysis in explaining your answer.

Information

The MAH is the largest headway that you will tolerate in a departing queue before the phase should terminate. The choice of this headway involves balancing two conflicting and competing issues: if the headway that you select is too small, then you run the risk of terminating the phase too soon and not serving all of the vehicles in the queue that formed during red. However, if the headway that you select is too large, then you run the risk of allowing the phase to extend too long, serving not just the initial queue but also vehicles that arrive after the initial queue has cleared. One problem comes in recognizing that some of the headways that you observe after the queue has cleared might be in the same range as those that you observed during the queue clearance. Conversely, a slowly reacting driver in a vehicle that is part of a departing queue will result in a headway that is longer than the normal saturation headway. The choice that you make in the value of the MAH will have some risk of both conditions: A Type 1 termination, or cycle failure, when not all of the initial queue is served or a Type 2 termination, an inefficient extension of the green, resulting in longer delays on the other approaches. While the ideal goal is to achieve a Type 3 termination (when the phase terminates just after the queue has cleared) each time a phase terminates, your challenge is to find a MAH that balances the risks of the Type 1 and Type 2 terminations.

So what is a phase termination analysis? A phase termination analysis is a tool that looks at the headway data generated by a simulation model from a stream of vehicles departing from an intersection and, given a value of the MAH, classifies each phase termination into one the three types described above. The first part of the analysis involves the stream of headways for the departing vehicles, with the stream separated into vehicles that were a part of the queue (noted by Q) and those that arrived after the queue had cleared (noted by NQ). An example of the headway data are shown in the first two columns of Figure 119. Note that when you collect your headway data in this activity, the passage time will be set to a very high value (5 seconds) so that you can observe a sufficient sample of both queued and non-queued vehicles and their resulting headways departing after the beginning of green. The high passage time provides you with enough vehicles (and resulting headways) to allow you to study different values of MAH and their effects on phase termination.

The next step in the analysis involves superimposing a value for the MAH that determines when the display would change from green to red. The display would continue as "green" as long as the headway in the data stream is less than the MAH, but would change to "red" (for simplicity, we've skipped "yellow" here) when a headway exceeded this value. The "Ideal Signal Display", column 3 in Figure 119, shows the display that would result from the MAH that has been selected: the green would be displayed until a headway occurs that is greater than the MAH. The "Change Occurs" column (column 4) shows when the change from green to red would occur.

The third part of the analysis, the "Termination Outcomes", determines whether this display change occurs when queued vehicles are still being served (a Type 1 termination) or whether the green would continue to be displayed even when non-queued vehicles would be served (a Type 2 termination). In this example, the two queued vehicles are served, but the first three non-queued vehicles are served as well. The display doesn't change to red until the sixth headway (7.79 seconds), which is the first headway in this example to exceed the MAH of 3.0 seconds.

In summary, the phase termination analysis subjects the simulated headway data stream to a range of values of the MAH to determine how effective each MAH value would be in serving these vehicles. We know whether each vehicle is a part of the queue or not, and we can determine for a given MAH if a vehicle would be served or not. In addition to the information described above, the phase termination analysis as shown in Figure 119 also provides:

- The Outcome Distribution for the value of the MAH selected; in this example there are three Type 1, five Type 3, and nine Type 2 terminations
- The Percentile is the percentage of vehicles with headways that are less than the MAH (in this example, 98.2%)
- The "Results" summarizes the Percentiles and the number of terminations for each type for four MAH values (ranging from 1.5 seconds to 3.0 seconds)

MAH	Percentile
3.0	98.2%

Headway (sec)		Signal Information		Termination Outcomes		
Queued	NonQueued	Ideal Signal Display	Change Occurs	Type 1	Type 3	Type 2
1.15		Green				
1.96		Green				
	1.15	Green				
	1.3	Green				
	1.91	Green				
	7.79	Red	Change			1
	8.03	Red				
	5.14	Red				

Outcome Distribution	
Type 1 Termination	3
Type 3 Termination	5
Type 2 Termination	9

Results of Phase Termination Analysis

Headway	1.5	2.0	2.5	3.0
Percentile	32.0%	76.4%	95.9%	98.2%
Type 1 Termination	17	14	6	3
Type 3 Termination	0	1	6	5
Type 2 Termination	0	2	5	9

Figure 119. Example phase termination analysis procedure

Task 1

Initial steps.

- Make a copy of the folder that includes your VISSIM data files from Activity #28. Name this new folder "a36". Use this VISSIM file as the basis for your analysis and design of the passage time.
- Select one approach on one of the major streets of your intersection for your headway study

- Increase the volume on this link to produce "beginning of green" queue lengths of 10 to 15 vehicles per cycle per lane
- Set the passage time to 5.0 seconds on the approach that you have selected
- Establish "data collection" points just downstream of the signal head for the two lanes on the approach that you selected previously. For instructions on how to add a Data Collector in VISSIM see the VISSIM tutorial.
- Select and configure the "Data Collection" evaluation file. To do this, make sure to check "raw data" in configuration box.

Task 2

Headway data.

- Run VISSIM
- Open and parse the MER file in tab 2 of your Excel template
- Copy the parsed data to tab 3 of your Excel file. Keep only three columns: CP, t(leave), and tQueue. Delete the other columns. Figure 120 shows the layout of the MER file after it has been opened in Excel and parsed.

Data Collection (Raw Data)

File: f:\sweet and us95 high volume\team1_lab3.inp
Comment:
Date: Monday, May 28, 2012 12:21:31 PM
VISSIM: 5.30-08 [29295]

Data Collection Point 1: Link 1 Lane 2 at 143.633 m, Length 0.000 m.
Data Collection Point 2: Link 1 Lane 1 at 143.610 m, Length 0.000 m.

Data C.P.	t(enter)	t(leave)	VehNo	Type	Line	v[m/s]	a[m/s²]	Occ	Pers	tQueue	VehLength[m]
2	16.42	-1	2	100	0	14.3	-0.29	0.08	1	0	4.5
2	-1	16.73	2	100	0	14.2	-0.29	0.03	1	0	4.5
1	66.77	-1	6	100	0	1.4	2.77	0.03	1	42.7	4.5
2	66.82	-1	7	100	0	1.5	2.93	0.08	1	42.6	4.5
1	-1	68.1	6	100	0	5.2	2.6	0	1	42.7	4.5
2	-1	68.13	7	100	0	5.2	2.55	0.03	1	42.6	4.5
1	69.57	-1	8	100	0	6.5	2.86	0.03	1	42.1	4.5
2	69.85	-1	14	100	0	6.9	2.95	0.05	1	30.8	4.5
1	-1	70.18	8	100	0	8.2	2.71	0.08	1	42.1	4.5
2	-1	70.44	14	100	0	8.5	2.82	0.04	1	30.8	4.5
1	71.62	-1	10	100	0	8.6	2.4	0.08	1	38.5	4.5

Keep three columns:
- Data C.P, designating the data collection point or lane
- t(leave), the time that the vehicle leaves the data collection point
- tQueue, the time that the vehicle spends in queue

Figure 120. VISSIM MER file in Excel

- Eliminate all rows in which t(leave) equals "–1" or in which t(leave) is less than 300
- Separate the data into two separate tables, one for each of the two data collection points (lanes 1 and 2)
- Add a new column for each table that identifies whether the vehicle was a part of the queue or not. This can be done by putting a Q or NQ in the cell adjacent to the tQueue data. Figure 121 shows an example of the determination of Q or NQ based on whether tQueue is zero or greater than zero using an Excel logic function.
- In a new column, compute the headway between each vehicle pair as shown in Figure 122

Data C.P.	t(leave)	VehNo	tQueue	Q/NQ
2	300.42	161	11.8	Q
2	301.41	164	10	Q
2	302.87	167	8.9	Q
2	304.76	171	4.5	Q
2	305.97	176	2	Q
2	315.15	184	0	NQ
2	318.14	187	0	NQ
2	324.84	190	0	NQ

If(tQueue>0,"Q","NQ")

Figure 121. Example determination whether a vehicle is in a queue or not (Q or NQ)

Data C.P.	t(leave)	VehNo	tQueue	Q/NQ	Headway
2	300.42	161	11.8	Q	
2	301.41	164	10	Q	1.0
2	302.87	167	8.9	Q	1.5
2	304.76	171	4.5	Q	1.9
2	305.97	176	2	Q	1.2
2	315.15	184	0	NQ	9.2
2	318.14	187	0	NQ	3.0
2	324.84	190	0	NQ	6.7

Figure 122. Example calculation of headway based on the "leave" times of two consecutive vehicles

Task 3

Headway distribution analysis.

- Using the headway data created in Task 2 for one of the lanes only, create frequency (histogram) and cumulative frequency plots using 0.25 second bins in tab 4 of the Excel template. Figure 123 shows an example of the data used to create the frequency and cumulative frequency plots including (for headway bins from 0.00 to 2.00 seconds) the frequency (number), the percent frequency, the cumulative frequency, and the percent cumulative frequency.

Bin	*Frequency*	*Freq%*	*CumFreq*	*CumFreq%*
0.00	0	0.000	0	0.000
0.25	0	0.000	0	0.000
0.50	0	0.000	0	0.000
0.75	0	0.000	0	0.000
1.00	0	0.000	0	0.000
1.25	16	0.136	16	0.090
1.50	42	0.356	58	0.328
1.75	36	0.305	94	0.531
2.00	42	0.356	136	0.768

Figure 123. Example frequency and cumulative frequency headway data

- Figure 124 and Figure 125 show the resulting frequency and cumulative frequency plots based on the frequency data shown in Figure 123

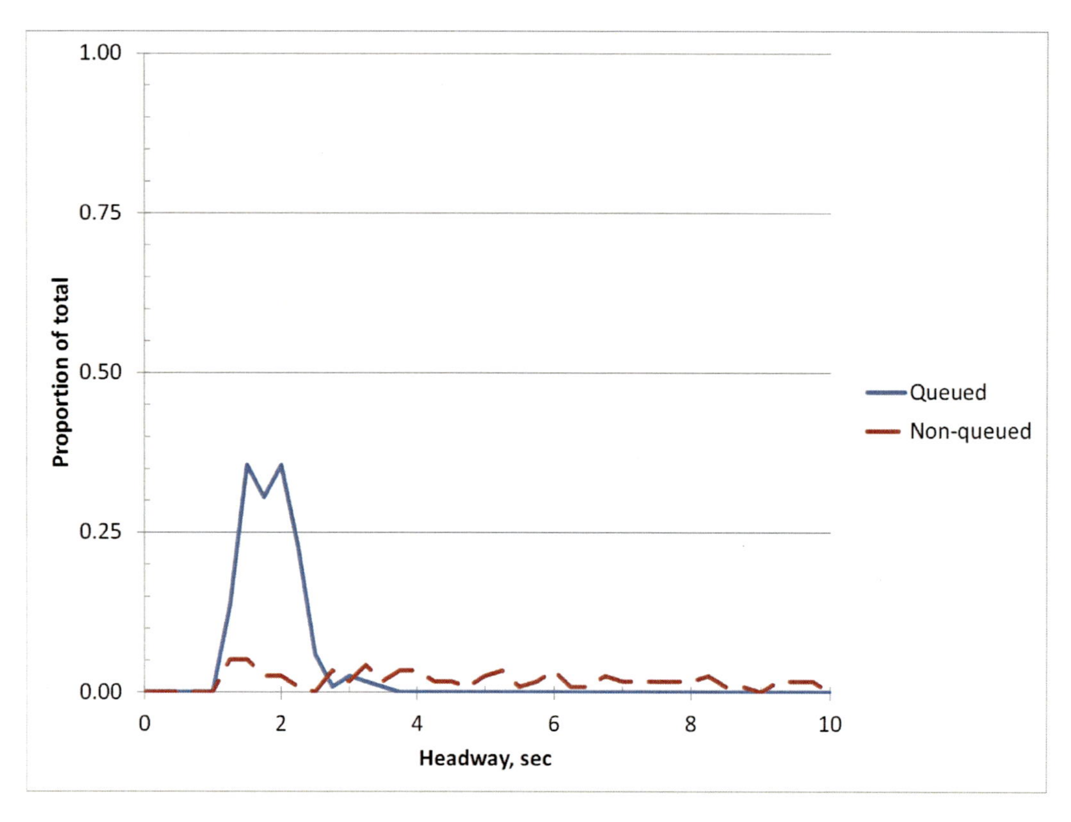

Figure 124. Headway frequency plot for queue and non-queued vehicles

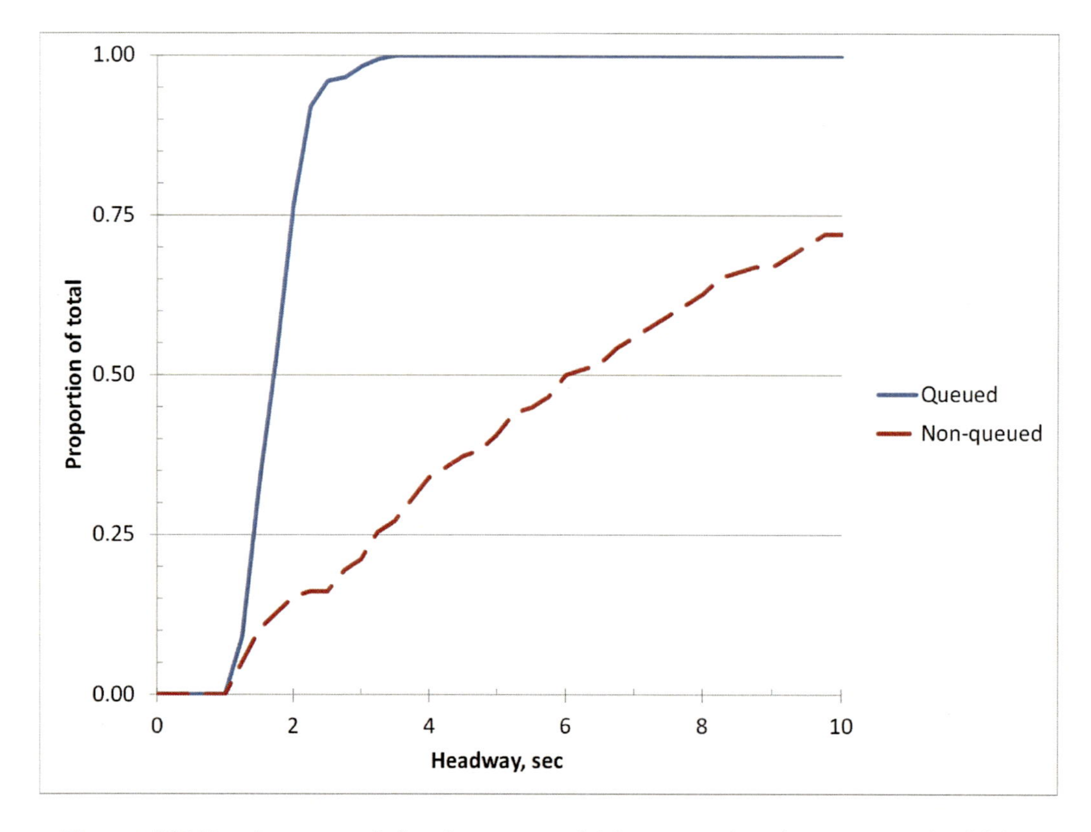

Figure 125 Headway cumulative frequency plot for queued and non-queued vehicles

TASK 4

Phase termination analysis.

- Create the data that you will need for the phases termination analysis in tab 5. Figure 126 shows two columns that have been created that place the queued and non-queued headway in the appropriate columns.

							Headway (sec)	
Data C.P.	t(leave)	VehNo	tQueue	Q/NQ	Headway		Queued	Nonqueued
1	300.18	163	9.6	Q				
1	301.33	165	9.1	Q	1.1		1.15	
1	303.29	169	5	Q	2.0		1.96	
1	304.44	174	0	NQ	1.1			1.15
1	305.74	177	0	NQ	1.3			1.3
1	307.65	179	0	NQ	1.9			1.91
1	315.44	186	0	NQ	7.8			7.79
1	323.47	189	0	NQ	8.0			8.03
1	328.61	193	0	NQ	5.1			5.14

Figure 126. Creation of input data for phase termination analysis

- Copy the headways for the queued and non-queued vehicles into the two input columns in the phase termination analysis spreadsheet template to tab 6 of your Excel template. The data should be pasted as "values." Figure 127 shows the headway data that have been pasted into the first two columns of the phase termination analysis template. The Signal Information and Termination Outcomes are calculated based on these input data.

Headway (sec)		Signal Information		Termination Outcomes		
Queued	NonQueued	Ideal Signal Display	Change Occurs	Type 1	Type 3	Type 2
1.15		Green				
1.96		Red	Change	1		
	1.15	Red				
	1.3	Red				
	1.91	Red				
	7.79	Red				
	8.03	Red				
	5.14	Red				

Figure 127. Outcomes for phase termination analysis showing type 1 termination

- Use the Phase Termination Analysis template to determine the Termination Outcomes for a range of possible MAH. Enter a value in the MAH cell and record the resulting number of Types 1, 2, and 3 terminations in the "results" section of the template. Each time you enter a new value for the MAH, a new set of outcomes will be calculated. Figure 128 shows the results of a phase termination analysis for four MAH cases, ranging from 1.5 seconds to 3.0 seconds. When the MAH was set to 1.5 seconds, the phase terminated before the queue had cleared for all 17 phases, and only 32 percent of the vehicles would be served. At the other extreme, if the MAH was set to 3.0 seconds, the phase would extend past the time the queue had cleared in 9 of the 17 cases (Type 2 terminations).

MAH	Percentile
3.0	98.2%

Headway (sec)		Signal Information		Termination Outcomes		
Queued	NonQueued	Ideal Signal Display	Change Occurs	Type 1	Type 3	Type 2
1.15		Green				
1.96		Green				
	1.15	Green				
	1.3	Green				
	1.91	Green				
	7.79	Red	Change			1
	8.03	Red				
	5.14	Red				

Outcome Distribution

Type 1 Termination	3
Type 3 Termination	5
Type 2 Termination	9

Results of Phase Termination Analysis

Headway	1.5	2.0	2.5	3.0
Percentile	32.0%	76.4%	95.9%	98.2%
Type 1 Termination	17	14	6	3
Type 3 Termination	0	1	6	5
Type 2 Termination	0	2	5	9

Figure 128. Phase termination analysis results for four example maximum allowable headways

Task 5

Select the MAH.

- Write a goal statement for the selection of the MAH based on your desired balance of type 1 and type 2 terminations
- Select your MAH using your goal statement and the analysis that you have completed as the basis for your selection

ACTIVITY

37 Determining the Passage Time

PURPOSE

The purpose of this activity is to give you the opportunity to select and justify design values for passage time.

LEARNING OBJECTIVES

- Compare headway distributions for one lane and two lane data
- Select passage time values for one lane and two lane approaches

REQUIRED RESOURCE

- Excel template and results from Activity #36

DELIVERABLE

- Excel spreadsheet that includes the selected design parameters (passage time), as well as justification for these values. This should be the same Excel file that you used for Activity #36 with the following tabs added:

  Tab 8: Headway analysis for two lane data

  Tab 9: Phase termination data

  Tab 10: Phase termination analysis for two lane data

  Tab 11: Performance data from VISSIM

  Tab 12: Analysis and summary based on Tasks 6, 7, and 8

CRITICAL THINKING QUESTIONS

1. What are the differences in the headway distributions that you prepared for the one lane approach and two lane approach conditions? Describe them in complete sentences.

2. Would the passage time for a two lane approach be lower or higher than for a one lane approach? Provide justification for your answer.

3. Can you think of a situation in which your answer to question #2 would change? If so, describe that situation.

INFORMATION

In Activity #36, you selected a value for the maximum allowable headway (MAH) that balanced early phase termination with inefficient extension of the phase for a one lane approach. In this activity you will consider the effect of a two lane approach, and how this might change the selection of the MAH.

Figure 129 and Figure 130 compare the headway distributions for one lane and two lane approaches for vehicles that were a part of a queue that existed at the beginning of the green. The headways are lower for the two lane case than for the one lane case since they are measured for consecutive vehicles that sometimes can depart from the approach at nearly the same time. In this latter case, a headway between two vehicles departing from the same approach but from different lanes could be near zero. For the case shown here, the mean headway for the two lane case is 0.8 seconds, while the mean for the one lane case is 1.7 seconds.

A phase termination analysis also shows the differences between the one lane and two lane cases, and why the MAH should be lower for the two lane case. Figure 131 and Figure 132 show the percent phase terminations for Types 1 and 2 terminations for the one lane and two lane data.

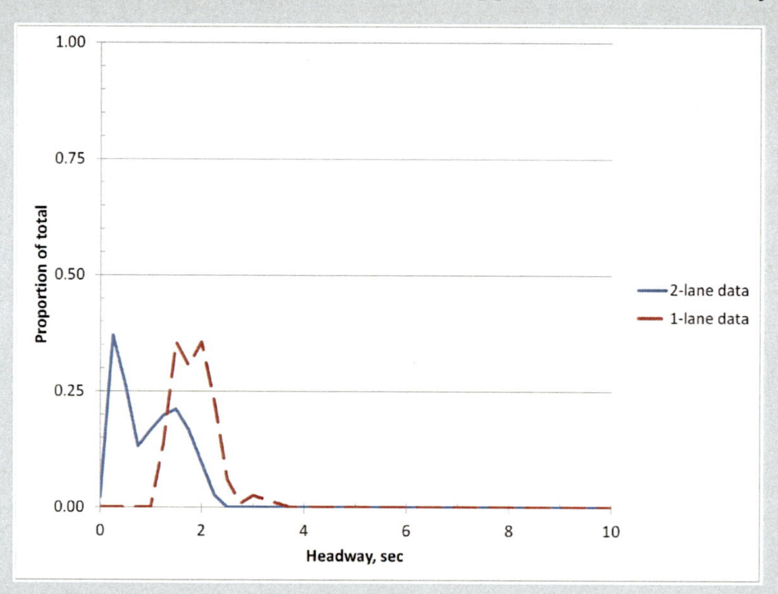

Figure 129. Headway density function, one lane and two lane data

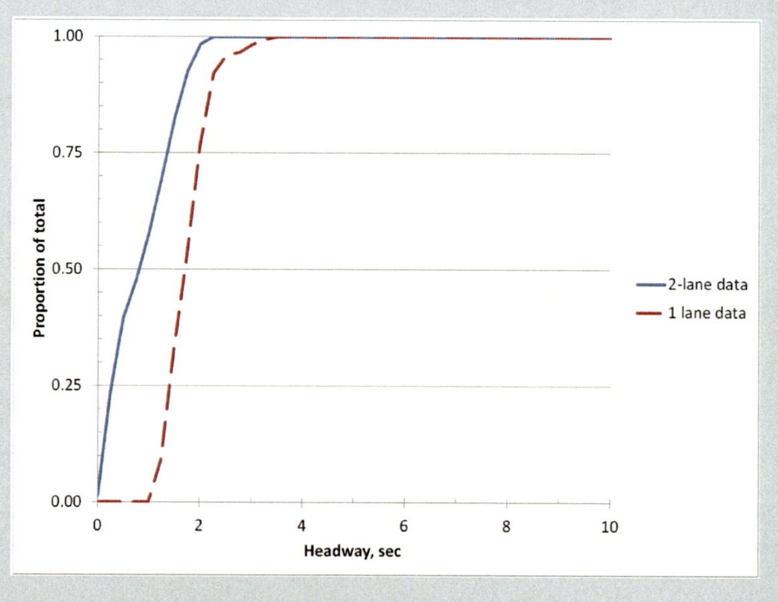

Figure 130. Cumulative headway density function, one lane and two lane data

In Figure 131 for example, the likelihood of a Type 1 termination (the phase terminates before the queue has cleared) for a given MAH is higher for the one lane case than for a two lane case. Stated another way, a smaller MAH can be used for a two lane case than for a one lane case to achieve the same result.

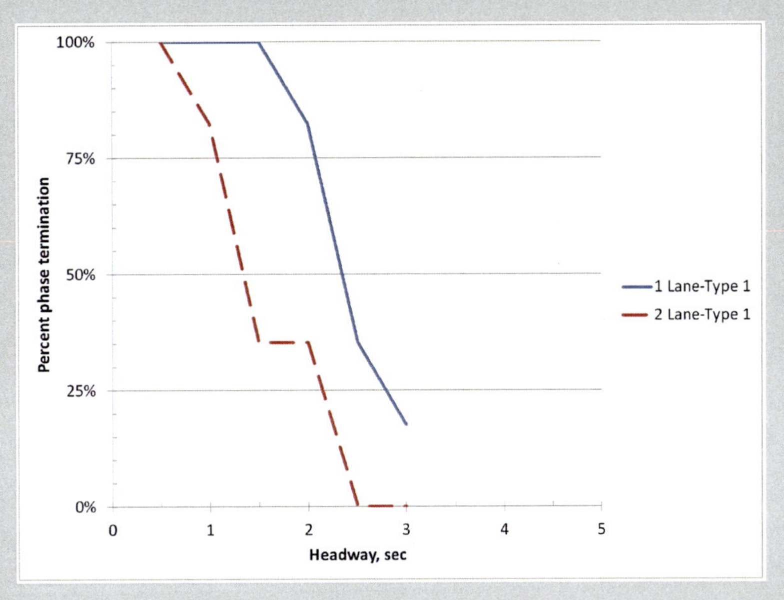

Figure 131. Frequency distribution for Type 1 terminations for the one lane and two lane cases

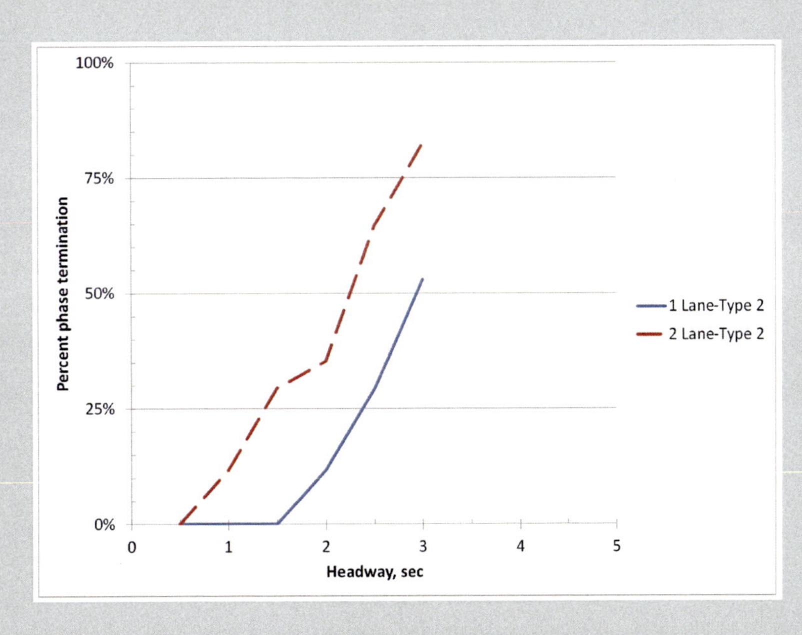

Figure 132. Frequency distribution for Type 2 terminations for the one lane and two lane cases.

Task 1

Initial steps.

- Open the Excel spreadsheet that you used in Activity #36
- Create new tabs for the following work:

  Tab 8: Headway analysis for two lane data

Tab 9: Phase termination data

Tab 10: Phase termination analysis for two lane data

Tab 11: Performance data from VISSIM

Tab 12: Analysis and summary based on Tasks 6, 7, and 8

TASK 2

Collect speed data.

- Review the speed data collected in the MER file as part of Activity #36
- Prepare a statistical analysis of the data, including a determination of the mean value and a density plot. Identify the speed data that represents vehicles traveling through the intersection after the queue begins to move after the start of green.
- Based on this analysis, select a speed value that you think is appropriate for the calculation of the passage time

TASK 3

Headway distributions for two lane data

- Using the data file created in Activity #36, create frequency (histogram) and cumulative frequency plots using 0.25 second bins based on the data for both lanes combined into one data set. These data should be placed in Tab 8 of your Excel file.
- Figure 133 and Figure 134 show examples of the resulting frequency and cumulative frequency plots for two lane data for both queued and non-queued vehicles

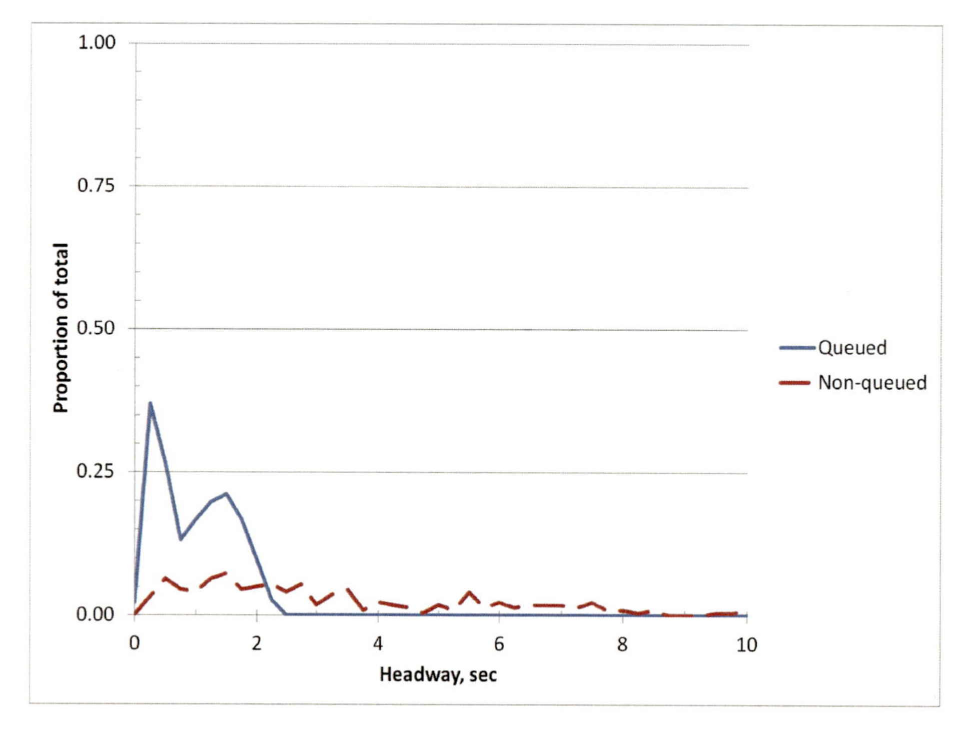

Figure 133. Headway frequency plot for queue and non-queued vehicles

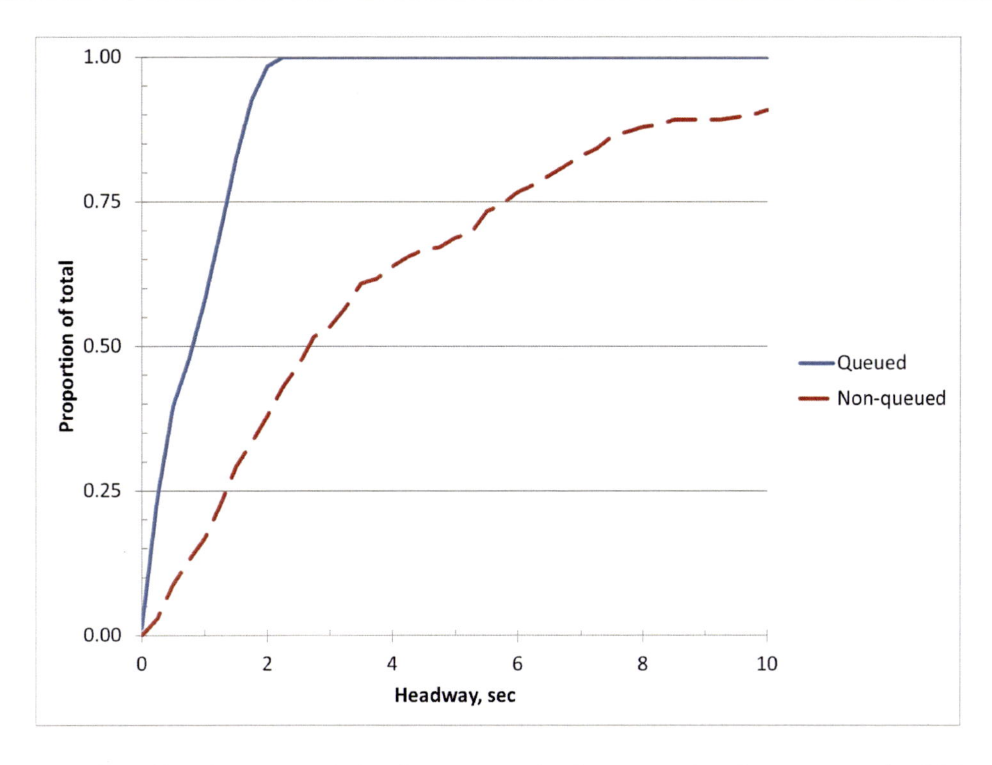

Figure 134. Headway cumulative frequency plot for queued and non-queued vehicles

Task 4

Phase termination analysis.

- Create the data that you will need for the two lane phases termination analysis in tab 9. Figure 135 shows two columns that have been created that place the queued and non-queued headways in the appropriate columns.

							Headway (sec)	
Data C.P.	t(leave)	VehNo	tQueue	Q/NQ	Headway		Queued	Nonqueued
1	300.18	163	9.6	Q				
2	300.42	161	11.8	Q	0.24		0.24	
1	301.33	165	9.1	Q	0.91		0.91	
2	301.41	164	10	Q	0.08		0.08	
2	302.87	167	8.9	Q	1.46		1.46	
1	303.29	169	5	Q	0.42		0.42	
1	304.44	174	0	NQ	1.15			1.15
2	304.76	171	4.5	Q	0.32		0.32	
1	305.74	177	0	NQ	0.98			0.98
2	305.97	176	2	Q	0.23		0.23	
1	307.65	179	0	NQ	1.68			1.68
2	315.15	184	0	NQ	7.5			7.5
1	315.44	186	0	NQ	0.29			0.29
2	318.14	187	0	NQ	2.7			2.7
1	323.47	189	0	NQ	5.33			5.33
2	324.84	190	0	NQ	1.37			1.37
1	328.61	193	0	NQ	3.77			3.77

Figure 135. Creation of input data for phase termination analysis

- Copy (and paste as "values") the headways for the queued and non-queued vehicles into the two input columns in the phase termination analysis spreadsheet template for the two lane analysis (tab 10 of your Excel file). Figure 136 shows the headway data that have been pasted into the first two columns of the spreadsheet. The Signal Information and Termination Outcomes are calculated based these input data.

Headway (sec)		Signal Information		Termination Outcomes		
Queued	NonQueued	Ideal Signal Display	Change Occurs	Type 1	Type 3	Type 2
0.24		Green				
0.91		Green				
0.08		Green				
1.46		Green				
0.42		Green				
	1.15	Green				
0.32		Green				
	0.98	Green				
0.23		Green				
	1.68	Green				
	7.5	Red	Change			1
	0.29	Red				
	2.7	Red				
	5.33	Red				
	1.37	Red				
	3.77	Red				

Figure 136. Outcomes for phase termination analysis showing type 1 termination

- Use the Phase Termination Analysis template to determine the Termination Outcomes for a range of possible values for the MAH for the two lane case. Enter a value in the MAH cell and record the resulting number of Types 1, 2, and 3 terminations. Figure 137 shows the results of a phase termination analysis for four MAH cases, ranging from 0.5 seconds to 2.0 seconds. When the MAH was set to 0.5 seconds for example, the phase terminated before the queue had cleared for all 17 phases, and only 39 percent of the vehicles would be served. At the other extreme, if the MAH was set to 2.0 seconds, the phase would extend past the time the queue had cleared in 6 of the 17 cases.

MAH	Percentile
3.0	#N/A

Headway (sec)		Signal Information		Termination Outcomes		
Queued	NonQueued	Ideal Signal Display	Change Occurs	Type 1	Type 3	Type 2
0.24		Green				
0.91		Green				
0.08		Green				
1.46		Green				
0.42		Green				
	1.15	Green				
0.32		Green				
	0.98	Green				
0.23		Green				
	1.68	Green				
	7.5	Red	Change			1
	0.29	Red				
	2.7	Red				
	5.33	Red				
	1.37	Red				
	3.77	Red				

Outcome Distribution	
Type 1 Termination	0
Type 3 Termination	3
Type 2 Termination	14

Results of Phase Termination Analysis				
Headway	0.5	1.0	1.5	2.0
Percentile	38.9%	57.0%	81.9%	98.1%
Type 1 Termination	17	14	6	6
Type 3 Termination	0	1	6	5
Type 2 Termination	0	2	5	6

Figure 137. Phase termination analysis results for four example maximum allowable headways

TASK 5

Select the MAH.

- Write a goal statement for the selection of the MAH based on your desired balance of type 1 and type 2 terminations for the two lane case
- Select the MAH for the two lane case using your goal statement and the analysis that you have completed as the basis for your selection

TASK 6

Compare one lane and two lane data.

- Compare the mean values of the headway distributions for the queued and non-queued vehicles for the one lane and two lane cases, and the frequency distributions for these two cases
- Compare the results of the phase termination analyses for both the one lane and two lane cases

TASK 7

Compute passage times.

- Compute the passage times for the one and two lane cases using the MAH values that you have selected and the speed value that you determined in Task 2. Assume an average vehicle length of 20 feet and a detection zone length of 22 feet.

TASK 8

Observe the VISSIM simulation using new values of passage time.

- Set the passage times that you selected in Task 7 into your VISSIM network for both the one lane and two lane approaches
- Set the detection zone lengths to 22 feet and the minimum green time to 5 seconds
- Gather the performance data (average delay and average queue length) and summarize these data for the individual approaches and for the intersection
- Observe the operation and make conclusions about what you see, and compare this with your performance data

Student Notes:

ACTIVITY

38 Actuated Traffic Control Processes

PURPOSE

The purpose of this activity is to give you the opportunity to compare the results of your design work (Activities #36 and #37) to the range of values discussed in the *Traffic Signal Timing Manual*.

LEARNING OBJECTIVE

- Contrast design values with those recommended in practice

REQUIRED RESOURCE

- *Traffic Signal Timing Manual*

DELIVERABLES

Prepare a document that includes

- Answers to the Critical Thinking Question
- Completed Concept Map

LINK TO PRACTICE

Read the section on "Phase Intervals and Basic Parameters" and "Actuating Timing Parameters" from the *Traffic Signal Timing Manual* as assigned by your instructor.

CRITICAL THINKING QUESTION

When you have completed the reading, prepare answers to the following question:

1. Describe the differences between your selected passage time value and the value ranges described in the *Traffic Signal Timing Manual*.

In My Practice...

by Tom Urbanik

In practice, you must resolve theoretical calculations with the realities of drivers and the technology that is deployed. The practical goal is efficient control without generating complaints or trouble calls due to an occasional short-timing of a phase. The activities that you have completed in this chapter included the issue of a slow truck causing a phase to gap out. In practice an inattentive driver could also fail to reset the passage timer in a detector design using small area (e.g., 6' by 6' loop) detection. This could occur if the detection zone was between two cars and the second car did not move over the detection zone before the passage timer expired. While this problem is largely overcome by using presence detection, it is still possible for the detection zone to be located between the two vehicles.

Although partially addressed in Activity #37 for two lanes, the complexity of multi-lane detection which sends the detector call to a single phase timer, makes selection of the passage time problematic. The passage time model you considered in Activity #36 was for a single lane. It does not account for calls on two or more lanes. Straggling cars in three lanes may look like closely spaced cars in a single lane. The partial solution is to adjust the passage time down (which could result in the phase occasionally running too short) or using a single lane value which can result in extending the phase even though the flow rate is much less than saturation flow.

So, the traffic signal timing engineer has to balance these practical issues in application of the model by making adjustments to the ideal passage setting or using advanced features. While not extensively used (unfortunately), there are advanced features to address these issues. Although these advanced features are not included in this course, they can be found in the *Traffic Signal Timing Manual*.

CONCEPT MAP

Terms and variables that should appear in your map are listed below.

call	occupancy time	maximum allowable headway	h	t_o
detection zone	recall		L_d	t_u
interval	unoccupancy time		L_v	v

Student Notes:

Timing Processes for the Intersection

PURPOSE

In Chapter 6, you studied the operation of one intersection approach and determined the value of the vehicle extension time that would extend the green for as long as a queue existed, but not longer. The focus was on the traffic flow and signal timing on one approach only, with no consideration of the effect of this value on the other intersection approaches. In this chapter, now that you have established a basic understanding of the timing processes for one approach of an intersection, we will consider how the timing parameters on one approach affect the operation on the other approaches and the intersection overall.

LEARNING OBJECTIVES

When you have completed the activities in this chapter, you will be able to

- Describe the function of the maximum green time
- Describe the effect of the maximum green time on the cycle length and delay
- Describe the maximum green time setting and timer process
- Determine the optimal maximum green time (based on the optimal cycle length) at a signalized intersection
- Determining the effect of the minor street vehicle extension setting on the efficiency of major street and intersection operations
- Describe the advantages and disadvantages of increasing maximum green time on intersection operations
- Set the maximum green time for both approaches of an intersection, balancing the performance of both the minor street and the major street
- Compare the maximum green times that you developed in your design with those used in practice

CHAPTER OVERVIEW

This chapter begins with a *Reading* (Activity #39) on the relationship between the maximum green time, the cycle length, and delay. The chapter also includes four activities including an assessment of your understanding of the effect of the maximum green time on intersection operations (Activity #40), the effect of long vehicle extension times on one approach on the operation of the other approaches (Activity #41), and the pros and cons of increasing the maximum green time in order to solve the problem of phase failure (Activity #42). You will complete a design activity (Activity #43) in which you will select the maximum green time for the design project with which you have been working. The chapter concludes with an *In Practice* activity (Activity #44) with background on the maximum green time, both recommended values for its setting and how this timing process works.

Activity List

Number and Title		Type
39	Maximum Green Time, Cycle Length, and Delay	*Reading*
40	What Do You Know About Maximum Green Time, Cycle Length, and Delay?	*Assessment*
41	Determining the Effect of the Minor Street Vehicle Extension Time on Intersection Operations	*Discovery*
42	Determining the Effect of the Maximum Green Time on Intersection Operations	*Discovery*
43	Setting the Maximum Green Timing Parameter for All Approaches of an Intersection	*Design*
44	Maximum Green Time	*In Practice*

Maximum Green Time, Cycle Length, and Delay

Purpose

The purpose of this activity is to give you the opportunity to learn how the setting of the maximum green time limits the delay experienced by vehicles traveling through the intersection.

Learning Objectives

- Describe the function of the maximum green time
- Describe the effect of the maximum green time on the cycle length and delay

Deliverables

- Define the terms and variables in the Glossary
- Prepare a document that includes answers to the Critical Thinking Questions

Glossary

Provide a definition for each of the following terms or variables. Paraphrasing a formal definition (as provided by your text, instructor, or another resource) demonstrates that you understand the meaning of the term or phrase.

cycle length	
maximum green time	
uniform delay	
C	
d_1	
g	

G	
g/C	
r	
s	
v	

CRITICAL THINKING QUESTIONS

When you have completed the reading, prepare answers to the following questions.

1. Why does delay increase as cycle length increases?

2. What is the function of the maximum green time?

3. What is the process followed by the maximum green timer?

INFORMATION

Overview

In Chapter 6 you learned how to set the passage time in conjunction with the length of the detection zone so that the phase continues to time as long as the queue is being served, and no longer. But what if the volumes are so high that the phase extends long enough to unfairly delay vehicles on the other intersection approaches? How long should a phase be allowed to time? In this chapter we will consider the effect of the maximum green time (and, as a result, the cycle length) on the delay experienced by all users of the intersection.

Uniform Delay Equation

The Highway Capacity Manual provides an equation for computing control delay (the delay attributed to the control device) for an approach at a signalized intersection. The equation includes three terms, one each for the following components of delay: uniform delay, incremental delay, and initial queue delay. For low or moderate traffic volumes, the first term of this equation (the uniform delay term) provides a reasonable estimate of delay, as a function of cycle length (C), green time (g), volume (v), and saturation flow rate (s). You considered this equation in Chapter 2 (Activity #8).

$$d_1 = \frac{0.5C(1-g/C)^2}{1-(v/s)}$$

Another view of the uniform delay term, substituting r/C for $(1-g/C)$, is given at right:

$$d_1 = \frac{0.5C(r/C)^2}{1-(v/s)}$$

In both formulations, we can see the effect of green time (g), red time (r), and cycle length (C) on delay. Delay increases as the red time increases, and thus as the cycle length increases.

Delay and Cycle Length

Consider an example intersection for which there are two intersecting one-lane one-way streets. Figure 138 shows the delay for one approach, assuming a green ratio (g/C) of 0.5, volume of 500 veh/hr, and a saturation flow rate of 1800 veh/hr/green based on the uniform delay equation shown above. As the cycle length increases, the delay increases in a linear manner.

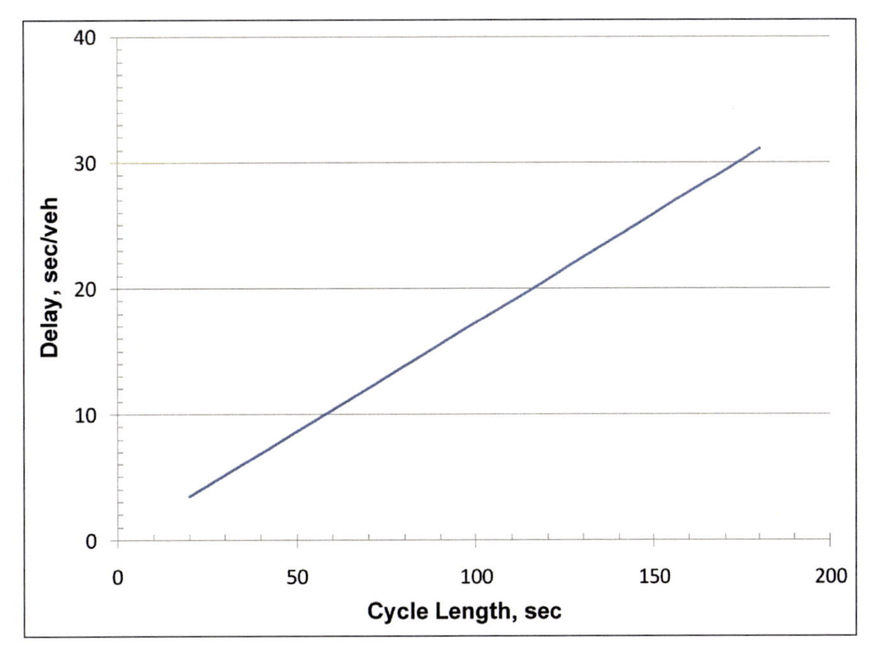

Figure 138. Delay vs. cycle length for one approach

How does this relate to efficient phase termination, particularly limiting the length of the cycle by setting a maximum green time for each approach? To understand this relationship, let's consider two cases, each case with different green interval durations. In case 1, the green time is half the length of the green time for case 2. In case 1, the red and green intervals are equal and their sum is the cycle length: $C_1 = r_1 + g_1$

For case 2, the red and green intervals are also the same, and the cycle length C_2 is twice the duration of C_1:

$$C_2 = 2C_1 = 2(r_1 + g_1) = r_2 + g_2$$

The timing for these two cases is illustrated in Figure 139.

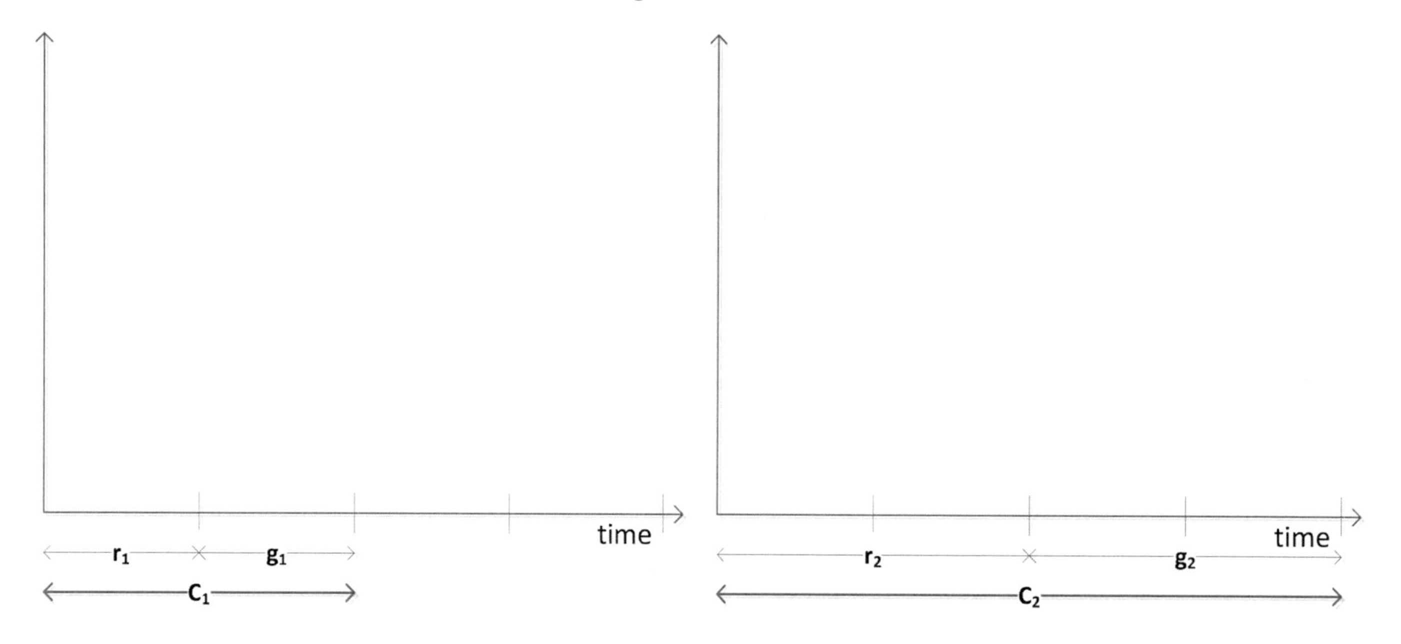

Figure 139. Two cases, with different green time durations

We will further assume that the arrival flow rates in both cases are the same and equal to v. We can illustrate this notion in the pair of cumulative vehicle diagrams shown in Figure 140, where the number of vehicles that have arrived at the intersection at the end of the second cycle in case 1 is equal to the number of vehicles that have arrived at the end of the first cycle in case 2. The slope of both lines is the arrival rate or volume, v.

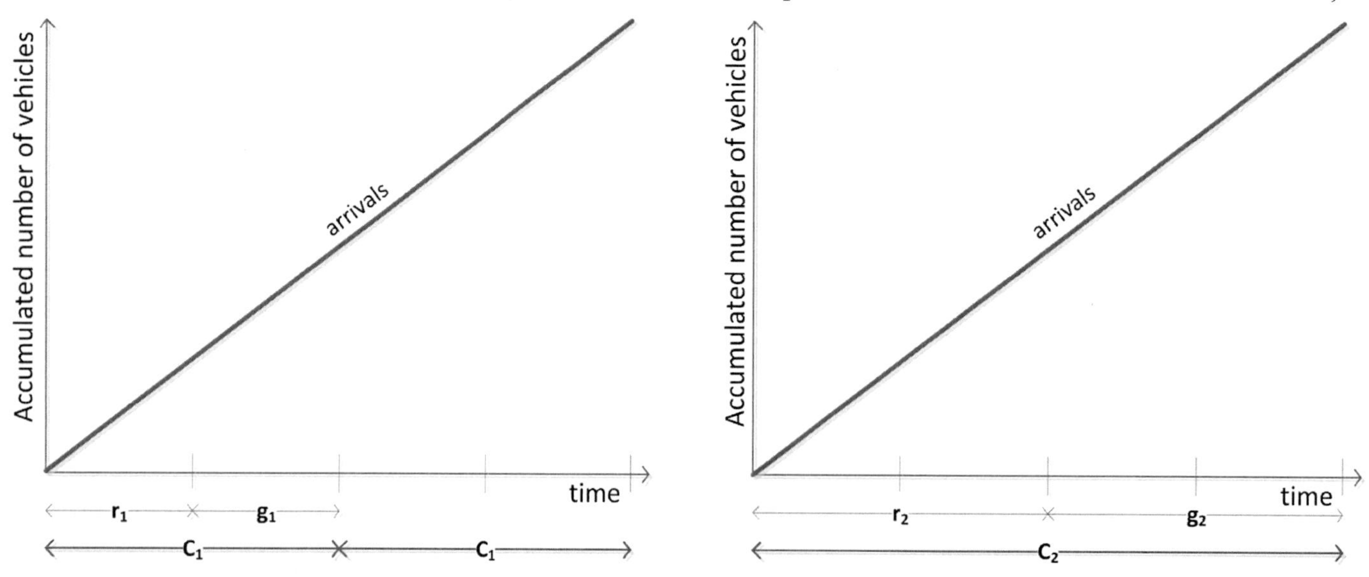

Figure 140. Cumulative vehicle diagrams for two cases

We will further assume that the green durations (g_1 and g_2) shown in each case are equal to the maximum green times for the cases and that the queues clear at the end of green. We can then show the departure curves for both cases, in Figure 141.

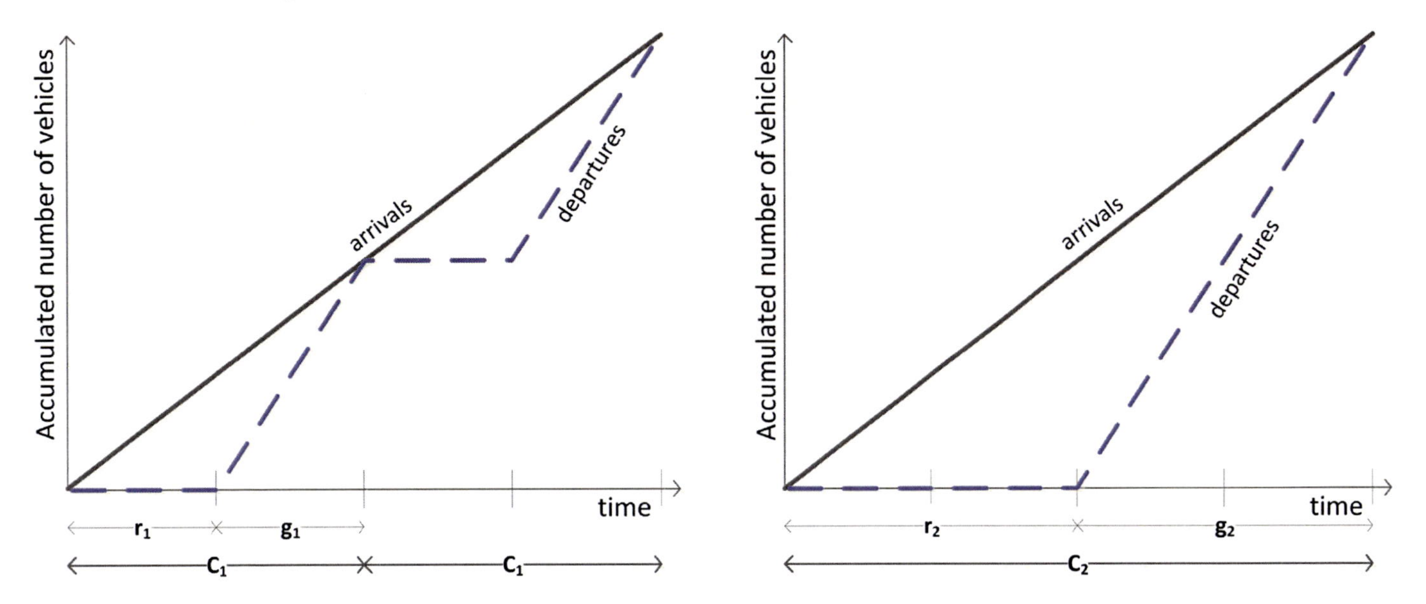

Figure 141. Cumulated Vehicle Diagrams, cases 1 and 2

As we saw in Chapter 2, another way of representing these flows is with a queue accumulation polygon. The vertical distance between the arrival and departure curves at any point in time is the queue length at that point in time. This vertical distance over time is shown in the queue accumulation polygons in Figure 142. In addition, the areas of the polygons (in this case, triangles) are equal to the total delay experienced by all users during that time interval.

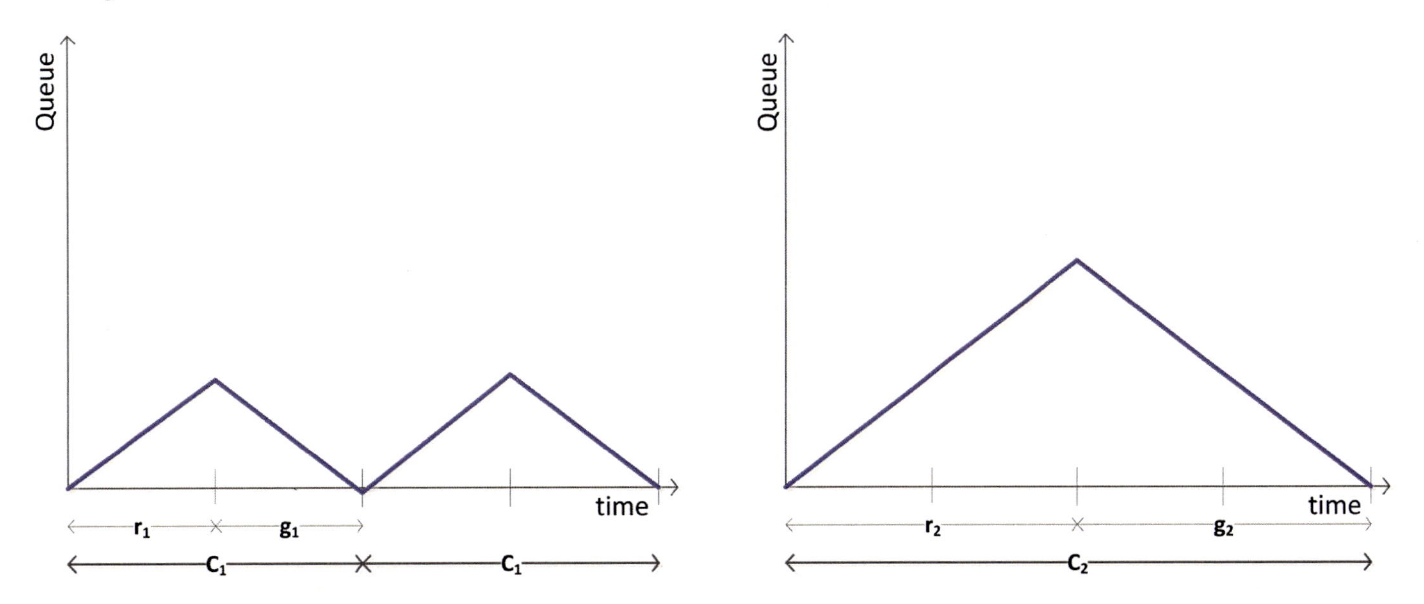

Figure 142. Queue accumulation polygons, cases 1 and 2

Figure 143 shows that the total delay for case 2, in which the maximum green time is twice as long as for case 1, is twice the delay for case 1.

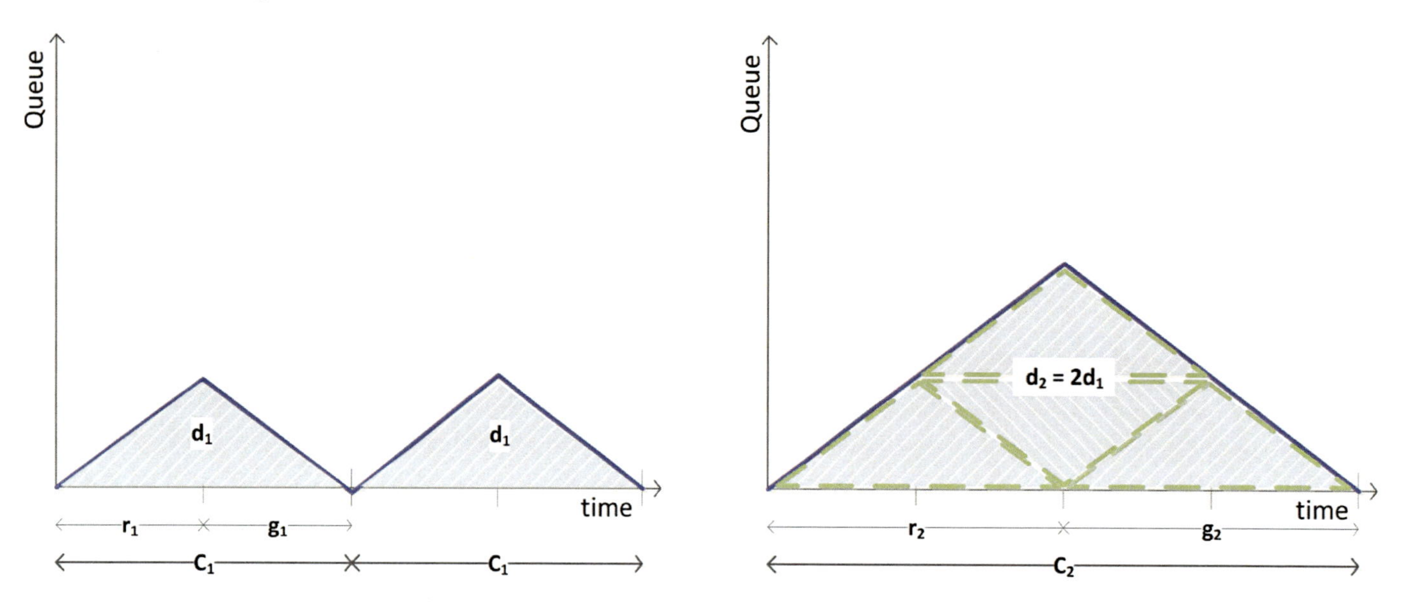

Figure 143. Total delay, cases 1 and 2

Example Calculation

Let's now look at a numerical example to validate what we just observed graphically. In this example, we'll show two cases, the first with a cycle length of 60 seconds, and then second with a cycle length of 120 seconds. In both cases the $g/C = 0.50$ and the arrival flow rate is 500 vehicles per hour.

For case 1, the area of the triangle, dt_1, is:

$$d_{t1} = 0.5(\text{base})(\text{height}) = 0.5C_1\ (vr_1) = 0.5(60)(4.2) = 125\ \text{veh} - \text{sec}$$

The average delay is equal to the total delay dt_1 divided by the number of vehicles that arrive during the cycle ($vC_1 = 8.3$ vehicles):

$$d_{a1} = \frac{d_{t1}}{vC_1} = \frac{125}{8.3} = 15\ \text{sec/veh}$$

For case 2, the total delay and the average delay are:

$$d_{t2} = 0.5(\text{base})(\text{height}) = 0.5C_1\ (vr_1) = 0.5(120)(8.3) = 500\ \text{veh} - \text{sec}$$

$$d_{a2} = \frac{d_{t2}}{vC_2} = \frac{500}{16.7} = 30\ \text{sec/veh}$$

So as we saw graphically, the delay doubles when the cycle length doubles.

Other Considerations

Much of the previous discussion in this reading focused on the effect of longer cycle length on increasing delay. But it is also important to note the impact of cycle length on intersection capacity. When the duration of the green is extended (through a longer maximum green time), the proportion of the cycle that is allocated to the change and clearance intervals (yellow and red clearance times) declines. A small but positive increase in intersection capacity results. Figure 144 shows this concept with three example cases, for cycle lengths of 30 seconds, 60 seconds, and 120 seconds. For each case, fixed yellow and red clearance times totaling 5 seconds (shown together as "red" in the figure) are assumed. For $C = 120$ seconds, the north-south movements would have 55 seconds of green (g_{NS}) followed by 5 seconds of yellow and red clearance times. The east-west movements would also have 55 seconds of green (g_{EW}) again followed by 5 seconds of yellow and red clearance times. This means that there is 110 seconds of green time available to serve the movements out of a cycle length of 120 seconds. Thus 92 percent of the cycle is available to serve traffic movements. However, for $C = 30$ seconds only 67 percent of the cycle is available for green time as the remainder is needed for the yellow and red clearance times.

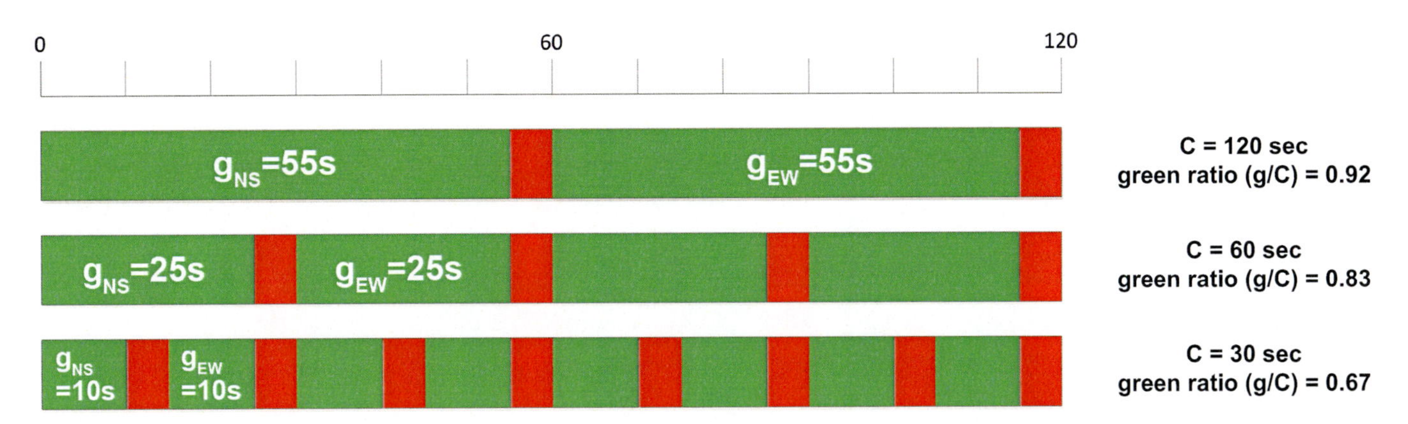

Figure 144. Proportion of green as a function of cycle length

Another consideration in setting the maximum green time is the impact on the way in which a phase terminates. For several reasons, most of which are beyond the scope of this book, it is preferable for a phase to terminate by gapping out. So, if a phase terminates primarily by maxing out, this may indicate that the maximum green time setting may be too low.

Conclusion

So, again we find that setting a signal timing parameter involves a trade-off. We want to set the maximum green time long enough so that in most cases the phase will terminate by gapping out. But we want to make sure that the phase doesn't time so long that delay becomes too high. Finding this balance is the challenge that you will face in the design of the maximum green time parameter later in this chapter. It is important to note that this balance doesn't mean trying to equalize the number of gap outs and max outs: it does mean trying to ensure that the phase gaps out as often as possible, except when volumes are high during peak periods.

Student Notes:

ACTIVITY **40**

What Do You Know About Maximum Green Time, Cycle Length, and Delay?

PURPOSE

The purpose of this activity is to help you understand the role of the maximum green time in providing efficient intersection operations. You will also validate your understanding of how to use the uniform delay equation to determine the relationship between the delay and cycle length (and thus the maximum green time).

LEARNING OBJECTIVES

- Describe the maximum green time setting and timer process
- Determine the optimal maximum green time (based on the optimal cycle length) at a signalized intersection

DELIVERABLE

- Prepare a spreadsheet with the following information:

  Tab 1: Title page with activity number and title, authors, and date completed

  Tab 2: Summary of the results of your analysis from Tasks 1 and 2 and your answers to the Critical Thinking Questions

TASK 1

A model for computing the average delay when traffic is arriving at a signalized intersection at a uniform rate was described in Activity #39. It is reproduced below. Using an Excel spreadsheet, develop a VBA function to compute average uniform delay as a function of red time, green time, cycle length, volume, and saturation flow rate. See the Excel Tutorial for assistance in creating a VBA function.

$$d_a = \frac{0.5C(1-g/C)^2}{1-v/s}$$

TASK 2

Assuming a volume from one of the major street approaches of your simulation network, compute the average uniform delay per vehicle as a function of cycle length, with a range of cycle lengths from 40 seconds to 100 seconds. Prepare a graph of delay vs. cycle length for this range of values. Assume $g/C = 0.5$ and $s = 1900$ vehicles per hour of green.

CRITICAL THINKING QUESTIONS

1. Prepare a brief discussion of the implications of your analysis for the maximum green time setting for your network. What limitations exist in this analysis that must be considered when you set the maximum green time? Include the discussion and answer in your spreadsheet.

2. The reading in Activity #39 emphasized the importance of keeping the cycle length (and thus the maximum green time) as low as possible. But what happens when the cycle length becomes too short? List one possible downside of a very short cycle length.

ACTIVITY 41

Determining the Effect of the Minor Street Vehicle Extension Time on Intersection Operations

PURPOSE

The purpose of this activity is to help you to understand how timing parameters on the minor street affect the major street and overall intersection operations.

LEARNING OBJECTIVE

- Determine the effect of the minor street vehicle extension setting on the efficiency of major street and intersection operations

REQUIRED RESOURCE

- Movie file: A41.wmv

DELIVERABLE

- Prepare a spreadsheet with the following information:

  Tab 1: Title page with activity number and title, authors, and date completed

  Tab 2: Answers to the Critical Thinking Questions and the data that you recorded in Table 20 through Table 24

CRITICAL THINKING QUESTIONS

1. How do the eastbound and southbound approach queue lengths vary given the two vehicle extension time values used for the southbound approach?

2. How does an increase in the southbound approach vehicle extension time affect the eastbound green interval duration?

3. How does the increase in the southbound approach vehicle extension time affect the cycle length?

4. What effect does the vehicle extension time have on the delay experienced for these two cases?

INFORMATION

In this activity you will observe the operation of both approaches at the intersection of State Highway 8 and Line Street. An aerial view of the intersection is shown in Figure 145. State Highway 8 has a five to six lane cross section, while Line Street has a three lane cross section. In this activity, both approaches have stop bar presence detection with a zone length of 22 feet. The volumes are moderate, with 1400 vehicles per hour on the eastbound approach and 600 vehicles per hour on the southbound approach.

The intent here is to vary the vehicle extension time on phase 4 serving the southbound Line Street (minor street) approach and to observe the effect of each setting on the queuing experienced by motorists on both the major and minor streets. The minimum green time has been set to 5 seconds and will not be varied in this activity.

You will consider two different settings of vehicle extension time: 2 seconds and 5 seconds. Both queue length and green time duration will be considered in evaluating the performance of these alternatives. You will also learn about the relationship of green time duration and cycle length on the delay experienced by motorists at the intersection.

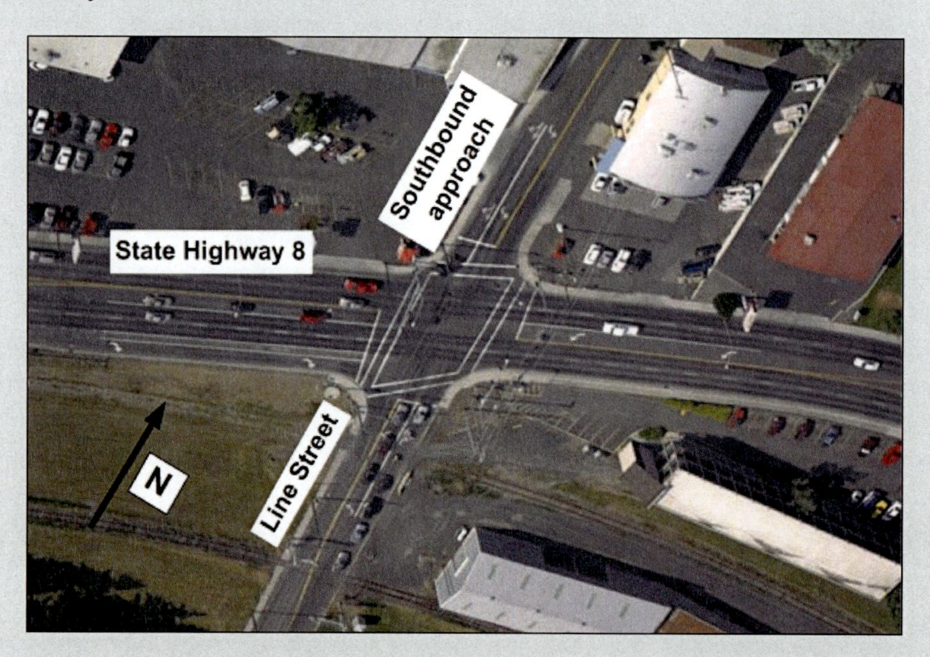

Figure 145. Aerial photo for State Highway 8 and Line Street

The movie (see Figure 146) shows side-by-side windows for the two different cases:

- Left window (case 1): the vehicle extension time for both the southbound approach and eastbound approach is 2 seconds
- Right window (case 2): the vehicle extension time for the southbound approach is set to 5 seconds whereas the vehicle extension time for the eastbound approach remains at 2 seconds

To assess the traffic operations quality in terms of queue length, duration of green time, and cycle length, you need to do the following:

- "Beginning of green" data collection: Once the signal indication for an approach turns green, pause the animation and record the length of the queue and the simulation time the signal indication turns green
- "End of green" data collection: Once the signal indication for the approach turns red, pause the animation and record the simulation time the signal indication turns red

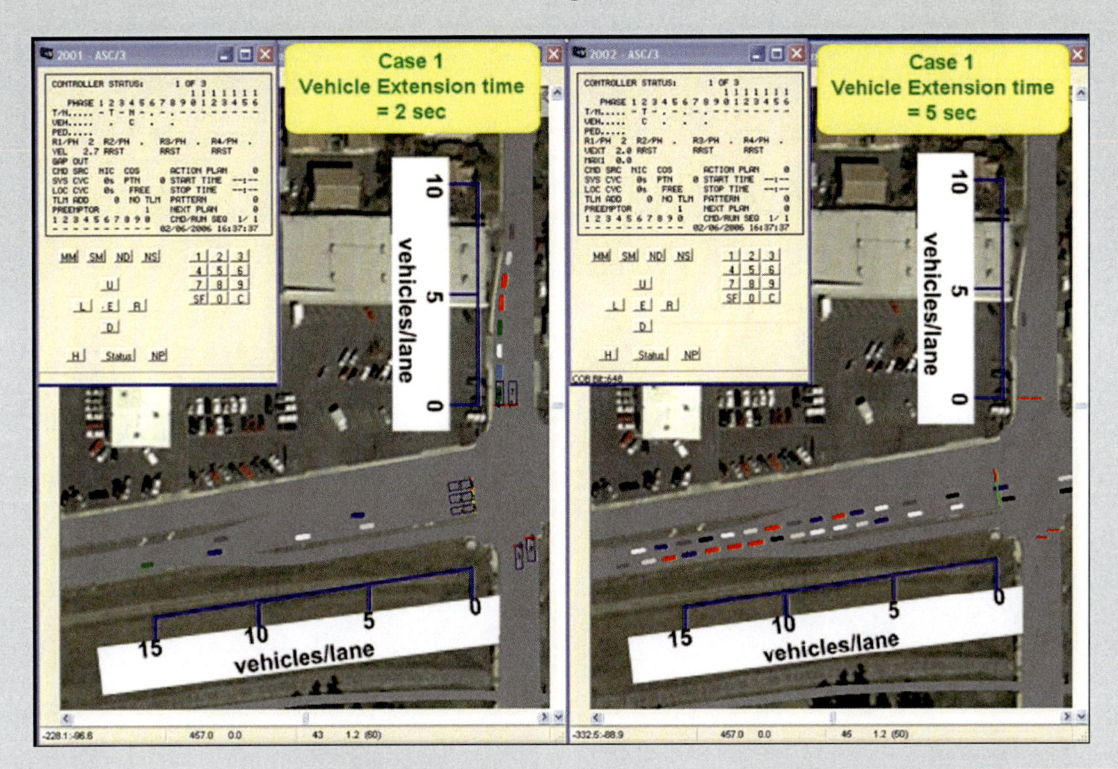

Figure 146. Side-by-side animation window

TASK 1

Open the file: "A41.wmv"

TASK 2

Observe the operation of both approaches at the intersection as well as the signal status data. Note that Case 1 is in the left window and case 2 is in the right window.

- Before starting the movie file, make sure you identify the simulation time clock in both windows (see circles in Figure 147). You will use this clock to record the beginning and ending of green. This animation starts at the simulation time of $t = 446.8$ seconds and ends at $t = 761.4$, a total time of a little more than five minutes.
- Figure 148 shows the side-by-side animation that you will observe. Scales have been provided to show the length of the queues (in vehicles) along the southbound and eastbound intersection approaches. The ASC/3 controller status windows are also shown so that you can follow the timing processes. Finally, notes will pop up periodically to point out things for you to observe. For example, for case 2, the note in Figure 148 shows that cycle 1 is timing, the eastbound green begins at 450.9 seconds, and there are 31 vehicles in the queue.

- Don't collect any data during this first observation. Just watch and observe. Note especially the differences that you see between case 1 (the left window) and case 2 (the right window). Make notes regarding your observations.

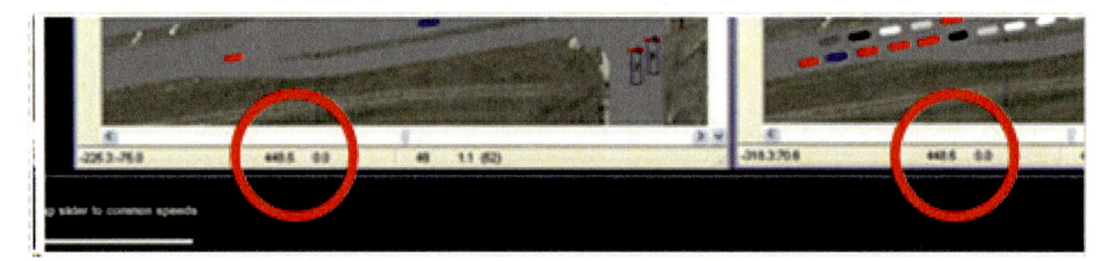

Figure 147. Simulation time in animation windows

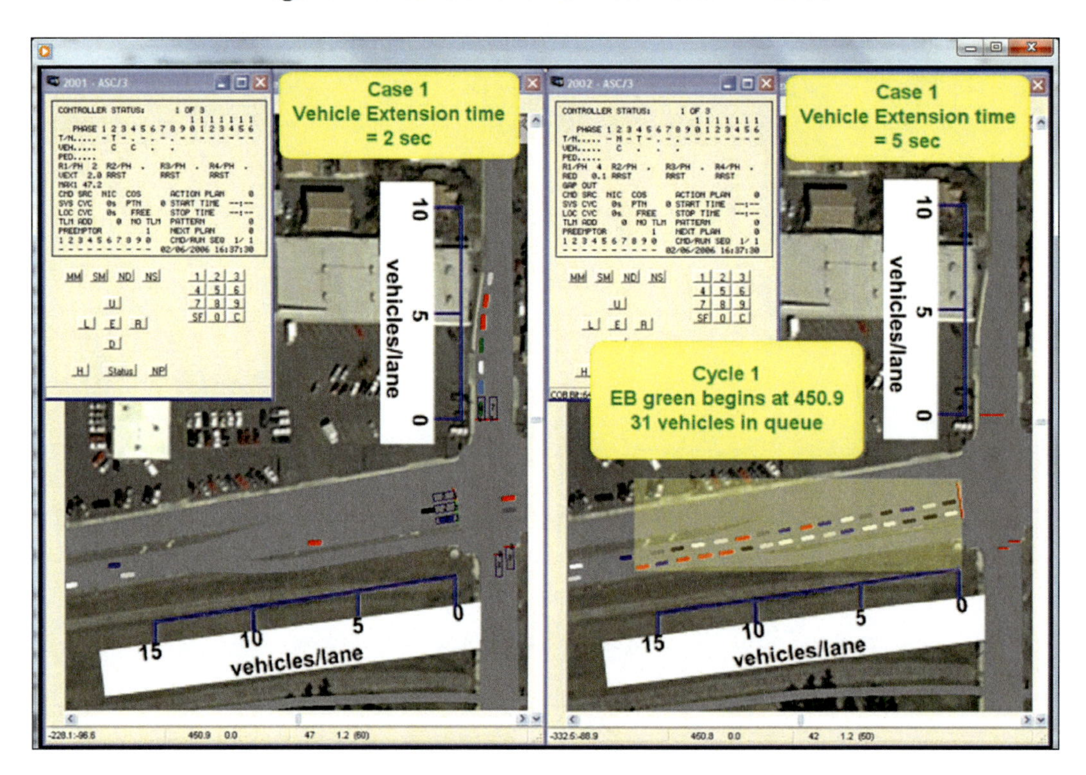

Figure 148. Case 1 and case 2 at t = 450.9

TASK 3

Observe and record the queue lengths and start/end of green intervals.

- Start the animation at the beginning. The VISSIM simulation time clock should read t = 446.8
- Record the queue length at the beginning of green and the simulation times for the beginning and ending of green for case 1 (in Table 20) and case 2 (in Table 21)
- For example, for case 2, the green indication for the eastbound approach turns green at t = 450.9 seconds (see Figure 148). If you pause the animation at this point you will see that there are 31 vehicles in the queue at the beginning of green. These two data points have been recorded for you in Table 3.
- Continue recording the queue length and the start and end of green time for each phase for both cases until the final SB phase ends at t = 752.8 seconds. You have space to record data for five cycles for case 1 and four cycles for case 2.
- Compute the green duration by taking the difference between the Green end and the Green begin times
- Compute mean values for the green duration and queue length and record these values in the last row of each table

Cycle	SB				EB			
	Green start, sec	Green end, sec	Vehicles in queue, start of green	Green duration, sec	Green start, sec	Green end, sec	Vehicles in queue, start of green	Green duration, sec
1								
2								
3								
4								
5								
		Mean →				Mean →		

Table 20. Data collection table for queue and display status for case 1 (SB vehicle extension time of 2.0 seconds) (Left window)

Cycle	SB				EB			
	Green start, sec	Green end, sec	Vehicles in queue, start of green	Green duration, sec	Green start, sec	Green end, sec	Vehicles in queue, start of green	Green duration, sec
1					450.9		31	
2								
3								
4								
5								
		Mean →				Mean →		

Table 21. Data collection table for queue and display status for case 2 (SB vehicle extension time of 5.0 seconds) (Right window)

Task 4

Summarize your data.

- Copy the "green start" data for the southbound and eastbound approaches from Table 20 and Table 21 into the appropriate cells in Table 22 and Table 23. Compute the length for each cycle, by taking the difference between the "green start" for each pair of consecutive cycles. Then compute the mean cycle length for each case.

- Based on the data that you recorded in Table 20, Table 21, Table 22, and Table 23, use Table 24 to summarize the average green duration, cycle length, and queue length for cases 1 and 2
- Study the results shown in the tables and prepare a summary of your observations

Cycle	Case 1		Case 2	
	Green start	Cycle length	Green start	Cycle length
1				
2				
3				
4				
5				
	Mean→		Mean→	

Table 22. Data summary, SB approach

Cycle	Case 1		Case 2	
	Green start	Cycle length	Green start	Cycle length
1				
2				
3				
4				
5				
	Mean→		Mean→	

Table 23. Data summary, EB approach

	SB		EB	
	Case 1	**Case 2**	**Case 1**	**Case 2**
Green duration, sec				
Cycle length, sec				
Queue length, vehicles				

Table 24. Mean values for Cases 1 and 2

Student Notes:

ACTIVITY

42 Determining the Effect of the Maximum Green Time on Intersection Operations

PURPOSE

The purpose of this activity is to help you to understand how the maximum green time settings affect intersection operations.

LEARNING OBJECTIVE

- Describe the advantages and disadvantages of increasing maximum green time on intersection operations

REQUIRED RESOURCES

- Movie files: A42-1.wmv A42-2.wmv

DELIVERABLE

- Prepare a document with your answers to the Critical Thinking Questions and your observations from Tasks 1 and 2

CRITICAL THINKING QUESTIONS

As you begin this experiment, consider the following questions. You will come back to these questions once you have completed the experiment.

1. Are all of the vehicles in the initial queue on the westbound approach served before the end of each green interval?

2. What is the mechanism for termination of the phase serving the westbound approach?

3. What are the advantages and disadvantages of the 40 second maximum green time for the operation of case 1?

4. What are the advantages and disadvantages of the 60 second maximum green time for the operation of case 2?

5. Which maximum green time setting would you select and why?

6. Identify the pros and cons of the two different maximum green time settings on the westbound approach and on the overall performance of the intersection. Consider what you observed and documented for both cases. Summarize the pros and cons of each case.

Information

In this activity, you will observe two cases, each focusing on the westbound approach of the major street, State Highway 8 (See Figure 149). In the first case, the maximum green time is set to 40 seconds. The demand is relatively high (1700 vehicles per hour across two lanes) on the westbound approach and the green time displayed is not sufficient to serve the demand. In the second case, the maximum green time is set at 60 seconds in an effort to serve more of the demand. But this change also has implications that must be considered for the operation of the intersection. Each case includes four cycles, focusing on the westbound approach. You will be asked to look at three things:

- What is the length of the queue at the beginning of each of the four green intervals on the westbound approach?
- How does the westbound phase terminate during each of the four cycles in each case?
- Are there vehicles from the westbound queue still unserved at the end of the green interval?

Figure 149. Activity 42 movie file

Task 1

Open the file: "A42-1.wmv." Observe the operation of case 1 and record your observations.

- Watch the entire video. It is nearly 3.5 minutes in length. Pay attention to the Critical Thinking Questions listed earlier.
- Record your observations

Task 2

Open the file: "A42-2.wmv." Observe the operation of case 2 and record your observations.

- Watch the entire video. It is nearly 4 minutes in length. Pay attention to the Critical Thinking Questions listed earlier.
- Record your observations

Student Notes:

ACTIVITY
43 Setting the Maximum Green Timing Parameter for All Approaches of an Intersection

PURPOSE

The purpose of this activity is to set the maximum green time for your intersection such that the delay is optimized for all approaches and for the intersection as a whole.

LEARNING OBJECTIVE

- Set the maximum green time for both approaches of an intersection, balancing the performance of both the minor street and the major street

REQUIRED RESOURCE

- VISSIM file from Activity #37

DELIVERABLE

- Prepare a spreadsheet that includes the analysis and reporting requirements listed in the tasks below:

  Tab 1: Title page with activity number and title, authors, and date completed

  Tab 2: Delay analysis for range of maximum green times

  Tab 3: Prepare a brief report that summarizes your conclusions, your recommended maximum green times, and the data that support your conclusions and recommendations. Include a plot of delay vs cycle length for the results that you generated.

INFORMATION

Consider this question: How do you set the timing parameters to balance the risks of early termination of green and inefficiently long green time? Consider the following criteria that could be used to produce efficient phase operations:

- The phase is not extended inefficiently for a very short queue
- The phase extends long enough to clear the standing queue
- The phase doesn't extend beyond the time that it takes for the queue to clear

In addition to these three criteria, the following criteria could also be considered to achieve intersection operational efficiency:

- The major street green time should be extended to serve vehicles arriving after the queue clears without causing excessive delay to the minor street traffic
- The maximum green time should be increased in case of phase failure when a phase consistently terminates by maxing out

Your objective in this activity is to determine the maximum green times such that the phases generally gap out (and not max out) balanced by making sure the cycle times are not excessive and long delay times are produced.

Task 1

Make a new copy of the folder that includes your VISSIM files from Activity #37. Rename this folder "a43". Use this VISSIM file as the basis for your analysis and design of the maximum green time.

Task 2

Set the maximum green time to 60 seconds for all approaches of your intersection. Use the settings for the minimum green time and the vehicle extension time that you determined in Activity #37. Collect delay and queue length data for each approach and for the intersection as a whole based on one simulation run of 3900 seconds (collecting data beginning at 300 seconds to account for network build-up). Observe the operation of the network for this time period and record the number of max outs and gap outs for each approach.

Task 3

Based on your results from Task 2, reduce the maximum green times by 10 seconds and run the simulation again. Again, collect the delay and queue length data for each approach and for the intersection as a whole based on one simulation run of 3900 seconds. Observe the operation of the network for this time period and record the number of max outs and gap outs for each approach.

Task 4

Continue iterating (reducing the maximum green by increments of 10 seconds and re-run the simulation) until you've reached a value of maximum green time that meets the objectives listed previously.

Task 5

Based on the results from Tasks 2, 3, and 4, select your design value for the maximum green time.

ACTIVITY

44 Maximum Green Time

PURPOSE

The purpose of this activity is to help you to learn how the maximum green time is set in practice.

LEARNING OBJECTIVE

- Compare the maximum green time that you selected with the range of values used in practice

REQUIRED RESOURCE

- *Traffic Signal Timing Manual*

DELIVERABLES

Prepare a document that includes

- Answers to the Critical Thinking Questions
- Completed Concept Map

LINK TO PRACTICE

Read the section from the *Traffic Signal Timing Manual* on maximum green times as assigned by your instructor.

CRITICAL THINKING QUESTIONS

When you have completed the reading, prepare answers to the following questions:

1. What is the function of the maximum green time?

2. What methods are used to set the maximum green time?

3. How do your design results from Activity #43 compare with the recommendations from the *Traffic Signal Timing Manual*?

IN MY PRACTICE...

by Tom Urbanik

The selection of the maximum green time starts with an understanding of the traffic volume, usually from a traffic count or traffic projection. These traffic volumes are only a snapshot of traffic conditions and may not reflect peak demand. Traffic at a school, for example, may have extreme peaking at the beginning or end of school, or following a special event like a football game. Typically, maximum green times will be increased to accommodate these extreme conditions. The main risk of larger than needed maximum green times includes extending the phase beyond saturation flow values if the passage time setting is large, thus driving the phase to maximum or in the case of detector failures, sending a continuous call which also extends the phase to maximum.

There are advanced controller features that can respond to fluctuations in traffic volumes. One example is the dynamic maximum which allows the controller to increase the maximum green time if the controller continues to "max out" rather than "gap out." The downside is you need cycle failures to increase the maximum green time.

Another feature which might not be thought of for increasing flexibility of fully actuated control to respond to fluctuations in volume is "soft recall." If the arterial is placed on soft recall, it only calls the arterial in the absence of calls on all other phases. So a large volume of traffic exiting a high school stadium is able to extend beyond its maximum green if there is no traffic on the arterial calling for service. Alternatively, if the arterial is on minimum recall, it will turn on the cross street maximum green timer every cycle even in the absence of traffic on the arterial, forcing the controller to cycle back to the arterial because of the recall.

CONCEPT MAP

Terms and variables that should appear in your map are listed below.

cycle length (C)

uniform delay (d_1)

volume (v)

maximum green time

green time (g)

Student Notes:

Left Turn Phasing—Permitted, Protected, or Both

Purpose

The safe separation of conflicting traffic movements was first discussed in Chapter 3. In Chapter 8, you will learn about the various ways in which left turns are served and some of the ways in which conflicts with the opposing traffic can be eliminated or reduced. Drivers can safely complete a left turn movement without waiting when a green arrow display is shown. Or, when a green ball or flashing yellow arrow is displayed, there is another level of instruction to the driver: you may proceed, but first you must judge that it is safe to do so, with a large enough gap in the opposing traffic that allows you to safely complete your left turn maneuver. You will complete a set of activities that describes the most commonly used methods of serving left turn movements, particularly the determination of whether a left turn should be protected (served by a green arrow) or permitted (served by a flashing yellow arrow or a green ball).

Learning Objectives

When you have completed the activities in this chapter, you will be able to

- Describe the methods of left turn phasing
- Describe the basic concepts of left turn phasing
- Determine the efficiency of permitted left turn operations under various opposing through traffic volumes
- Show that protected left turn phasing is more efficient than permitted left turn phasing under some conditions
- Describe the trade-offs and relative efficiencies between protected/permitted and protected left turn phasing
- Select optimal left turn phasing treatment based on analysis of performance data and observation of simulation conditions

Chapter Overview

This chapter begins with a *Reading* (Activity #45) on left turn phasing. The chapter also includes five activities, including an assessment of your understanding of left turn phasing (Activity #46) and three activities (Activities #47, #48, and #49) in which you observe the operation and performance of various left turn phasing options. In a design activity (Activity #50) you will test various left turn options for your intersection and select the best option for the conditions that you have observed. The chapter concludes with an *In Practice* activity (Activity #51) in which you will compare your left turn phasing plan with information used in practice from the *Traffic Signal Timing Manual*.

Activity List

Number and Title	Type
45 Left Turn Phasing	*Reading*
46 What Do You Know About Left Turn Phasing?	*Assessment*

Number and Title	Type
47 Permitted Left Turn Operations	*Discovery*
48 Comparing Permitted and Protected Left Turn Phasing	*Discovery*
49 Comparing Protected/Permitted and Protected Left Turn Phasing	*Discovery*
50 Analysis and Design of Left Turn Treatment	*Design*
51 Left Turn Phasing Options	*In Practice*

PURPOSE

The purpose of this activity is to give you the chance to learn more about left turn phasing and the various options for serving left turn movements.

LEARNING OBJECTIVE

- Describe the methods of left turn phasing

DELIVERABLES

- Define the terms and variables in the Glossary
- Prepare a document that includes answers to the Critical Thinking Questions

GLOSSARY

Provide a definition for each of the following terms or variables. Paraphrasing a formal definition (as provided by your text, instructor, or another resource) demonstrates that you understand the meaning of the term or phrase.

lagging left turns	
leading left turns	
left turn phasing	
permitted left turns	
protected left turns	

Critical Thinking Questions

When you have completed the reading, prepare answers to the following questions.

1. What performance measures can you extract from the graphical representations of the three queuing models presented in the reading?

2. Why should permitted left turn phasing always be considered as a phasing plan option?

3. When should protected left turn phasing be considered?

Information

The ring barrier diagram establishes the sequence of phases to be served at an actuated signalized intersection. For a standard intersection with four approaches, the movements and the numbering scheme used to identify the movements are shown in Figure 150. For leading protected left turns, the eight phases (their sequencing and the movements that they control) are represented in the ring barrier diagram shown in Figure 151.

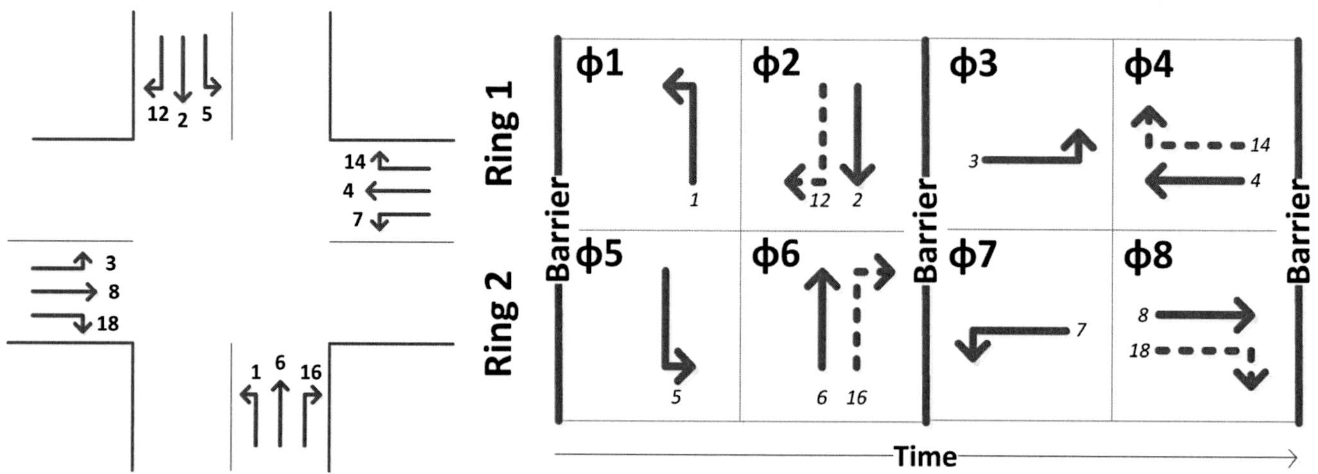

Figure 150. Movement numbers

Figure 151. Ring barrier diagram for leading protected left turns

In ring 1, phase 1 (which controls the northbound left turn movement) must occur before phase 2 (which controls the opposing southbound through movement) since the left turn movement is protected: it does not have any opposing movement to contend with while phase 1 is timing. This complete time separation between the opposing left turn and through movements provides a high level of safety, especially when volumes are high.

The arrival and departure flow patterns for a protected left turn can be represented by a flow profile diagram, a cumulative vehicle diagram, and a queue accumulation polygon as originally discussed in Chapter 2. These three diagrams for a protected left turn movement are shown in Figure 152, Figure 153, and Figure 154. The left turn movement flows at the saturation flow rate during the green, without opposing flows, and the phase terminates when the flow has been served. Figure 153 and Figure 154 show the left turn queue building during red and clearing at the end of green.

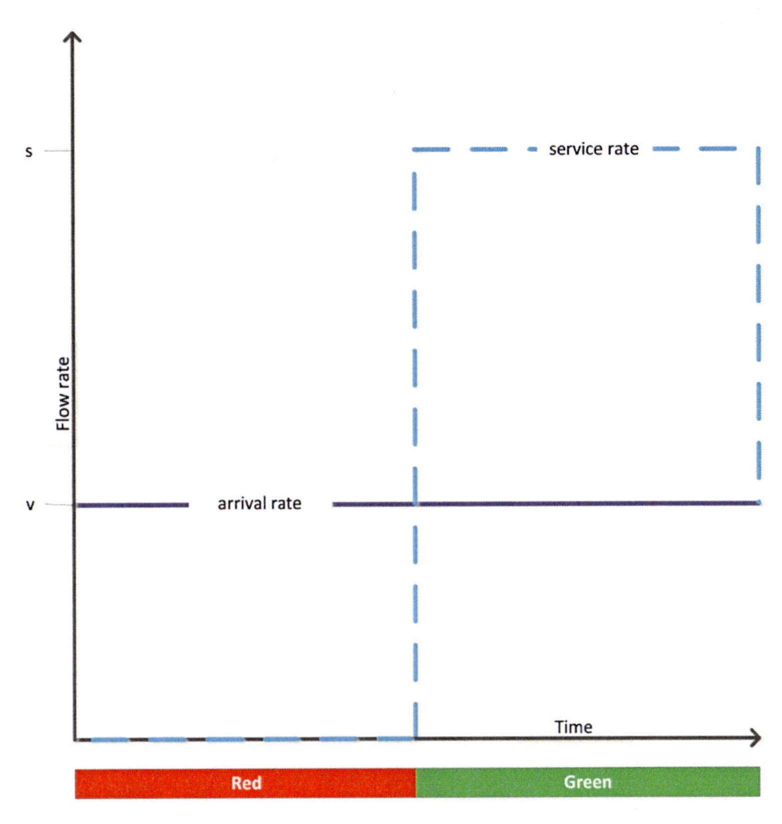

Figure 152. Flow profile diagram, protected left turn movement

But there are conditions, particularly when the left turn and/or opposing volumes are lower, where the driver can be given an option in which he or she can exercise safe judgment and accept or reject a gap in the opposing traffic stream. Examples of this gap acceptance behavior exist for other traffic facilities. Vehicles entering a

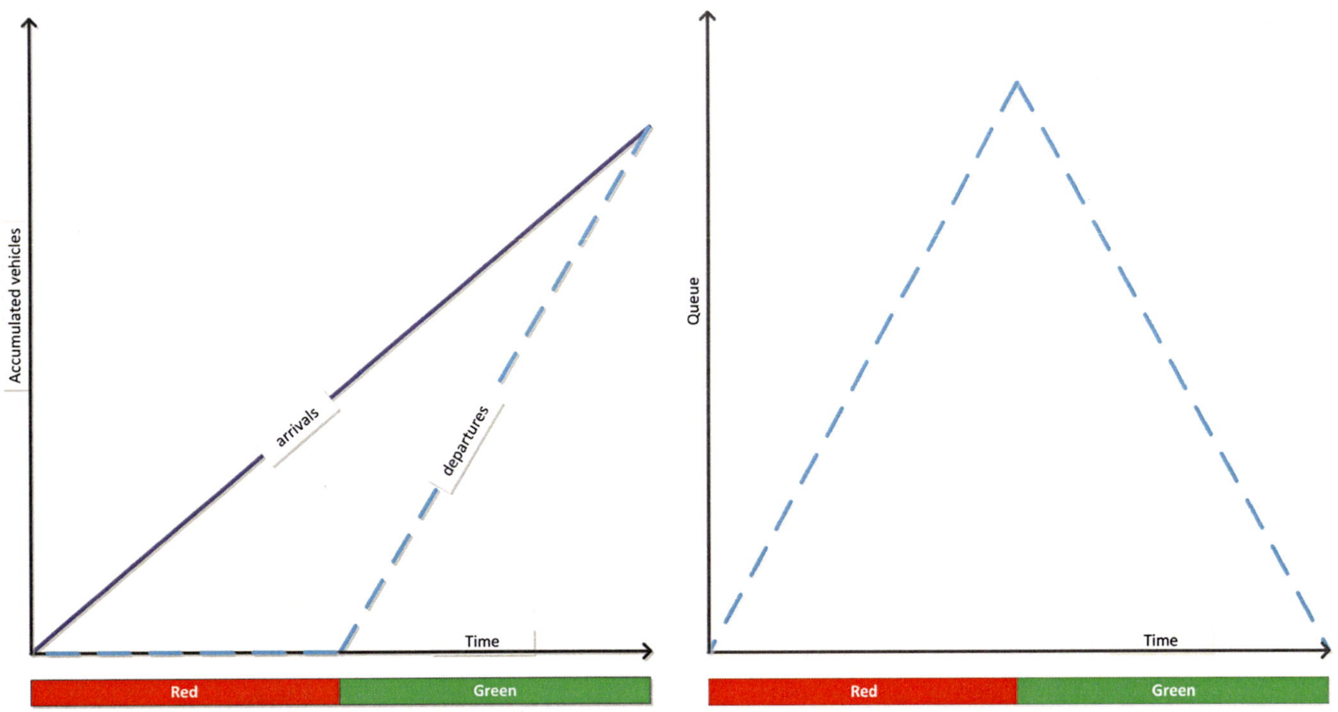

Figure 153. Cumulative vehicle diagram, protected left turn movement

Figure 154. Queue accumulation polygon, protected left turn movement

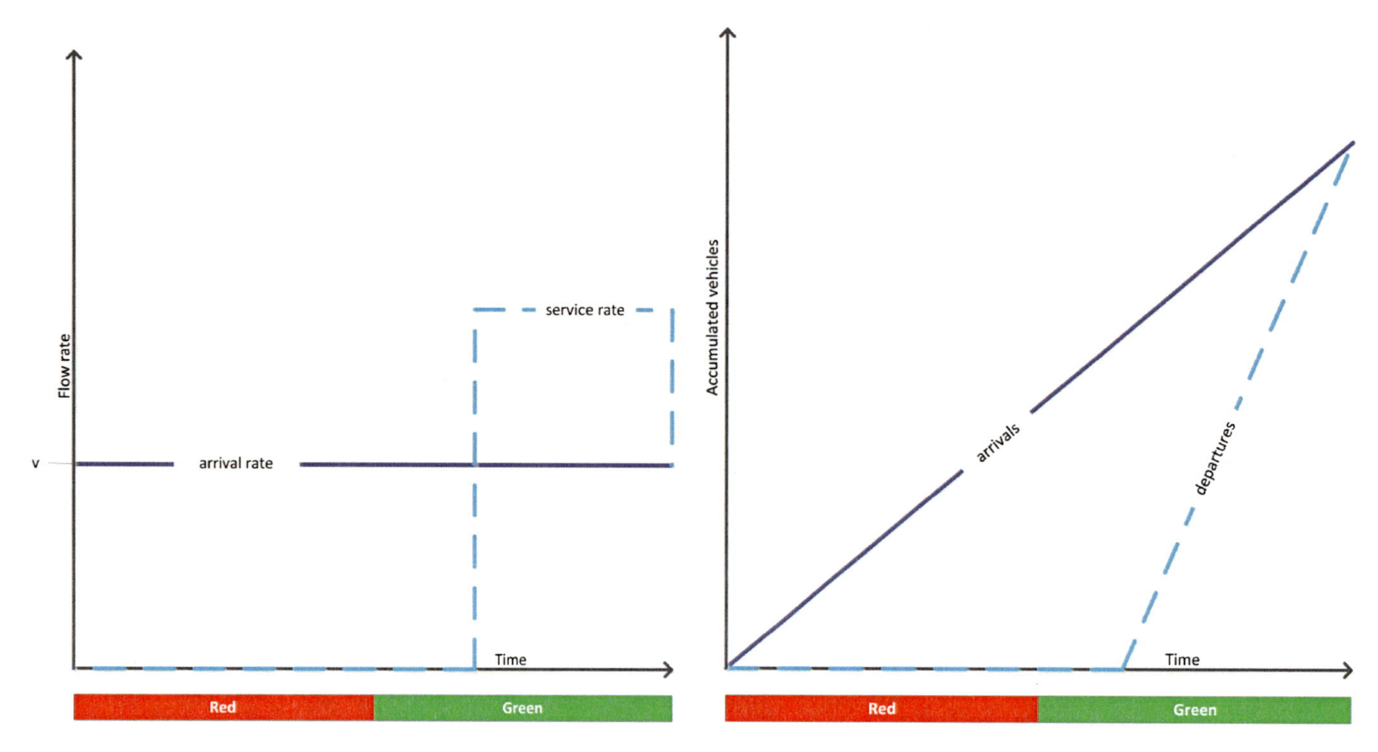

Figure 155. Flow profile diagram for permitted left turn movement

Figure 156. Cumulative vehicle diagram for permitted left turn movement

freeway have to look for a safe gap in the traffic already on the freeway before they enter the mainline. And, drivers on the minor street approach of a two-way stop-controlled intersection must find a large enough gap in the major stream before they either cross or merge into the major street. A similar option exists for permitted left turn maneuvers. The driver is shown either a solid green ball or a flashing yellow arrow, displays that indicate that a driver can proceed if, in their judgment, there is a large enough gap in the opposing traffic through which they can safely complete their turning maneuver.

Figure 155 shows the flow profile diagram for a permitted left turn movement. Here, the departure flow for the initial period of green is zero, as the opposing through movement queue is clearing. Once that opposing queue has cleared, the left turn vehicles can filter their way through the opposing gaps, when these gaps are large enough to be useful. Figure 156 and Figure 157 show the resulting cumulative vehicle diagram and queue accumulation polygon for the permitted left turn movement. As before, the total delay experienced by the left turn movement can be calculated as the area of either of these two figures.

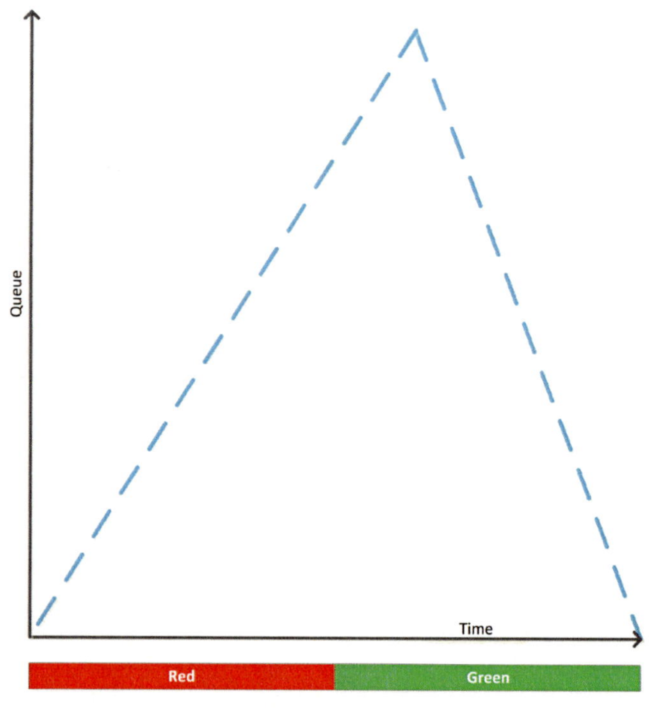

Figure 157. Queue accumulation polygon, permitted left turn movement

The ring barrier diagram for permitted left turn phasing is shown in Figure 158. Only four phases are needed, since the even numbered phases control both the left turn and through movements.

In some cases, a combination, or protected plus permitted, phasing can be used. Here at least the initial portion of the left turn queue is served with protected phasing while the remainder of the queue and any additional left turn vehicles that arrive can be served during the permitted phase. The queuing diagrams for this phasing option are shown in Figure 159, Figure 160, and Figure 161. Figure 159 shows that vehicles depart at the saturation flow rate during the protected phase. During the initial part of the permitted phase, the departing flow is zero when the opposing queue is clearing. Once the opposing queue has cleared, the departure rate is greater than zero but less than the saturation flow rate. Figure 160 and Figure 161 both show that the left turn queue is partially or completely served during the protected phase when vehicles are able to depart at the saturation flow rate (depending on the left turn flow rate and the amount of green time provided). During the permitted phase, the queue grows again when the left turn flow is zero during the period when the opposing through movement queue is clearing. Once this opposing queue has cleared, the queue of permitted left turn vehicles decreases to zero as left turn vehicles are able to filter through the opposing gaps. In this example, the queue clears at the end of green.

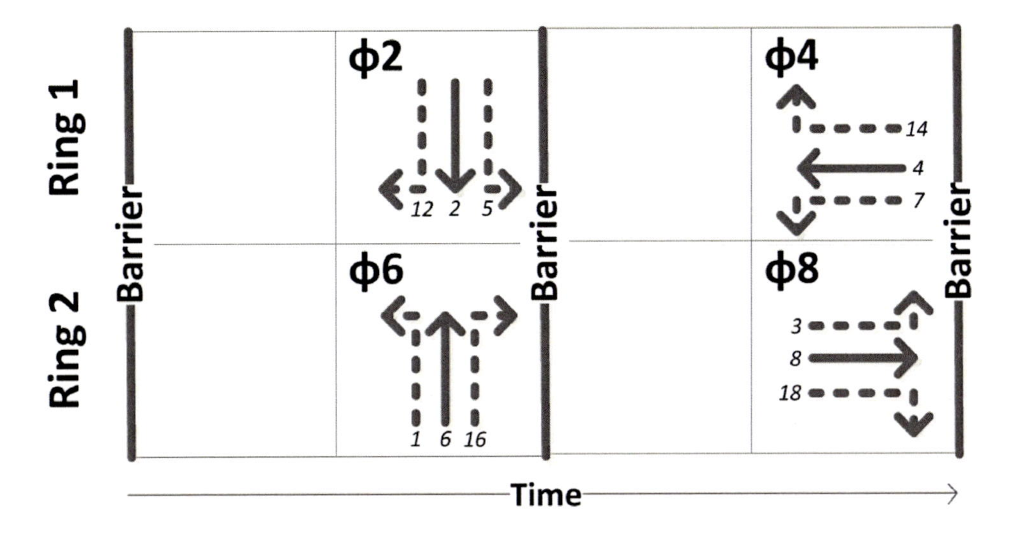

Figure 158. Ring barrier diagram, permitted left turn phasing

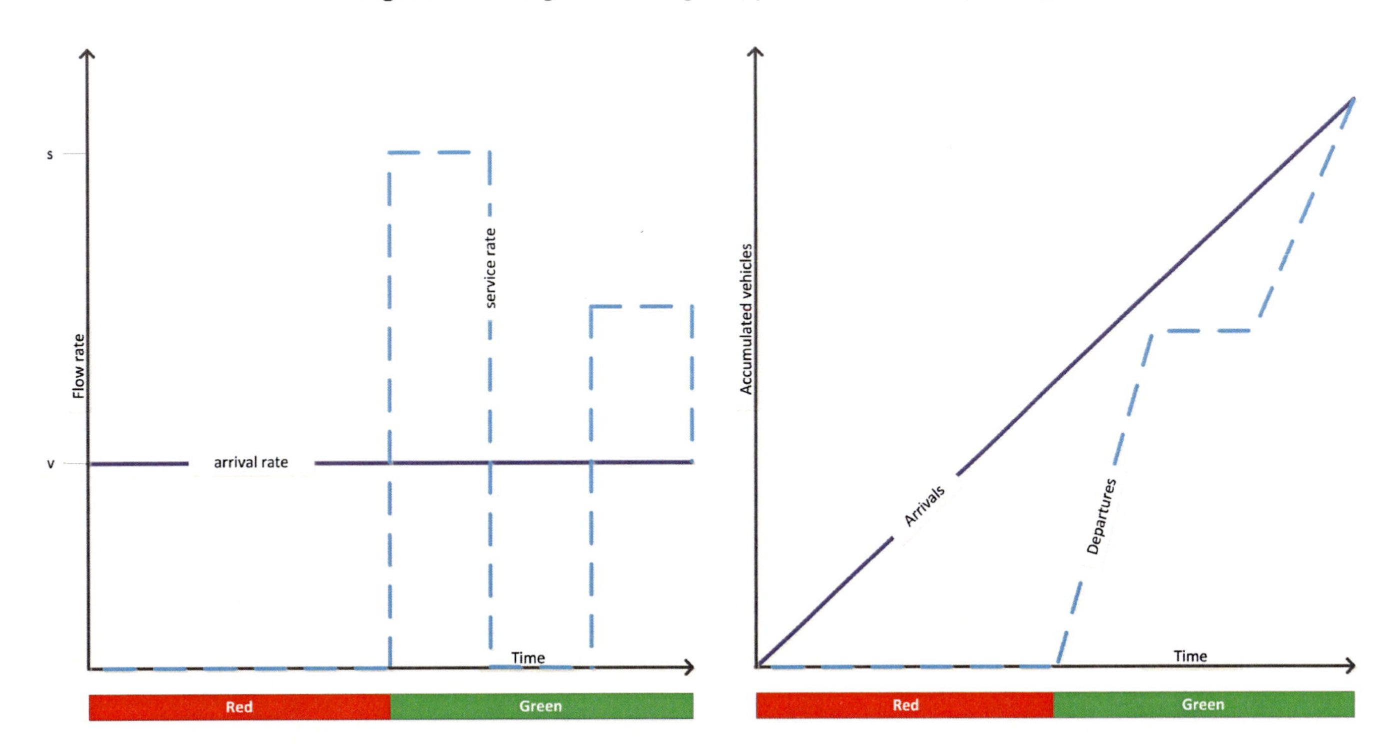

Figure 159. Flow profile diagram, protected plus permitted left turn phasing

Figure 160. Cumulative vehicle diagram, protected plus permitted left turn phasing

The ring barrier diagram for protected permitted phasing is shown in Figure 162. The left turn movement is shown as a solid line during the protected phase and as a dashed line during the permitted phase.

So how do you determine which left turn phasing option is best for a particular situation? While there are a number of considerations in practice, we will consider two points that will help you in the design activity that you will complete later in this chapter. The first point is to have as few phases as possible during the cycle as this reduces the number of yellow-red clearance transitions and allows for more green time to be available to serve the traffic demand. Permitted left turn phasing means a fewer number of phases than required for protected left turn phasing. Figure 158 shows that all movements can be served with four phases requiring only two transitions per cycle. This potential efficiency means that permitted phasing should at least be considered as an option when possible.

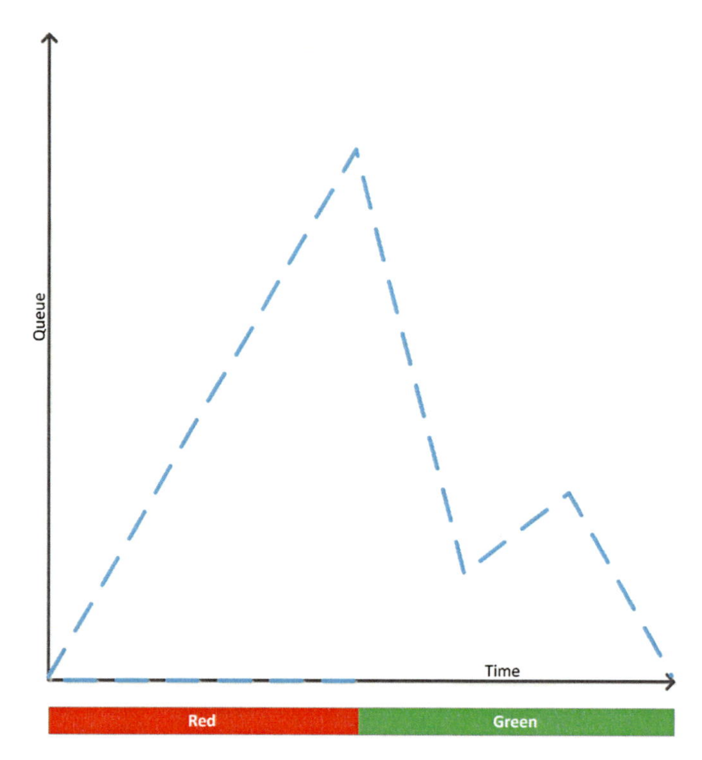

Figure 161. Queue accumulation polygon, protected plus permitted left turn phasing

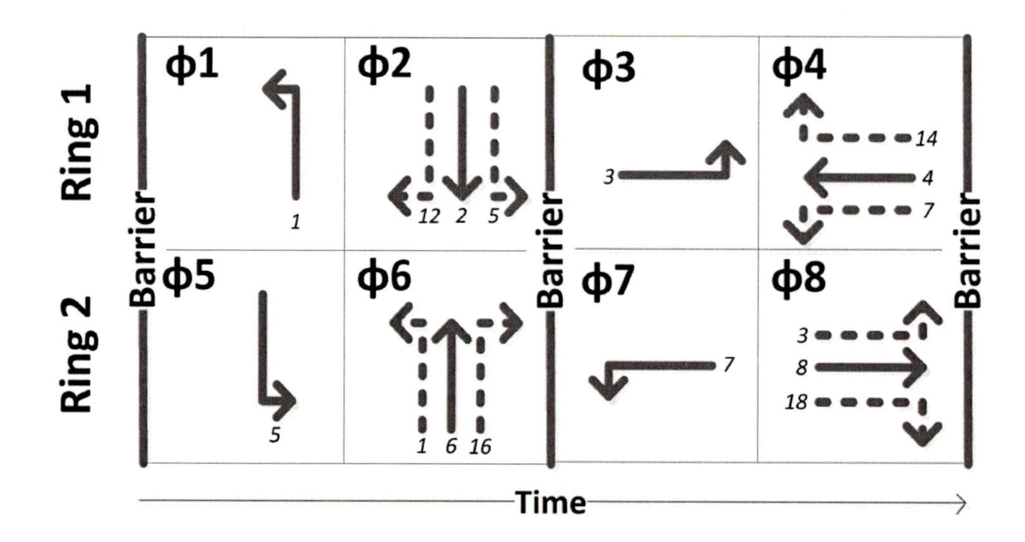

Figure 162. Ring barrier diagram, protected plus permitted left turn phasing

The second point is to make sure that there is sufficient capacity for the left turn movements. This means that if the combination of the left turn flow rate and the opposing through flow rate is high enough, there will only be sufficient left turn capacity if the phase serving the left turn movement is protected.

As you complete the activities to follow, you will see specific examples of both points when you observe the interaction of the left turn and opposing through traffic movements.

Student Notes:

What Do You Know About Left Turn Phasing?

Purpose

The purpose of this activity is to give you the chance to assess your understanding of left turn phasing options.

Learning Objective

- Describe the basic concepts of left turn phasing

Deliverable

- Prepare a document with your answers to the Critical Thinking Questions

Critical Thinking Questions

1. What experiment could you construct to determine the capacity limit of the left turn and opposing through volumes for a given intersection? Describe your experiment.

2. What calculation could you make to determine the reduction in capacity that would occur in the through movements if the phasing plan was changed from permitted left turn phasing to protected left turn phasing? Describe your calculation.

Student Notes:

Permitted Left Turn Operations

PURPOSE

The purpose of this activity is to give you the opportunity to increase your understanding of permitted left turn phasing.

LEARNING OBJECTIVE

- Determine the efficiency of permitted left turn operations under various opposing through traffic volumes

REQUIRED RESOURCE

- Movie file: A47.wmv

DELIVERABLE

- Prepare a document that includes your answers to the Critical Thinking Questions

CRITICAL THINKING QUESTIONS

As you begin this activity, consider the following questions. You will come back to these questions at the end of the activity.

1. How does the opposing volume affect the quality of the left turn permitted operation for each of the two cases?

2. What change to the phasing plan would you consider, if any, to improve the quality of the operation for case 2?

3. Do the two cases that you observed conform to the queuing model diagrams described in the reading (Activity #45)? Explain your answer.

4. Prepare a brief summary of the performance of the left turn movements for each case. Consider the relative size of the queues that form and the relative delay experienced by the left turn movements.

INFORMATION

In this activity you will observe the operation of State Highway 8 and Line Street, focusing on the left turn operations on State Highway 8. An aerial view of the intersection is shown in Figure 163. State Highway 8 has two through lanes in each direction, while Line Street has one through lane in each direction.

The left turn phasing that you will observe is called "permitted," since the left turn traffic is allowed or permitted to complete their turning maneuver only if there is a safe or acceptable gap in the opposing through traffic. If an inadequate number of gaps in the opposing through traffic present themselves, the quality of the left turn operations will deteriorate.

Two cases will be considered here, each with different opposing through volumes. In the first case, the opposing through movement is 800 vehicles per hour. In the second case, the opposing through movement is 1450 vehicles per hour. In both cases, the left turn movements are 100 vehicles per hour. The minor street movements (northbound and southbound through movements) have the same volume, 600 vehicles per hour.

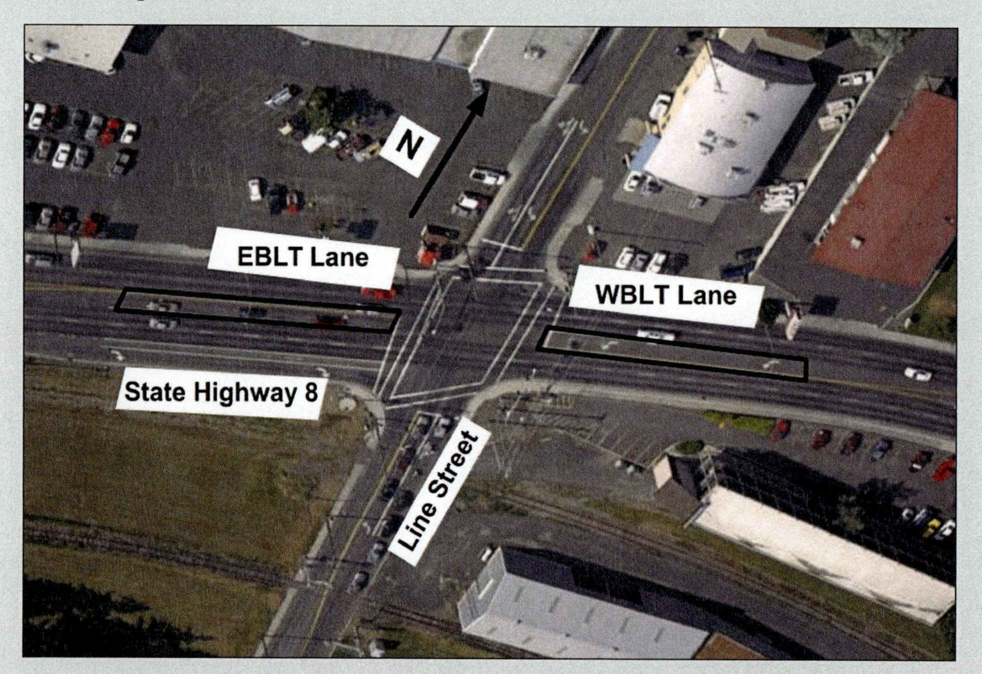

Figure 163. Aerial photograph, State Highway 8 and Line Street

Figure 164 shows the ring diagram for both cases. The permitted left turn movements are indicated with dashed lines.

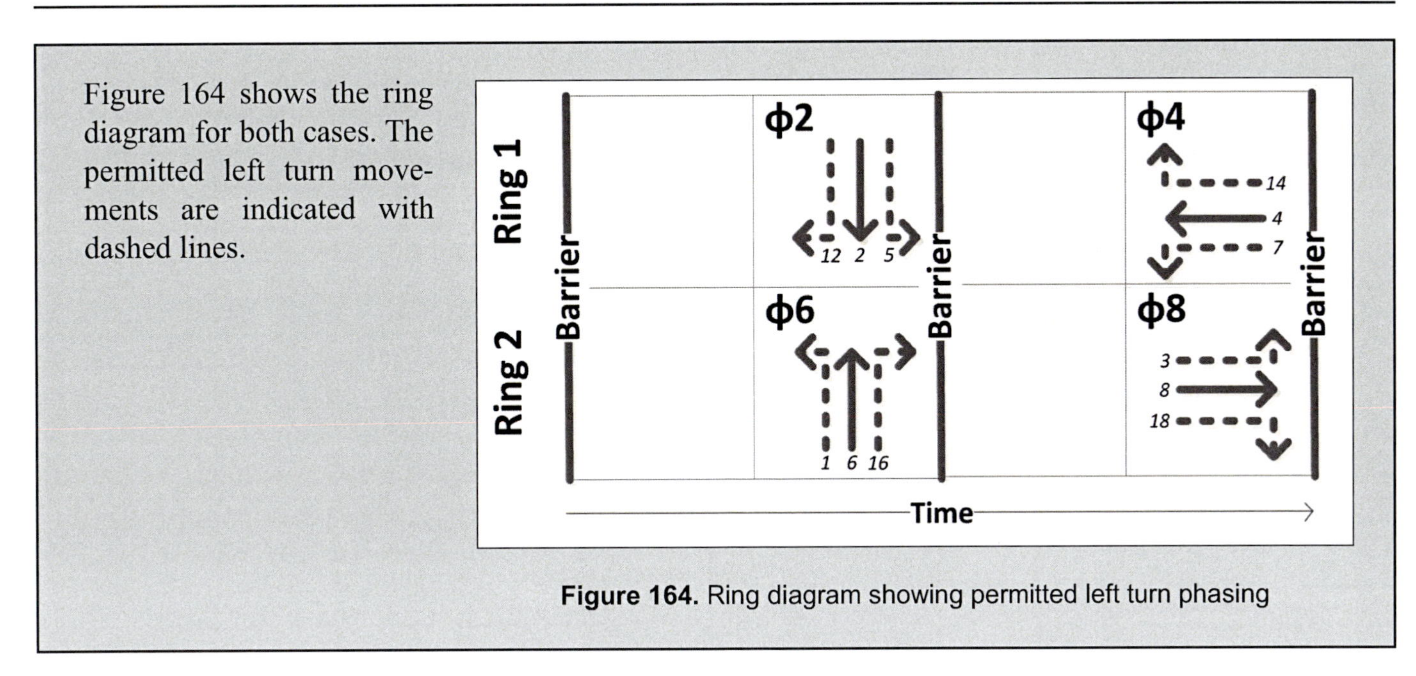

Figure 164. Ring diagram showing permitted left turn phasing

Task 1

Open the file: "A47.wmv."

Task 2

Observe the operation of the two cases. (See Figure 165.)

- Observe the relative size of the gaps in the through traffic on State Highway 8
- Observe the eastbound left turn and westbound left turn vehicles as they first wait, and then accept gaps in the opposing through traffic. Note the relative size of the queues that form in both cases.

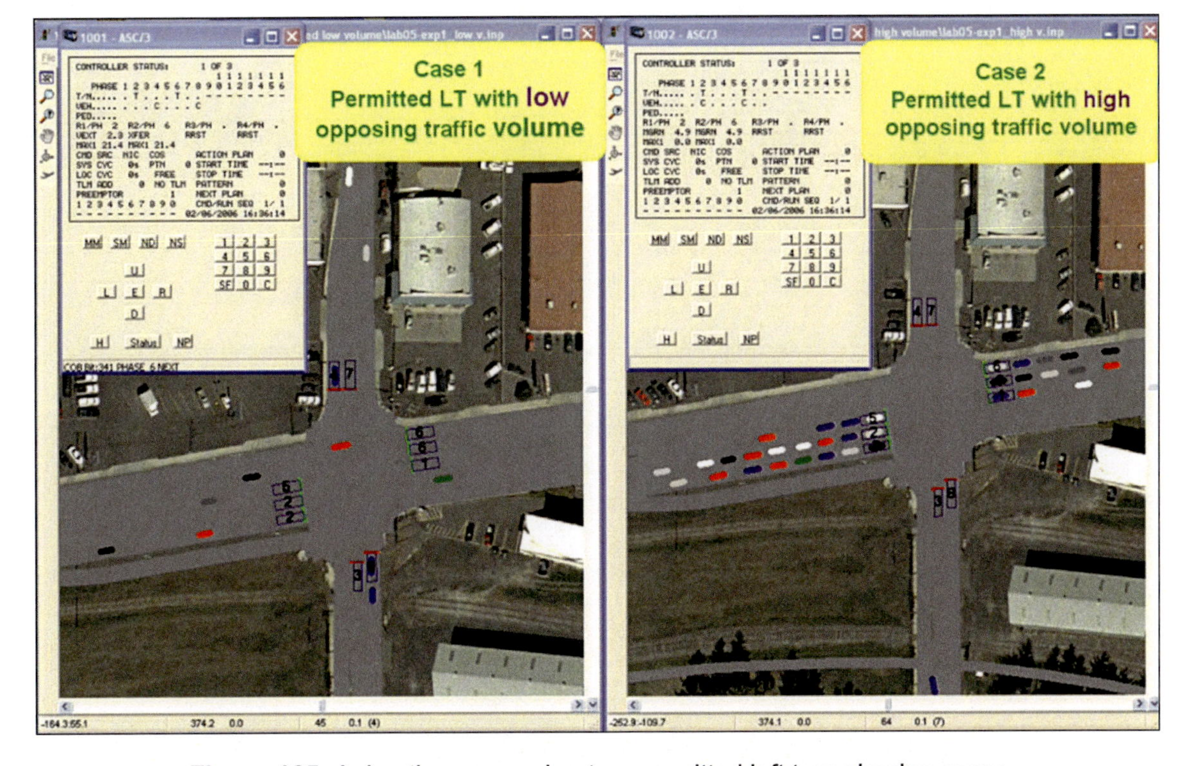

Figure 165. Animation comparing two permitted left turn phasing cases

Student Notes:

Comparing Permitted and Protected Left Turn Phasing

Purpose

The purpose of this activity is to give you the opportunity to compare permitted and protected left turn phasing.

Learning Objective

- Show that protected left turn phasing is more efficient than permitted left turn phasing under some conditions

Required Resource

- Movie file: A48.wmv

Deliverable

- Prepare a document that includes your answer to the Critical Thinking Question

Critical Thinking Question

As you begin this activity, consider the following question. You will come back to this question when you have completed your observations.

1. How does changing from permitted to protected left turn phasing affect the left turn operation and the operation of the entire intersection?

Information

In the previous activity, you considered the efficiency of permitted left turn operations. You saw that high opposing through volumes could seriously degrade the quality of permitted left turn operations. One option to improve the left turn operation is to change the phasing from "permitted" to "protected." Figure 166 illustrates the ring barrier diagram for full left turn protection.

In this activity you will again observe the left turn operation on State Highway 8. Both cases that you will observe have through volumes of 1450 vehicles per hour per lane and left turn volumes of 100 vehicles per hour. The only difference is in the left turn phasing. Case 1 is permitted left turn phasing (similar to case 2 in the previous activity) while case 2 is protected left turn phasing.

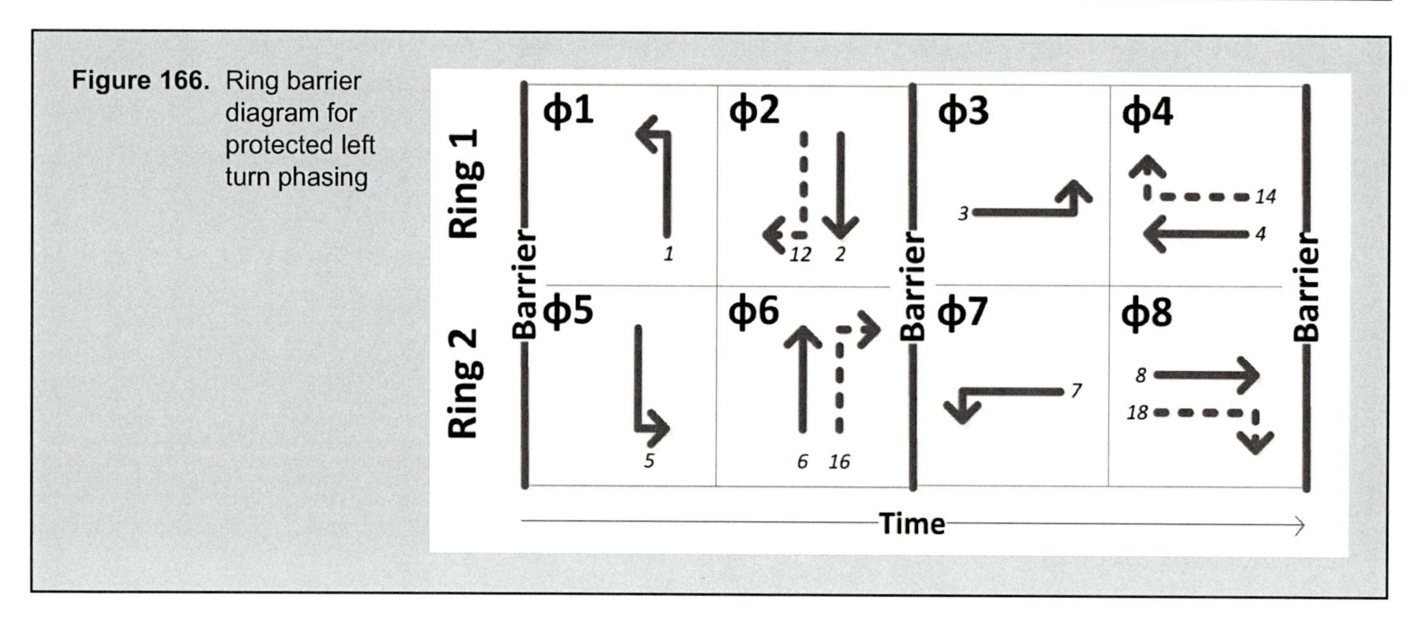

Figure 166. Ring barrier diagram for protected left turn phasing

Task 1

Open the file: "A48.wmv"

Task 2

Observe the operation of the two cases.

- Observe the left turn vehicles on the eastbound and westbound approaches for case 1 (permitted left turn) and case 2 (protected left turn). Observe the queue length for the eastbound left turn and westbound left turn movements for case 1 and the waiting time for those vehicles. Observe the same vehicles in case 2 and notice how all vehicles are served during the protected left turn. (See Figure 167.)
- Summarize your observations

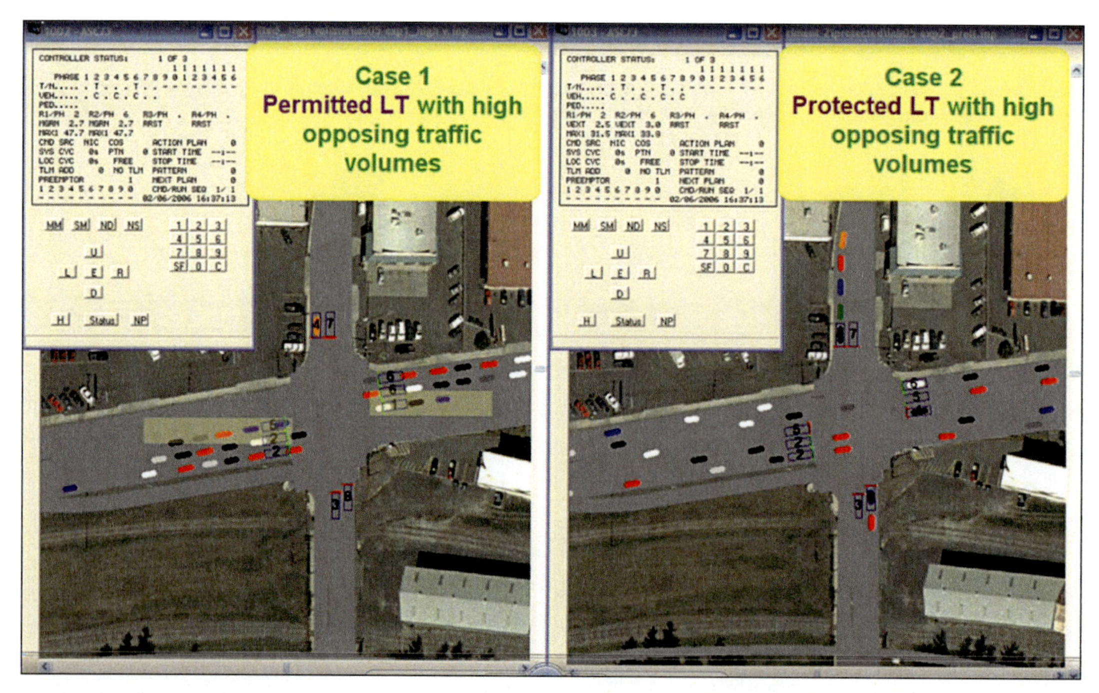

Figure 167. Animation comparing permitted and protected left turn phasing

ACTIVITY

49 Comparing Protected/Permitted and Protected Left Turn Phasing

PURPOSE

The purpose of this activity is to give you the opportunity to learn about protected plus permitted left turn phasing.

LEARNING OBJECTIVE

- Describe the trade-offs and relative efficiencies between protected plus permitted and protected left turn phasing

REQUIRED RESOURCE

- Movie file: A49.wmv

DELIVERABLE

- Prepare a document that includes your answer to the Critical Thinking Question

CRITICAL THINKING QUESTION

As you begin this activity, consider the following question. You will come back to this question when you have completed the experiment.

1. Why do the eastbound left turn and westbound left turn movements have lower delay when they are operating as protected/permitted phasing as compared to the protected left turn case?

INFORMATION

In the previous activity, you considered permitted and protected left turn phasing. Protected left turn phasing offers some benefits over permitted left turn operations, such as reduced left turn delay when opposing through volumes are high, but at the expense of increasing delay for other movements. In this activity you will consider another type of left turn treatment, protected plus permitted phasing. In this type of treatment, left turn movements have two separate green intervals, protected operations followed by permitted operations.

Protected plus permitted phasing is shown in the ring barrier diagram in Figure 168.

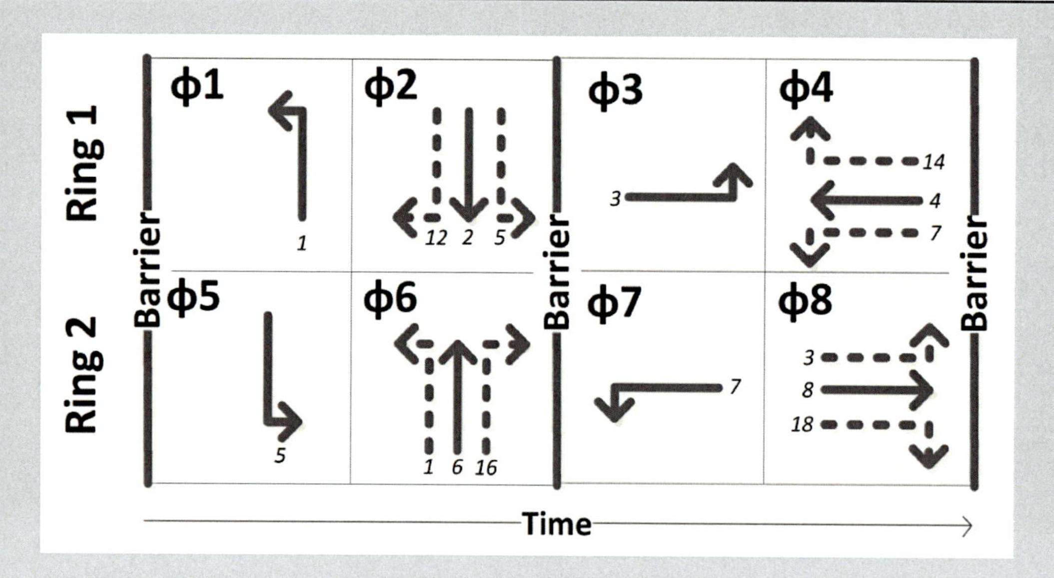

Figure 168. Ring barrier diagram for protected plus permitted left turn phasing

In this activity, you will perform tasks similar to what you did in Activity #48. You will observe the same intersection, State Highway 8 and Line Street, this time with protected and protected plus permitted left turn phasing.

Traffic volumes for all movements are the same as for the previous experiment except for eastbound left turn and westbound left turn.

- Eastbound through and westbound through: 1450 vph
- Eastbound left turn and westbound left turn: 200 vph

Task 1

Open the file: "A49.wmv."

Task 2

Observe the operation of both simulations.

- Observe the left turn vehicles on the eastbound and westbound approaches for case 1 (protected left turn) and case 2 (protected plus permitted left turn) (See Figure 169.)
- Observe vehicles that are served during the permitted phase in case 2 but are still waiting for the protected phase in case 1
- Observe the queue length that resulted for both cases
- Summarize your observations

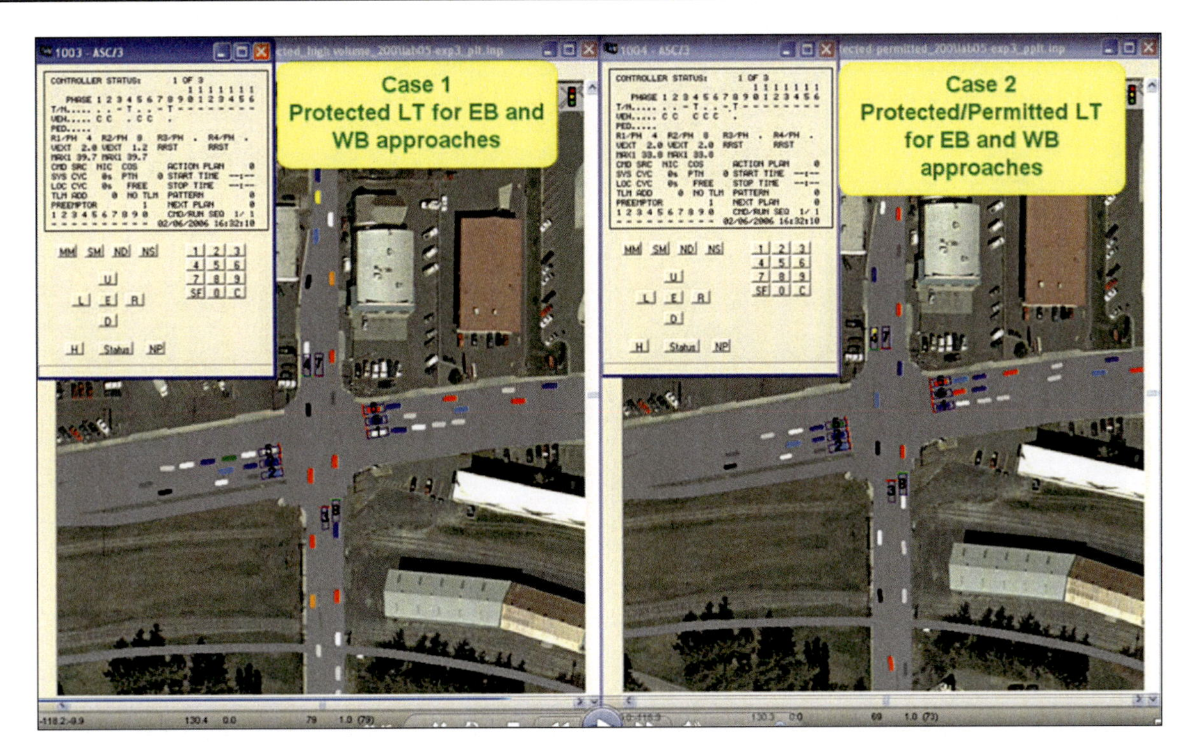

Figure 169. Animation comparing protected and protected plus permitted left turn phasing

Student Notes:

ACTIVITY

50 Analysis and Design of Left Turn Treatment

Purpose

The purpose of this activity is to give you the chance to compare protected and permitted left turn phasing treatments for your design problem and to select the most appropriate phasing treatment for the intersection.

Learning Objective

- Select optimal left turn phasing treatment based on an analysis of performance data and observation of simulation conditions

Required Resources

- VISSIM input file created in Activity #43

Deliverables

- Prepare a spreadsheet with required data, analysis, and conclusions as per Tasks 2 through 5:

  Tab 1: Title page with activity number and title, authors, and date completed

  Tab 2: Performance data comparing both left turn treatments

  Tab 3: Ring barrier diagram for your recommended phasing plan

Information

> How will you balance the relative advantages of permitted left turn phasing compared to protected left turn phasing? Completing the following tasks will help you make this decision.

Task 1

Make a copy of the folder that includes your VISSIM files from Activity #43. Name this new folder "a50". Use this VISSIM file as the basis for your analysis and design of your left turn treatments.

Task 2

Change the phasing plan to "permitted left turn" operation. See the VISSIM tutorial for help in making these changes. Collect data for delay and queue length, as in previous design activities.

Task 3

Observe the simulation for the two left turn options. Make notes on your observations on the operation and performance of both simulations.

Task 4

Compare the performance data and visual observation notes for both permitted and protected left turn treatment. Based on this comparison, make a determination of your recommended left turn treatment option.

Prepare a ring barrier diagram for the phasing plan that you recommend.

ACTIVITY

51 Left Turn Phasing Options

Purpose

The purpose of this activity is to give you the opportunity to learn how left turn phasing decisions are made in practice.

Learning Objectives

- Describe the process for selecting left turn phasing
- Contrast the advantages and disadvantages of each left turn phasing option

Required Resource

- *Traffic Signal Timing Manual*

Deliverables

Prepare a document that includes

- Answers to the Critical Thinking Question
- Completed Concept Map

Link to Practice

Read the sections from the *Traffic Signal Timing Manual* on left turn phasing as assigned by your instructor.

Critical Thinking Questions

When you have completed the reading, prepare answers to the following questions:

1. What are the advantages and disadvantages of the common left turn phasing options?

2. Describe the process followed in practice to select an appropriate left turn phasing plan.

3. Based on your reading, would you change the left turn phasing plan that you developed in Activity #50? Explain your answer.

IN MY PRACTICE...

by Tom Urbanik

Two issues have a profound effect on left turn operation and therefore phasing considerations. First is available storage which may be constrained due to closely spaced intersections. If the cycle length is long and the volume is high, consideration may need to be given to running the left turn twice per cycle. While one might think it is less efficient with two clearance intervals per cycle, a green indication with through traffic blocking the left turning traffic or left turning traffic blocking through traffic is more inefficient.

The second issue is when does left turning traffic arrive at the left turn lane. Again, with closely spaced intersections, left turning traffic may arrive too late to be served on the current cycle, causing increased delay for left turning traffic. Lagging rather than leading the left turn phase may provide reduced delay. This situation is very common at diamond interchanges where lagging the left turn to the ramp is often the preferred sequence.

CONCEPT MAP

Terms and variables that should appear in your map are listed below.

lagging left turns

leading left turns

left turn phasing

permitted left turns

protected left turns

Student Notes: ____________________

Right of Way Change: Change and Clearance Intervals

Purpose

In Chapter 9, you will learn about the vehicle change and clearance intervals, the timing intervals during which yellow and red, respectively, are displayed. Signalized intersections clearly and unambiguously assign right-of-way to specific movements in such a way that green is never displayed for conflicting movements at the same time. Thus, the transition from the green for one movement to another is an important process, one that must be done in way that maximizes safety for both the drivers that are currently being served and those about to be served. This is accomplished through the clearance and change intervals.

Like many parts of the signal timing process, technology and human factors must be considered together. How fast do drivers respond to a newly displayed yellow indication? What is the variation of this response among drivers? How does the variation in the approach speed of the drivers, and the distance that they are upstream from the intersection, affect their response? How much of a safety factor should be built into this change and clearance process? You will explore some of these questions in the activities that follow in this chapter.

Learning Objectives

When you have completed the activities in this chapter, you will be able to

- Describe the process for setting the yellow and red clearance times
- Describe the purpose and method of calculation of the vehicle change and clearance intervals
- Describe the different responses of drivers to the yellow indication based on their location upstream of the intersection
- Compare field data with the theoretical basis of stopping
- Describe driver behavior at the onset of the yellow interval
- Determine the vehicle change and clearance intervals
- Describe the factors considered when the yellow and red clearance intervals are set by practicing traffic engineers

Chapter Overview

This chapter begins with a *Reading* (Activity #52) on the change and clearance intervals. The chapter then proceeds to three activities including an assessment of your understanding of the change and clearance intervals (Activity #53) and activities in which you observe drivers responding to the yellow and red displays using field data (Activities #54 and #55). A design activity (Activity #56) follows in which you determine the change and clearance intervals for your design problem. The chapter concludes with an *In Practice* activity (Activity #57) in which you compare the results of your design work with the *Traffic Signal Timing Manual*.

Activity List

Number and Title	Type
52 The Theoretical Basis of the Yellow and Red Clearance Intervals	*Reading*
53 What Do You Know About the Change and Clearance Intervals?	*Assessment*
54 Drivers Responding to Yellow and Red Indications	*Discovery*
55 Vehicle Response to Displays at End of Green	*Field*
56 Determining Vehicle Change and Clearance Intervals	*Design*
57 Yellow and Red Clearance Intervals	*In Practice*

ACTIVITY

52 The Theoretical Basis of the Yellow and Red Clearance Intervals

PURPOSE

The purpose of this activity is to show you some of the issues that must be considered when setting the yellow and red clearance intervals.

LEARNING OBJECTIVE

- Describe the process for setting the yellow and red clearance intervals

DELIVERABLES

- Define the terms and variables in the Glossary
- Prepare a document that includes answers to the Critical Thinking Questions

GLOSSARY

Provide a definition for each of the following terms or variables. Paraphrasing a formal definition (as provided by your text, instructor, or another resource) demonstrates that you understand the meaning of the term or phrase.

change interval	
clearance interval	
perception-reaction time (δ)	
stopping distance (x_s)	
v	

a	
w	
L	

CRITICAL THINKING QUESTIONS

When you have completed the reading, prepare answers to the following questions.

1. What are the two factors that make up the stopping distance?

2. What are the two conditions that must be true in order for a driver to be able to safely stop or safely clear the intersection when the yellow is first displayed?

3. The value of t derived in the reading is the minimum value necessary to ensure that a driver will either be able to safely stop or safely clear the intersection when the yellow is displayed. What happens if the value is set higher than t just to provide an extra margin of safety? What would the trade-offs be in this decision?

INFORMATION

The ring barrier diagram defines which phases can time concurrently (those in different rings, on the same side of the barrier) and which must be timed sequentially (those in the same ring). Safety considerations require that a time separation be placed between the phases that must be timed sequentially. This time period consists of the change interval which is indicated by the yellow display and the clearance interval which is indicated by the red display. The theory that supports the determination of these intervals requires us to consider two factors:

- How long does it take a driver to perceive the need to stop and then to actually brake to a stop?
- How long does it take a driver to safely and completely clear the intersection?

As you will see, this theory makes several assumptions: vehicles arrive at the intersection at the same (and constant) speed, each vehicle has the same length, and all drivers have the same response characteristics to a change in the display from green to yellow. In reality, none of these assumptions is true: there is some variability in each of these three factors. So we will first develop the theory as a base for understanding how the change and clearance timing intervals are set, and then we will look at how the variability in these three parameters complicates the selection of the yellow time (change interval) and the red clearance time (clearance interval). Finally, we will consider the effects on driver behavior (how the driver responds to the change from green to yellow) when we increase or decrease the duration of the yellow indication.

We will first define what we will call the "choice point." It is the point upstream of which the driver will be able to safely stop at the intersection stop bar should he or she choose to do so. If the driver is any closer to the intersection than this choice point, he or she would not be able to safely come to a stop when the signal display changes from green to yellow. We can determine the location of this choice point by considering both the process that the driver must undergo to begin the stopping maneuver as well as the braking process. When the yellow is first displayed, it takes some time for the driver to perceive and react to this information. We call this the perception-reaction time (δ). If the driver decides to stop, he or she will apply the brakes and begin the deceleration process. The minimum stopping distance (x_s) is computed as the sum of (1) the distance traveled during the driver's perception-reaction time and (2) the braking distance (calculated using the basic equations of motion or kinematics).

$$x_s = v\delta + \frac{v^2}{2a}$$

where: v = initial velocity of the vehicle at the onset of yellow (ft/sec)

a = maximum comfortable acceleration rate (ft/sec/sec)

δ = perception-reaction time (sec)

Now, suppose that instead of stopping in response to the yellow indication, the driver at the choice point decides to continue through the intersection. The distance that the vehicle would have to travel from the choice point to the point where the rear bumper of the vehicle clears the far side of the intersection is given by the clearance distance (x_c):

$$x_c = x_s + w + L$$

where: x_s = minimum stopping distance as defined previously (ft)

w = intersection width (ft)

L = vehicle length (ft)

The time that it takes for a vehicle to travel the clearance distance (x_c) must equal the sum of the yellow (Y) and red clearance (RC) times:

$$Y + RC = \frac{x_c}{v} = \frac{x_s + w + L}{v}$$

where v is the speed of the vehicles approaching the intersection.

Why must this be true? In theory, our goal for a vehicle that continues through the intersection without stopping is to provide sufficient yellow time (Y) for the vehicle to travel from the choice point to the stop bar and sufficient red clearance time (RC) for the vehicle to travel from the stop bar to clearing its rear bumper through far side of the intersection. Thus:

$$Y = \frac{x_s}{v} \qquad RC = \frac{w + L}{v}$$

To illustrate this model, let's consider an intersection that is 40 feet wide. The location of the choice point is calculated using the equation for the minimum stopping distance (x_s). Suppose that drivers approaching the intersection travel at 35 miles per hour (51.33 feet per second) and have a perception-reaction time of 1 second. The minimum comfortable deceleration rate of 10 feet per second per second is also assumed.

$$x_s = v\delta + \frac{v^2}{2a} \qquad x_s = (51.33)(1) + \frac{(51.33)^2}{2(10)} = 184\ ft$$

In this case the choice point is located 184 feet upstream of the intersection as shown in Figure 170.

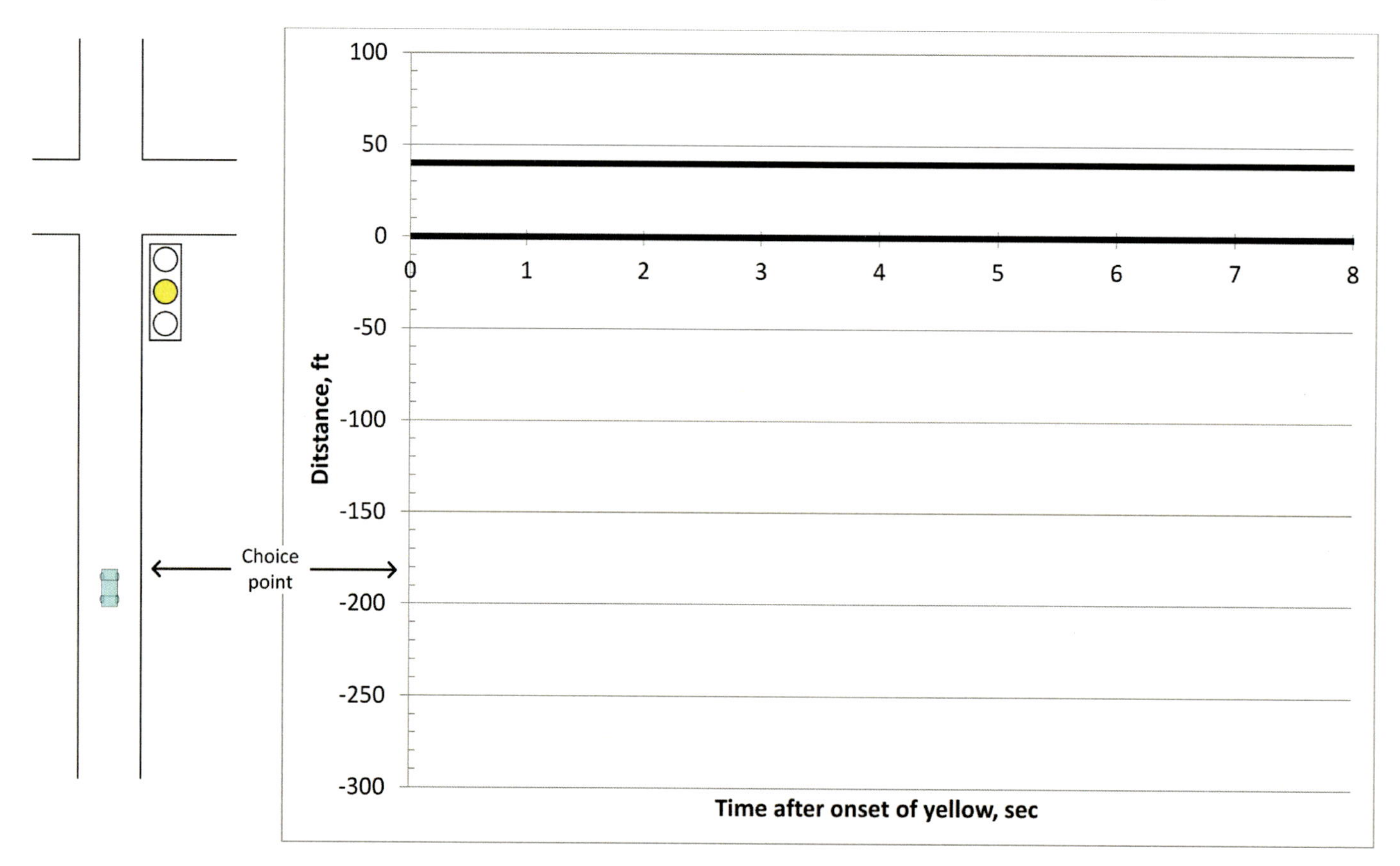

Figure 170. Vehicle position at onset of yellow indication

The time required for a vehicle to stop (t_s) when it is at the choice point and yellow is first displayed is the sum of the perception-reaction time and the braking time, in this case:

$$t_s = \delta + \frac{v}{a} = 1 + \frac{(51.33\ ft/sec)}{10\ ft/sec^2} = 6.2\,sec$$

The trajectory of this vehicle is shown in Figure 171. The trajectory is linear during the one second of the perception-reaction process, and curvilinear during the braking time.

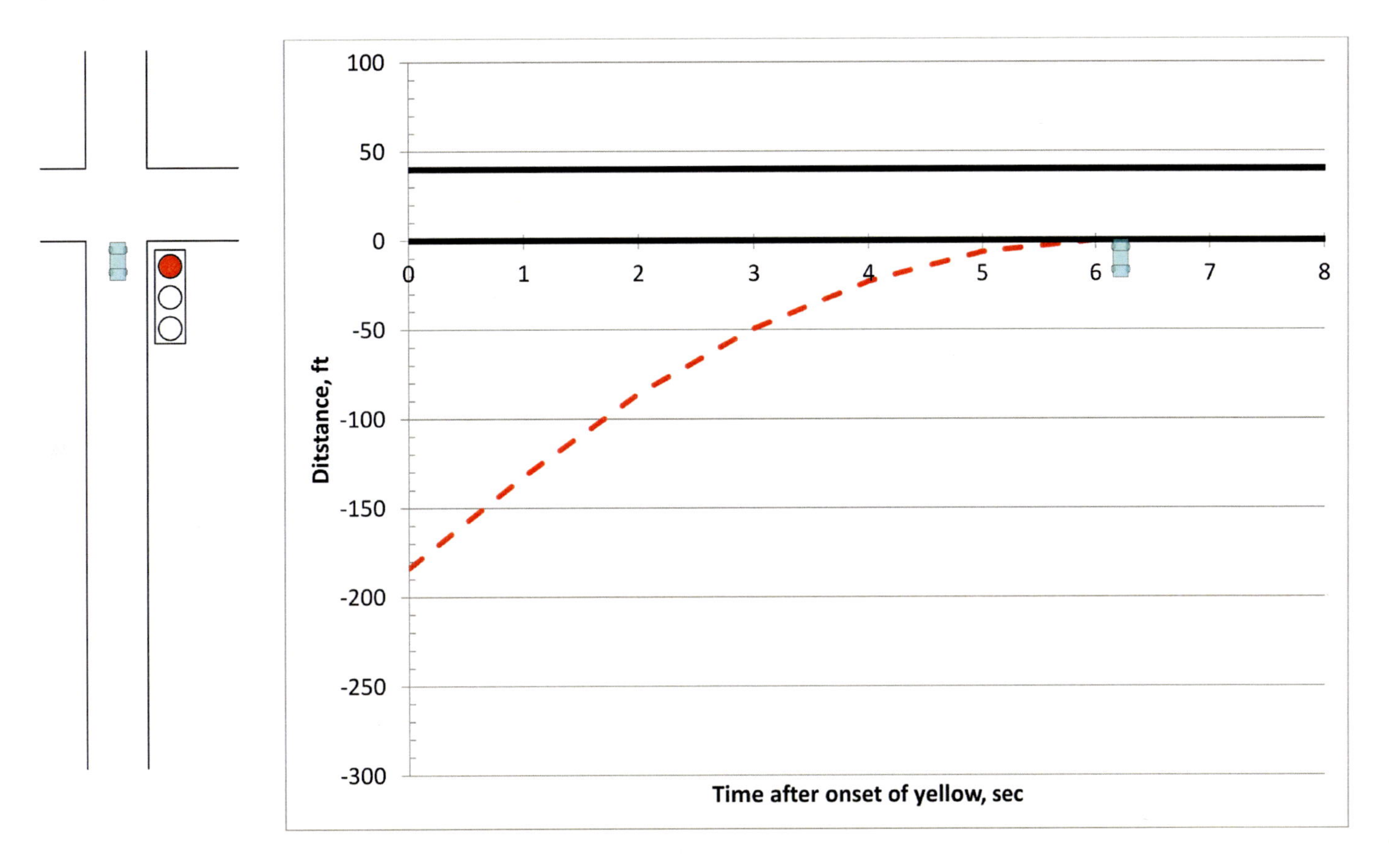

Figure 171. Vehicle trajectory stopping in response to yellow indication

Now let's look at the trajectory of a vehicle starting at this choice point (184 feet upstream of the intersection), and continuing through the intersection until its rear bumper clears the far side of the intersection (as shown in Figure 172). This is the clearance distance (x_c) and is equal to sum of the stopping distance (x_s), the width of the intersection (w), and the length of the vehicle (L). We know from the discussion above that the sum of the yellow and red clearance times must be equal to the time that it takes to travel the clearance distance (x_c). Assuming in this example that the intersection width is 40 feet and the vehicle length is 20 feet, then:

$$Y + RC = \frac{x_s + w + L}{v} = \frac{184\,ft + 40\,ft + 20\,ft}{(51.33\,ft/sec)} = \frac{248\,ft}{51.33\,ft/sec} = 4.7\,sec$$

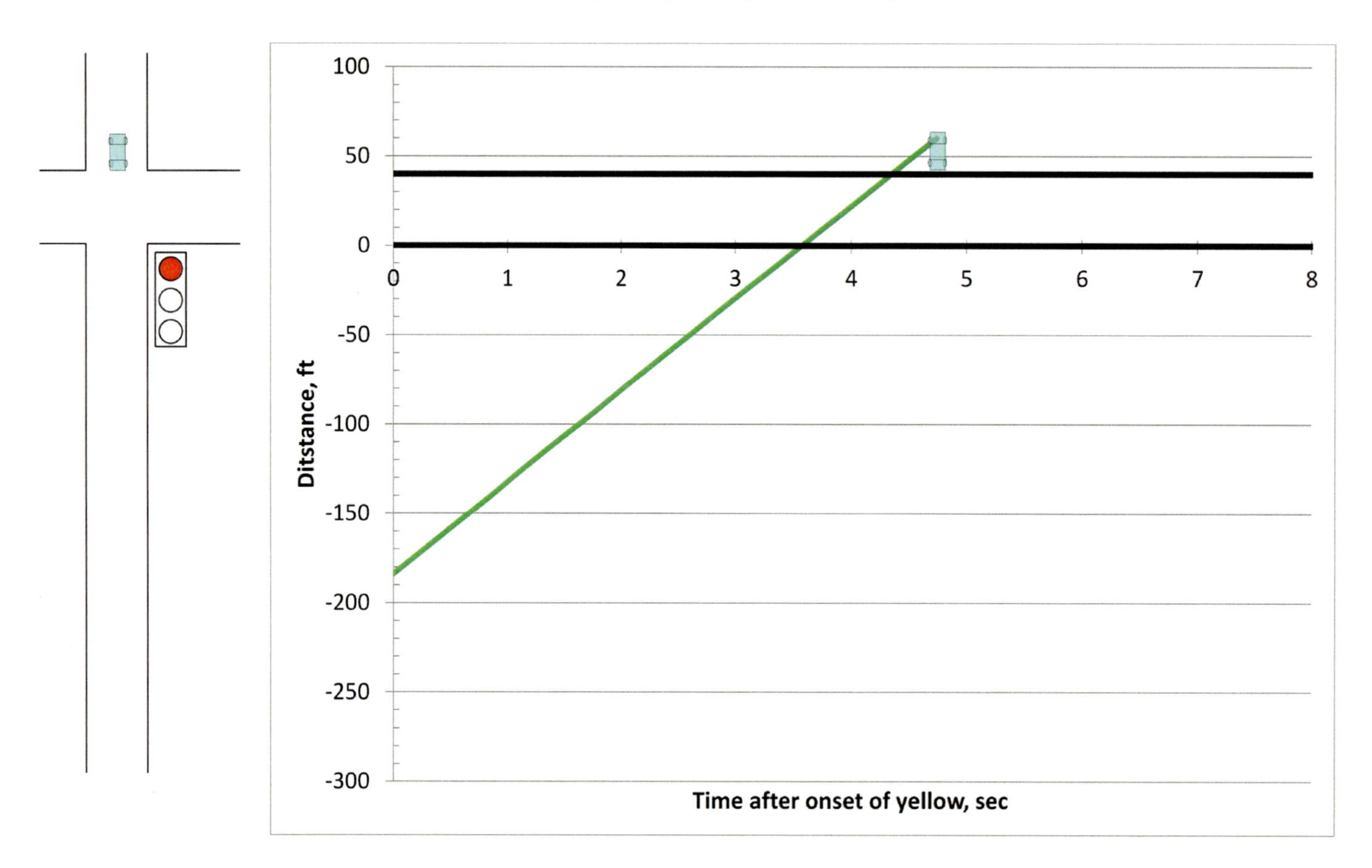

Figure 172. Vehicle trajectory safely passing through the intersection

After reviewing the trajectories from Figure 171 and Figure 172, we can see that it makes sense to allocate the sum of the yellow and red clearance times calculated above (4.7 seconds) into two parts. The yellow time (3.6 seconds in this example) is the travel time from the choice point to the stop bar for a driver that chooses not to stop, a result that puts the vehicle at the entry point to the intersection (at the stop bar) just as the display changes from yellow to red. The red clearance time (1.1 seconds) is the travel time from the entry of this vehicle into the intersection until its rear bumper clears the far side of the intersection.

$$Y = \frac{x_s}{v} = \frac{184\,ft}{(51.33\,ft/sec)} = 3.6\,sec$$

$$RC = \frac{w + L}{v} = \frac{40\,ft + 20\,ft}{(51.33\,ft/sec)} = \frac{60\,ft}{51.33\,ft/sec} = 1.1\,sec$$

Figure 173 shows a compilation of these results. If the sum of the yellow and red clearance times is set to 4.7 seconds, a vehicle (vehicle 3 in the figure) will be able to safely clear the intersection if it is at the choice point (or closer to the stop bar) when yellow is first displayed. And, if a vehicle is at the choice point (or further upstream from the stop bar), it will be able to safely stop when the yellow is displayed (vehicle 2).

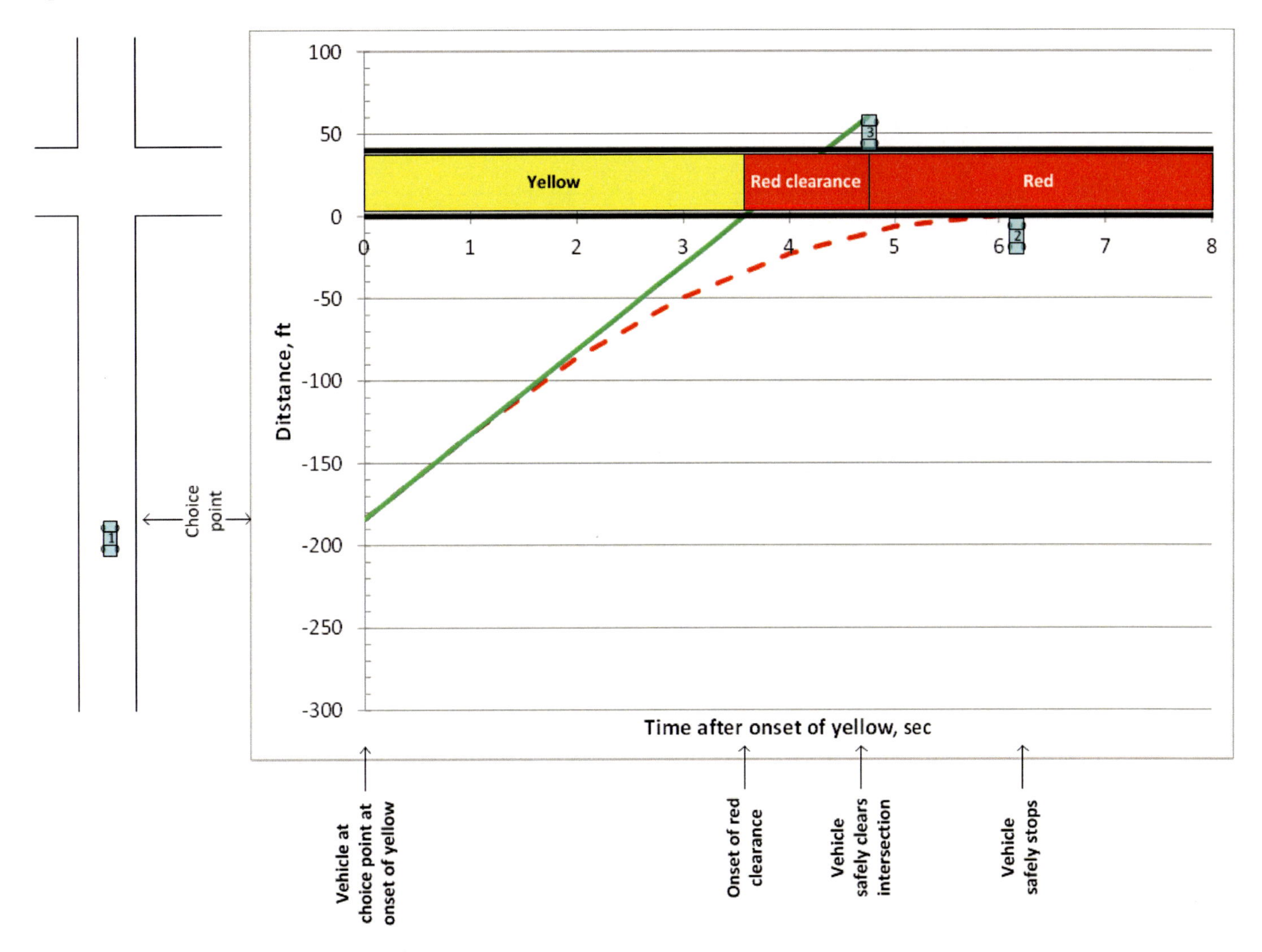

Figure 173. Vehicle trajectories with yellow and red clearance times

But as we noted at the beginning of this reading, there is a big step from the theoretical calculations presented here and the reality of traffic conditions in the field. What if a driver takes longer than one second to perceive and react to the display change from green to yellow? What if the driver's speed is lower (or higher) than the assumed approach speed? How does this model apply to longer vehicles such as trucks or buses? You will have the chance to work through some of these issues as you develop your design for the change and clearance intervals in the activities that follow.

Student Notes:

ACTIVITY
53 What Do You Know About the Change and Clearance Intervals?

PURPOSE

The purpose of this activity is to help you build your base of understanding of the change and clearance intervals.

LEARNING OBJECTIVE

- Describe the purpose and method of calculation of the vehicle change and clearance intervals

DELIVERABLE

- Prepare a completed spreadsheet that includes the following:

  Tab 1: Title page with activity number and title, authors, and date completed

  Tab 2: Calculations and results for Tasks 1 and 2

  Tab 3: Answers to the Critical Thinking Questions

CRITICAL THINKING QUESTIONS

1. In addition to the values assumed in the example in the reading ($v = 35$ miles per hour, $L = 20$ feet, $a = 10$ feet per second per second) for passenger cars, what are the implications in setting the yellow and red clearance times if the traffic stream also includes trucks with $L = 58$ feet and $a = 6.4$ feet per second per second? What values for these two timing intervals would you recommend and why?

2. Experience should tell you that there is likely to be a variation in the speeds of vehicles and the perception-reaction times of their drivers arriving at an intersection. Describe and complete a sensitivity analysis that you would perform to test the implications of the variation in perception/reaction times and in actual approach speeds. What impact does this analysis have on your conclusions about the duration of the yellow and red clearance times?

TASK 1

How long does it take a vehicle to stop, given the following information? Prepare a time distance plot showing your results.

- Vehicle length = 15 feet
- Approach speed = 25 miles per hour
- Perception-reaction time = 1 second
- Deceleration rate = 10 feet per second per second

TASK 2

What is your recommendation for the yellow and red clearance times given the conditions in Task 1?

Drivers Responding to Yellow and Red Indications

PURPOSE

The purpose of this activity is to provide you with the opportunity to learn about the variability of how drivers respond to the yellow indication, including how distance and time from the intersection at the onset of yellow affects the driver's likelihood of stopping (or not).

LEARNING OBJECTIVES

- Describe the different responses of drivers to the yellow indication based on their location upstream of the intersection
- Compare field data with the theoretical basis of stopping

REQUIRED RESOURCE

- Data file: A54.xlsx.

DELIVERABLE

- Prepare a spreadsheet that includes the following information:

  Tab 1: Title page with activity number and title, authors, and date completed

  Tab 2: Raw data for stopped and not-stopped vehicles

  Tab 3: Statistical summaries as required in Task 1

  Tab 4: Plot from Task 2

  Tab 5: Plot as required in Task 3

  Tab 6: Answers to the Critical Thinking Questions

CRITICAL THINKING QUESTIONS

As you begin this activity, consider the following questions. You will come back to these questions once you have completed the activity.

1. Based on the statistical summaries that you prepared in Task 1, how are the characteristics the same or different for vehicles that stop in response to yellow and those that don't?

2. What conclusions about driver behavior (the decision to stop or not) can you make based on the plots that you prepared in Tasks 2 and 3?

Information

It is important to learn about the connections (and, sometimes, differences) between theory and how drivers actually behave in the field. In this activity, you will work with a data set from the NGSIM project that includes observations of drivers along Lankershim Blvd. in Los Angeles (described earlier in Activities #3 and #10). This data set contains 303 records of vehicles responding to the onset of the yellow indication; each record includes whether the vehicle stopped or continued through the intersection, and, at the time of yellow onset, (1) how far the vehicle was from the stop bar, (2) its speed, and (3) how long it would take to reach the stop bar at this speed and from this distance. You will see that whether drivers decide to stop or not when the yellow is displayed, even when a set of drivers are the same distance or travel time upstream of the intersection, is a probabilistic outcome. When a driver is closer to the intersection, there is a higher probability that he or she will continue through the intersection without stopping; the farther away the driver is from the intersection, the probability increases that he or she will stop in response to the yellow indication.

Task 1

Prepare statistical summaries of both the "stopped" and "didn't stop" data sets, including mean values of and frequency distributions for the distance and time from the intersection at the onset of yellow. Prepare a table showing the probabilities of vehicles stopping or not stopping in 50 foot "distance from the stop bar" bins.

Task 2

Prepare a chart of "time from the stop bar" (x-axis) and "distance from the stop bar" (y-axis) for both data sets (vehicles that stop and those that don't). Include a vertical line on the chart that represents the onset of yellow (at $t = 2.9$ seconds).

Task 3

Compute the trajectories for two vehicles that respond to the yellow indication at $t = 2.9$ seconds. Assume that one vehicle continues to travel at 25 miles per hour through the intersection. Assume that the second vehicle, after a one second perception-reaction time, begins to decelerate at 10 feet per second per second and comes to a stop at the stop line. Add these two lines to the chart that you prepared in Task 2.

Vehicle Response to Onset of Yellow

Purpose

The purpose of this activity is for you to observe how vehicles in the field respond to a change in signal displays and to describe this behavior.

Learning Objective

- Describe driver behavior at the onset of the yellow interval

Required Resource

- Data file: A54.xlsx

Deliverable

- Prepare a spreadsheet that includes the following information:

  Tab 1: Title page with activity number and title, authors, and date completed

  Tab 2: Answer to the Critical Thinking Question

  Tab 3: Aerial photo of intersection approach and distance grid

  Tab 4: Field data and calculations from Table 25

  Tab 5: Plot from Task 5

  Tab 6: Probability analysis from Task 5

Critical Thinking Question

1. What conclusions can you make about the differences or similarities between the data that you collected in Activity #54 and this current activity?

TASK 1

Select one major street approach on your intersection. Using Google Earth (or another similar mapping tool), select an aerial view of the approach and identify points up to 300 feet upstream of the stop bar for that approach, in 50 foot increments. Print this aerial view with the "50 foot increment" points. An example of this aerial view with increments marked is shown in Figure 174.

Figure 174. Example intersection approach showing 50 foot intervals upstream of the stop bar

TASK 2

Observe the operation of the traffic stream and signal displays on this approach for five minutes, giving particular attention to the "onset of yellow" period when vehicles will be making decisions to stop or not.

TASK 3

Using the aerial map that you prepared in Task 1, record the location of 20 vehicles that you observe at the beginning of the yellow interval by placing a number on the map corresponding to the location of each of the vehicles. When you select these 20 vehicles, make sure that there is no vehicle between them and the stop bar at the onset of yellow. Also, record (in Table 25) the response of the driver to the yellow display (either "Go" or "Stop").

TASK 4

For each of these vehicles, record the following information in Table 25:

- Your estimate of the distance upstream from the stop bar
- The response of the driver to the yellow display (Go or Stop)
- Your estimate of the travel time from the observed location to the stop bar based on your estimated distance and the posted travel speed

TASK 5

Prepare a plot of the "time from the stop bar" (x-axis) and "distance from the stop bar" (y-axis) at the onset of yellow segregating the data according to whether the vehicle stopped or continued through the intersection. Compare this plot with the equivalent plot that you made in Task 2 of Activity #54. Again, compute the probability of "stopping or not stopping", segregating the data into 50 foot intervals.

Table 25. Field observations and calculations

Vehicle number	Distance of vehicle from stop bar at onset of yellow	Response of driver to the yellow display (Go/Stop)	Estimated time for vehicle to travel to stop bar at onset of yellow
1			
2			
3			
4			
5			
6			
7			
8			
9			
10			
11			
12			
13			
14			
15			
16			
17			
18			
19			
20			

Student Notes:

ACTIVITY

56 Determining Vehicle Change and Clearance Intervals

PURPOSE

The purpose of this activity is to give you the opportunity to determine the vehicle change and clearance intervals for your design.

LEARNING OBJECTIVES

- Determine the vehicle change and clearance intervals
- Describe how the variability of vehicle approach speeds affects the determination of the vehicle change and clearance intervals

REQUIRED RESOURCE

- VISSIM input file from Activity #50

DELIVERABLE

- Prepare a report using an Excel spreadsheet that includes your simulation data, your analysis, and your answers to the Critical Thinking Questions.

  Tab 1: Title page with activity number and title, authors, and date completed

  Tab 2: Data from FZP file

  Tab 3: Parsed data for selected approach

  Tab 4: Chart showing speed vs. distance upstream from the stop bar

  Tab 5: Calculations for the yellow and red clearance times

  Tab 6: Table of measures of effectiveness for each approach and for the intersection

  Tab 7: Answers to Critical Thinking Questions

CRITICAL THINKING QUESTIONS

1. What effect did the range in speeds that you observed for vehicles approaching the intersection have on the final result of the values of the yellow and red clearance times? Discuss your answer.

2. What effect would trucks have on your final result?

3. What are the implications of your work in Activities #54 and #55 on your final selection of the yellow and red clearance times? Discuss your answer.

Information

The reading in Activity #52 described the model used to calculate the yellow time and the red clearance time based on the width of the intersection, the length of the vehicle, and the speed of the vehicle as it approached the intersection.

$$Y = \delta + \frac{v}{2a} \qquad RC = \frac{w + L}{v}$$

In this activity you will select your yellow and red clearance times based on this model but recognizing the variability of the speeds of vehicles approaching the intersection.

Task 1

Make a copy of the folder that includes your VISSIM files from Activity #50. Name this new folder "a56". Use this VISSIM file as the basis for the analysis and design of your change and clearance intervals. Select one of the approaches on the major street of this intersection for this activity, preferably the longest approach.

Task 2

Start VISSIM and open your simulation file. Set the simulation resolution and speed:

- Select "Simulation," then "Parameters"
- Set the "Simulation resolution" to 1 time step per second
- Set the "Simulation speed" to maximum speed
- Set the "Period" to 900 seconds

Task 3

Select "Evaluation," then "Files," then "Vehicle record" to create the file to store the vehicle trajectory data.

- From "Configuration", select vehicle number, simulation time, link number, "x" world coordinates (or "y" if your major street is oriented north-south), and speed (mph)
- From "Filter," select 0 to 900 seconds
- These data will be written to the output file with the extension "fzp"

Task 4

Complete the simulation run.

Task 5

Open and parse the FZP file into tab 2 of an Excel worksheet. Copy the parsed data to tab 3 of your worksheet. In tab 3, keep only the data for the link that corresponds to the approach that you selected in Task 1.

Task 6

Identify the relevant world coordinate location of the stop bar for your selected approach. Using the data from tab 5 (for your selected approach), prepare a plot of vehicle speed (y-axis) and the relevant world coordinate (x-axis) with the x-axis showing the range from the stop bar to 300 feet upstream of the stop bar. Visually identify those vehicles that continue through the intersection without stopping (noted by their relatively horizontal trajectory on your speed-distance plot. Based on this visual inspection, select an approach speed that you believe represents the 85th percentile of speeds for vehicles that continue through the intersection without stopping.

Task 7

Compute the yellow and red clearance times based on the speed that you determined in Task 6, the average vehicle length, and the width of your intersection.

Task 8

Using your final design values for yellow and red clearance times, run VISSIM to produce estimates of delay and queue length. Prepare a table of these measures of effectiveness for each approach and for the intersection.

Student Notes:

Yellow and Red Clearance Intervals

Purpose

The purpose of this activity is to give you the opportunity to learn how the yellow and red clearance intervals are set in practice.

Learning Objective

- Describe the factors considered when the yellow and red clearance intervals are set by practicing traffic engineers

Required Resource

- *Traffic Signal Timing Manual*

Deliverables

Prepare a document that includes

- Answers to the Critical Thinking Question
- Completed Concept Map

Link to Practice

Read the section of the *Traffic Signal Timing Manual* relating to Vehicular Change and Clearance Intervals as assigned by your instructor.

The *Traffic Signal Timing Manual* notes that:

> "The intent of the vehicle phase change and clearance intervals is to provide a safe transition between two conflicting phases. It consists of a yellow change interval and, optionally, a red clearance interval. The intent of the yellow change interval is to warn drivers of the impending change in right-of-way assignment. The red clearance interval is used when there is some benefit to providing additional time before conflicting movements receive a green indication."

The yellow display warns the drivers that the right-of-way is about to change, while the red clearance display allows drivers to safely clear the intersection.

Critical Thinking Questions

When you have completed the reading, prepare answers to the following questions:

1. Consider the method that you used in Activity #56 to set the yellow and red clearance intervals, as well as your results. How do they compare and contrast with the methods and recommendations described in the *Traffic Signal Timing Manual*?

2. How does your own driving experience compare with the material from the *Traffic Signal Timing Manual* relating to the yellow and red clearance intervals?

IN MY PRACTICE...

by Tom Urbanik

Clearance interval practice differs significantly among agencies. The result produced by the model that you read about in Activity #52, while being based on the laws of physics, is considered to be larger than necessary by some practitioners, resulting in a wide variability of clearance interval times used in practice. One issue is the interpretation of the meaning of yellow which is defined differently based on state laws. The intent is that yellow is notice to stop unless it is not possible. However, some drivers, due to many reasons, have over the years tended to use more and more of the yellow to avoid stopping. Adding to this the practical definition of violating the red as red light running, results in changed behavior of some to see the yellow as the amount of time to enter the intersection. Furthermore, the introduction of photo enforcement further complicates the issues by also defining red light running as entering on red. The end result has been longer clearance times and widespread use of red clearance time to account for drivers trying to maximize the use of the yellow to enter the intersection.

So, in practice, local clearance interval timings are often driven by historical local practices. As a signal timing engineer you need to resolve local practice against other agency practices in your area. Desirably, signal timing should be consistent in a local area as drivers in one jurisdiction may develop habits based on their agency's practice. This could become an issue if another local jurisdiction has different practices. Finally, the addition of red light running cameras may put pressure on adjusting local practices.

Concept Map

Terms and variables that should appear in your map are listed below.

change interval	perception-reaction time (δ)	v	w
clearance interval	stopping distance (x_s)	a	L

Student Notes:

CHAPTER 10 Your Final Design: Putting It All Together

Purpose

In Chapter 10, you will complete your signal timing design and present your results. In the previous nine chapters of this book, you have learned how an actuated traffic control system at an isolated intersection works and the process for designing the phasing and timing parameters for a given traffic demand and geometric design. In this chapter you will assemble the various signal timing components that you have designed and evaluated, and integrate them into a final design. When you have completed the activities in this chapter, you will have completed and presented a report of the work that you have done!

Learning Objectives

When you have completed the activities in this chapter, you will be able to

- Integrate information from previous work
- Justify design choices
- Communicate results
- Prepare a timing plan for an isolated actuated signalized intersection based on an analysis of traffic flow quality and intersection performance for a range of timing parameter values and phasing alternatives
- Synthesize ideas from a professional engineering report
- Compare your design results with values recommended for practice
- Integrate information into a professional style report and presentation
- Clearly communicate the timing plan design for an isolated actuated signalized intersection based on an analysis of traffic flow quality and intersection performance for a range of timing parameter values and phasing alternatives
- Provide effective feedback to others

Chapter Overview

This chapter begins with a *Reading* (Activity #58) that describes some of the issues that you will face in putting together your final report. A discovery activity (Activity #59) follows in which you will assemble the elements of your signal timing design. You will compare your work with two links to professional practice: a review of a signal timing design report (Activity #60) and the *Traffic Signal Timing Manual* (Activity #61). The chapter concludes with two design activities on the completion and presentation of your design (Activities #62 and #63).

Activity List

Number and Title	Type
58 Integrating Information, Justifying Choices, and Communicating Results	*Reading*
59 Assembling Information For Your Timing Plan Design	*Discovery*
60 What Do You Know About the Signal Timing Process?	*In Practice*
61 Signal Timing Design in Practice	*In Practice*
62 Design Report	*Design*
63 Design Evaluations and Assessments	*Design*

ACTIVITY
58 Integrating Information, Justifying Choices, and Communicating Results

PURPOSE

The purpose of this activity is to give you the opportunity to consider issues relating to your final design report.

LEARNING OBJECTIVES

- Integrate information from previous work
- Justify design choices
- Communicate results

DELIVERABLES

- Define the terms in the Glossary
- Prepare a document that includes answers to the Critical Thinking Questions

GLOSSARY

Provide a definition for each of the following terms. Paraphrasing a formal definition (as provided by your text, instructor, or another resource) demonstrates that you understand the meaning of the term or phrase.

level of aggregation	
measures of effectiveness	

CRITICAL THINKING QUESTIONS

When you have completed the reading, prepare answers to the following questions.

1. When we talk about integrating information, what do we mean? Provide an example from everyday life.

2. Identify a kind of information that you often see presented. Describe two ineffective ways of presenting that information. Why are those ways ineffective?

3. What criteria make for an effective oral presentation or report? Name at least five, describing how each contributes to the effectiveness of the presentation or report.

4. Why is there not one "right" answer to a problem that you might observe that will apply to all traffic conditions (for example: short or long queues, low or high volumes)? Explain.

Information

The Design Process

What is engineering design? More specifically, what is traffic signal timing design? While you may not have thought about the design process explicitly as you have completed the preceding 57 activities in this book, design is exactly what you have been doing. You have defined a problem, analyzed the performance of an existing signal timing plan, and evaluated the performance of your final design elements. Let's step back for a minute and look back to what you have considered, from several perspectives: the components of the traffic control system and how these components interact, the inputs to the system that you have access to and control of, and the way the system performs based on changes that you make to it.

As you learned in Chapter 1, the traffic signal control system can be represented by four components: users, detectors, controllers, and displays. The users arrive at the intersection, and their arrivals are sensed by detectors. The detectors transmit what they have sensed to the controller. The controller determines which users to serve, in what sequence, and for how long, and specifies which users are currently being served (and those that are not) by driving the displays. The displays provide information on what actions are appropriate for the users to take. There is a clear process of interaction and influence among and between these four components (see Figure 175).

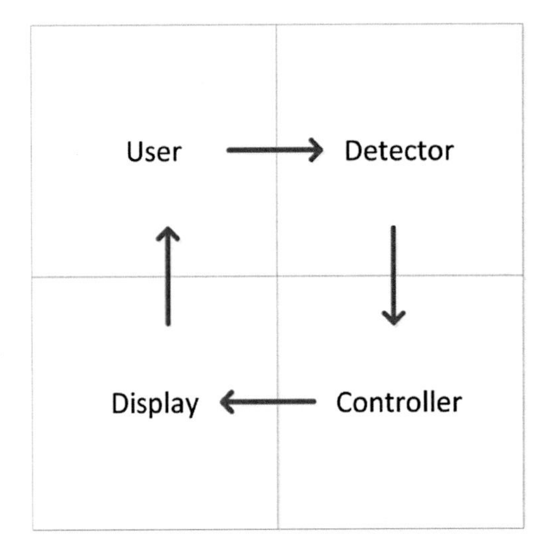

Figure 175. Traffic control process model

But which of these components can you influence in this design process and what are the results of the design choices that you make? We will consider these issues in the list below.

1. You have little or no influence on the number of users that arrive at the intersection during a given interval of time. So the user demand is generally a "given" that you must assume. You also have little influence over the types of users that arrive at the intersection, though you might give priority to certain types of users over others. For our work, the number of users has been specified as a flow rate (in vehicles per hour), and we have focused on users who drive cars or trucks.

2. You do have direct influence on the detection component. You can specify the technology (for example, inductive loops, video, or microwave), where the technology is located, and the manner in which it senses the arrival or presence of users. We have focused on one technology in this project: inductive loops that are embedded in the pavement and that sense the presence of vehicles that have arrived at the intersection. We have assumed that the loops are located at the stop bar and that they are 22 feet long. So while you as a traffic engineer do have influence on the detector design, as part of this project you were given a completely specified detection component to assume.

3. You have direct influence on the controller component. You can specify the controller type (NEMA, 2070, other), the sequence in which the phases are served, the timing processes and durations of the timing intervals, and whether the intersection operates in isolation or as part of a coordinated system. Through a series of experiments and observations that encompassed the previous activities, you determined the phasing plan (through specifying the ring barrier diagram) and you specified the duration of the minimum green, passage time, maximum green, yellow, and red clearance timing intervals for each phase.

4. You have direct influence on the display component. Within guidelines from the *Manual of Uniform Traffic Control Devices* and the *Traffic Signal Timing Manual*, you can specify the types and locations of the displays and indications. However, the specification of the displays and indications were beyond the scope of your work in this design project and you assumed standard vehicle displays with green, yellow, and red indications.

This information is summarized in Table 26, where each component is described, how much influence or control the designer in general has over each component, and what your role was for each component as part of this design project.

Component (attributes)	Degree of influence or control of component by designer	Your degree of influence or control of component in your design project
User (automobiles, trucks, pedestrians, transit, trains)	None or little	None (user volume assumed)
Detectors (technology, detection area, location)	High (technology type, detection area, location)	None (assumed inductance loops, presence detection, located at stop bar, 22 feet in length)
Controller	High (phase sequence, timing processes, timing durations)	High (selected/determined phase sequence, basic actuated timers, timer durations, yellow interval, red clearance interval)
Display (indications)	High (type, location)	None (display configuration assumed)

Table 26. Your design role in the traffic signal control system

So, how did the system perform, given the geometric layout of the intersection, the volume and type of users, the specification of the detection system, the controller timing plan, and the vehicle displays that you assumed? You used several methods to describe the performance of the system including (1) description of what you observed while watching the simulation and (2) analysis of the data that VISSIM generated including standard performance measures (delay and queue length). In this reading, we will explore some of these issues in more detail, including integrating different kinds of information, using various kinds of measures of effectiveness, presenting data, using experimental results to make design choices, and communicating your results.

Integrating Different Kinds of Information

In Chapter 5, we introduced the notion of "learning to see." What we meant was this: a traffic engineer should spend time in the field observing the flow of traffic and, based on these observations, determine if traffic is flowing well or if the quality of flow could be improved. For example, we could observe that a queue spills out of a left turn pocket impeding the flow of traffic in the adjacent through lanes. Or we could observe that a short queue that forms on the through lanes clears quickly after the beginning of green. Both of these observations are statements about the quality of the flow of traffic that might lead us to, in the first example, ideas that change the sequence of phases or the duration of one or more timing intervals, or, in the second example, to do nothing because the system is already performing in an acceptable manner.

We can quantify these observations by adding numeric performance data that we collect during field observations or from a simulation run. While measures of effectiveness will be discussed in the next section of this Reading, let's first explore two examples that illustrate the integration of visual observations with numeric performance data.

Example #1: Queue spillback from left turn pocket

Consider the two cases of the operation of a left turn lane shown in Figure 176. Figure 176a shows a queue that spills out of the left turn lane and into the adjacent through lanes. By contrast, in Figure 176b, the left turn queue is served without impeding the through traffic flow. Both observations are valuable and lead us to make different conclusions about the quality of flow. But can we supplement these observations with performance data that we collect from the field or from a simulation run? The answer is yes, as shown in Table 27, where the delay and queue length data for both conditions are shown. In the first example, there is an average of four vehicles in the queue while in the second example the average is only one vehicle in the queue.

a. LT lane queue spillback

b. No LT spillback

Figure 176. Queue spillback from left turn pocket

Case	Average delay (sec/ veh)	Average queue length (vehicles)
Queue spillback	27	4
No queue spillback	5	1

Table 27. Comparison of numeric performance data for different flow conditions (Example #1)

Example #2: Queue doesn't clear before end of green

Now consider two cases of the operation of a through lane shown in Figure 177. Figure 177a shows a queue that doesn't clear before the end of green. By contrast, in Figure 177b, the queue does clear before the end of green. As in the example #1, Table 28 shows the delay and queue length data for both conditions, data that can be integrated with the visual observations made from Figure 177.

a. Queue at end of green

b. Short queue that will clear before end of green

Figure 177. Queue in through lane

Case	Average delay (sec/veh)	Average queue length (vehicles)
Queue doesn't clear	35	14
Queue clears	12	2

Table 28. Comparison of numeric performance data for different flow conditions (Example #2)

Measures of Effectiveness

A measure of effectiveness or MOE is a parameter that describes the performance of the system, or how effective the system is in meeting the needs of its users. For an isolated signalized intersection, two measures are commonly used for this purpose.

- The average delay per vehicle is the average additional travel time experienced by users if they weren't impeded by other vehicles or the control system. Average delay includes users that experience delay as well as those that do not. Theoretically, if no user travels at less than his or her desired speed, the average delay would be zero.
- Average queue length is the average length of the queue during the period of measurement or observation. If any vehicle arrives during red, and must stop or queue, the average queue length would be non-zero.

Both of these measures can be used by the traffic engineer to describe intersection performance. But since users can also directly experience delay and observe the length of the queue on an intersection approach, these MOEs have value over those that can't be directly experienced by the user. An example of the latter is the degree of saturation or volume/capacity ratio. The traffic engineer can measure the volume on an intersection approach, calculate the capacity of the approach, and then determine the volume/capacity ratio to help in the evaluation of the performance of the intersection. However, the user cannot directly experience this parameter.

It is also useful to look at MOEs at different levels of aggregation. For example, we can measure delay for vehicles in one lane, for all of the lanes on a given approach, and for all vehicles traveling through the intersection. When we want to understand performance at a detailed level (in order to identify and develop solutions for a specific problem), the lane or approach is the appropriate level or view. If, however, we want to broadly compare the performance of an intersection under two different signal timing designs, we could use the average delay for each of the designs. Each level of aggregation is important and tells a different part of the "performance story." Three figures from Chapter 5 (Figure 178, Figure 179, and Figure 180) are repeated here to illustrate these three levels of aggregation.

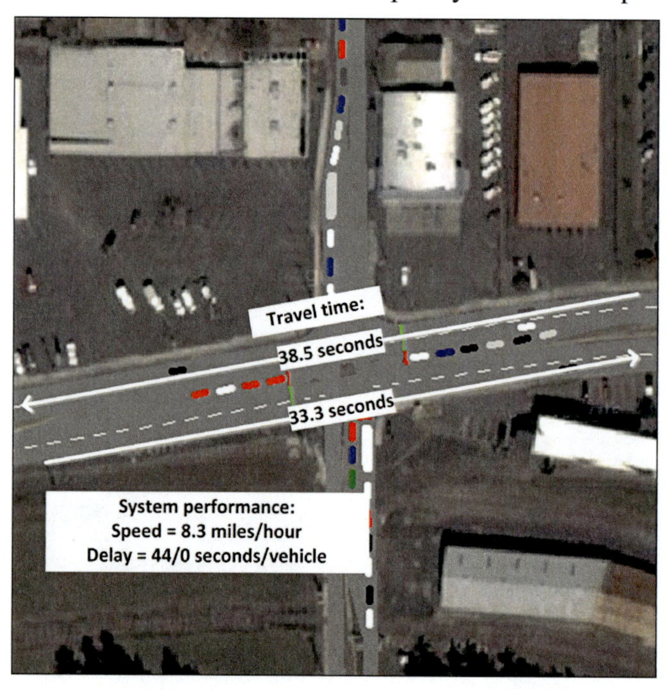

Figure 178. System performance data

Presenting Data

Tables and charts are potentially a good way to show your data and the story that your data can tell. We say "potentially" because tables and charts can also obscure your results and get in the way of the story you want to tell about your data. In this section we will look at both good and bad examples of tables and charts. We will also consider examples of "before and after" comparisons, precision and accuracy, and differences, both statistically and operationally significant. Read through the examples and see how each one might be of value as you prepare your final report. Each one is from a report previously completed by a university student. A quote from Edward Tufte (from *The Visual Display of Quantitative Information*) says it clearly:

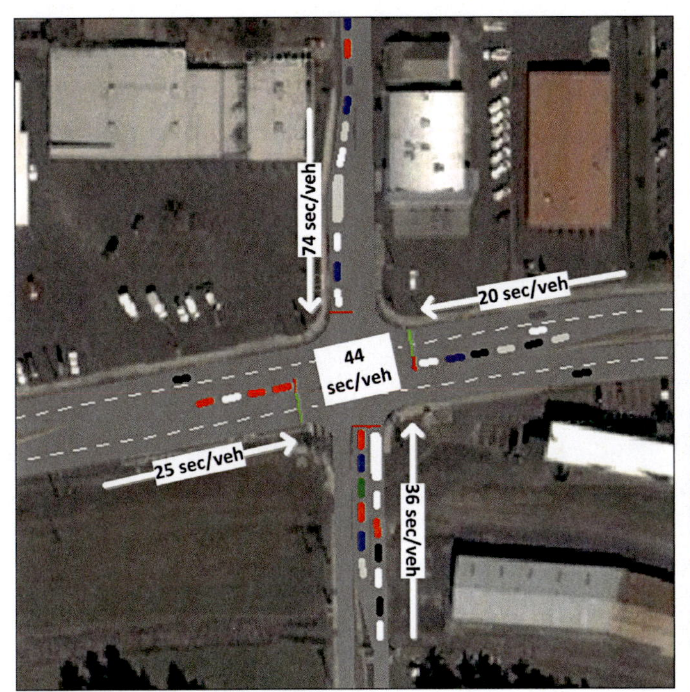

Figure 179. Intersection and approach performance data (average delay)

Figure 180. Movement performance data (average delay)

"What is to be sought in designs for the display of information is the clear portrayal of complexity. Not the complication of the simple; rather the task of the designer is to give visual access to the subtle and the difficult – that is, the revelation of the complex."

Figure 181 shows the duration of the mean green times for each of the eight phases at a signalized intersection. Much of the chart is good: the bar charts clearly show the green time values and it is easy to see the differences: phases 2 and 6 (generally the phases that control major street through movements) have the longest mean green durations, and the phases controlling the left turn movements (phases 1, 3, 5, and 7) have the lowest mean green time durations. However, the title is incorrect since this isn't really a histogram or frequency diagram.

Table 29 compares four attributes for a set of simulation runs where the maximum green time setting has been varied from 100 seconds down to 20 seconds. The number of gap outs and max outs vary as expected, as the maximum green time varies: for lower maximum green times, the phase is more likely to max out while for higher maximum green times the phase is more likely to gap out. Both are clearly shown in the table. The delay varies, and decreases as the maximum green time is decreased, which is what we would expect from theory (higher cycle lengths generally results in higher delay). This table is a good example of presenting and comparing results for the analysis of one timing parameter. But the table could be improved by giving the units for the average queue (feet) or delay (sec/veh).

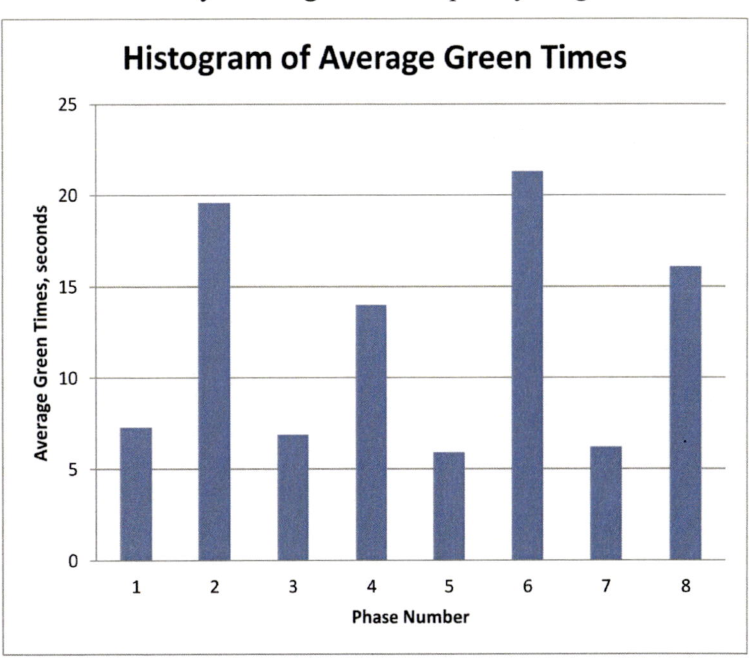

Figure 181. Example chart with incorrect title

Measure	Maximum Green Time (sec)									
	10	20	30	40	50	60	70	80	90	100
Gap outs	0	129	138	154	175	180	178	190	198	205
Max outs	209	80	71	55	33	30	22	11	4	4
Average queue	37.9	34.6	41.2	43.6	53.2	49.3	57.4	50.9	52.2	48.4
Delay	24.0	29.0	32.0	35.2	41.1	39.0	43.2	41.6	43.1	41.9

Table 29. Effective comparison of the variation in the number of gap outs and max outs by maximum green time

Figure 182 shows a cumulative frequency chart for two sets of data, NQ and Q. We most likely know that Q is for "queued vehicles" and NQ is for "non-queued vehicles." However the reader should never have to guess what these abbreviations are. Another element in the chart that could be improved is the labeling of the y-axis range: the maximum should be 1.0 (the actual maximum value for a cumulative frequency chart) and not 1.20. In addition, only one decimal place is needed and not the 2 that are included in the figure. Similar comments can be made about the labeling of the x-axis: a spacing of 1 is sufficient; 0.25 results in too many numbers displayed on the x-axis.

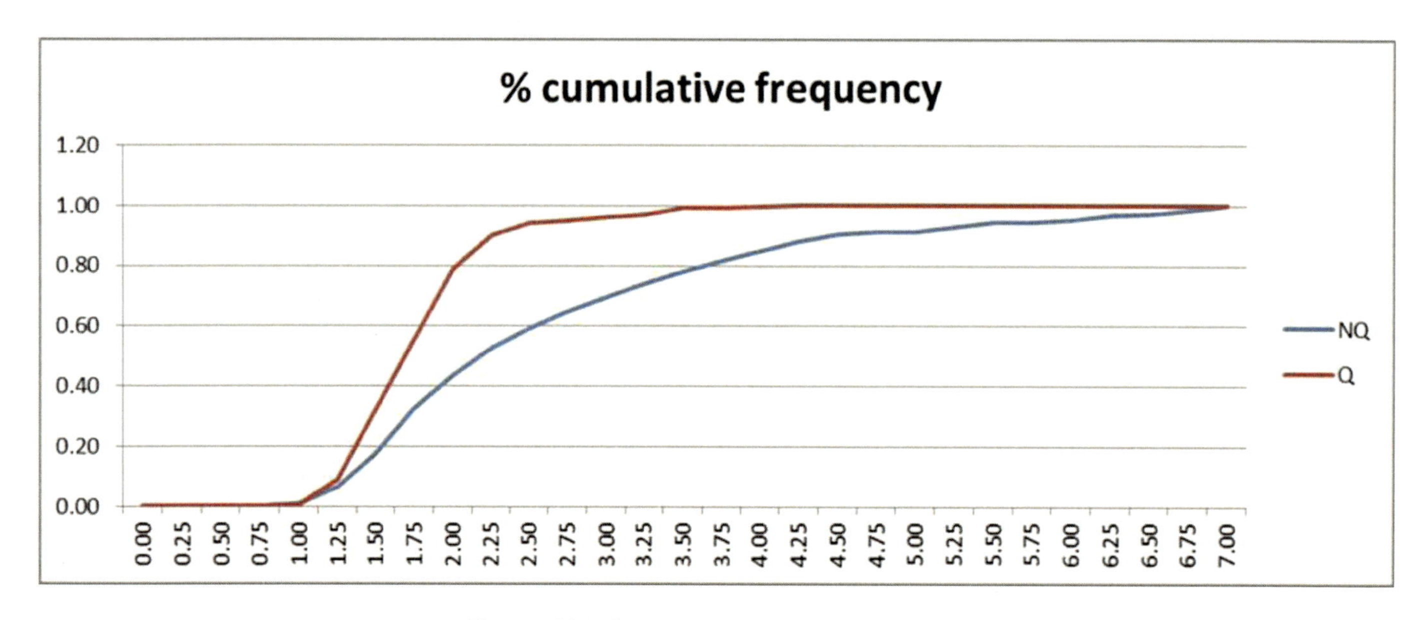

Figure 182. Poor example of graph labeling

Figure 183 is a screen capture from a VISSIM simulation that shows the traffic flow on one approach at a signalized intersection. This figure clearly captures the problem of traffic spilling out of the left turn pocket, impeding the flow of traffic on the two through lanes, and is a good way to illustrate a "traffic problem."

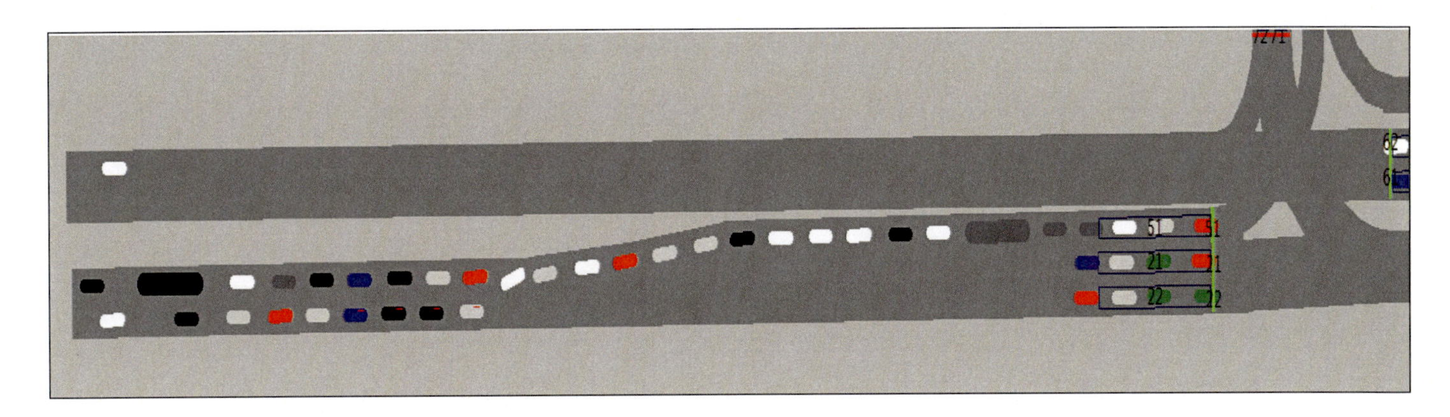

Figure 183. Good example of VISSIM screen capture showing left turn queue problem

Figure 184 shows a cumulative frequency plot for non-queued vehicles. While we might assume that it shows the distribution for headways, we don't know this for sure since the x-axis is not labeled. The numbers shown on the y-axis have two decimals places; whole numbers would be sufficient here since there is no decimal information conveyed. In addition, spacing intervals of 20 percent is sufficient for this range and not every 10 percent as shown in the chart. The x-axis also has too many numbers; spacing intervals of 1 or 2 would be sufficient as an interval.

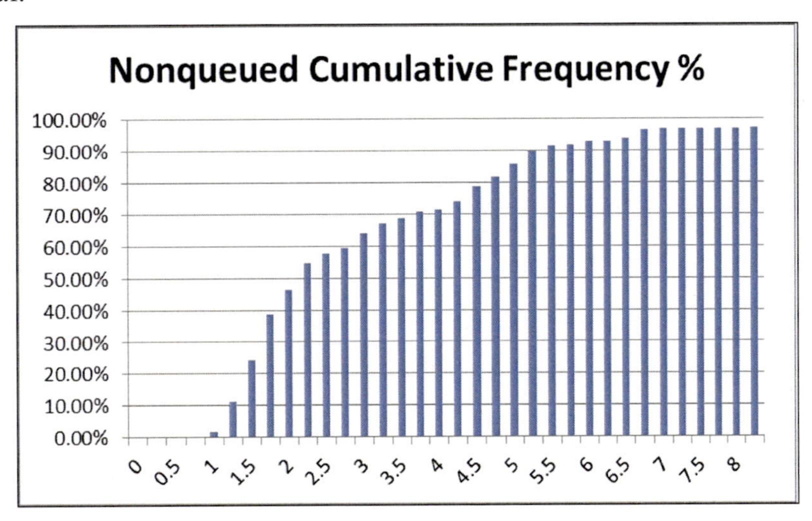

Figure 184. Poor example of cumulative frequency chart

Table 30 shows the before and after ("base case" compared to "final design") data for average queue length, total delay, and level of service. The table clearly makes the comparison between both cases, and we can see that the final design results in significant (observably significant) differences for both average queue length and total delay. The table also refers to level of service, a concept from the *Highway Capacity Manual*. The HCM provides a range of delay values for a set of grades from A to F. While these categories are useful for comparing two or more alternatives, it should be kept in mind that these category ranges are somewhat arbitrary, backed by little human factors research.

Measure	Before (base case)	After (final design)
Average queue (vehicles)	31.6	12.9
Total delay (seconds)	25.1	15.4
Level of service	C	B

Table 30. Before and after comparison

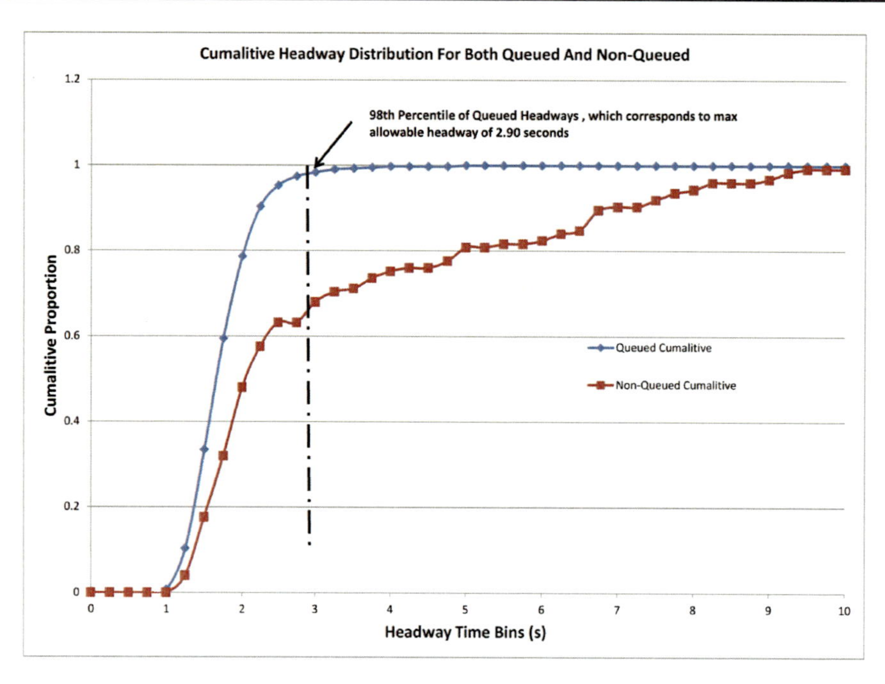

Figure 185. Example headway distribution for queued and non-queued vehicles (original)

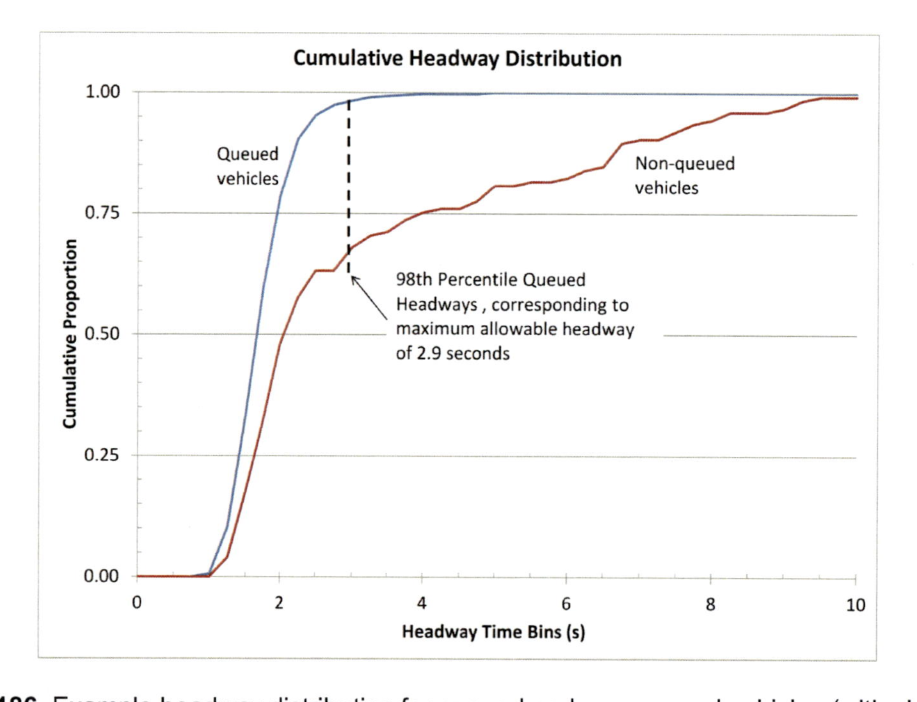

Figure 186. Example headway distribution for queued and non-queued vehicles (with changes)

Figure 185 shows a cumulative headway distribution for both queued and non-queued vehicles. The key parts of the chart are labeled so the reader can "see" the amount of difference in the headways between these two vehicle categories: the queued vehicles have lower headways (expected since they vary about the theoretical saturation headway), while the non-queued vehicles have a much wider range of values. However, there are improvements that can be made to the presentation of this information

Figure 186 shows the same information but with improvements. The type size is larger, making the chart easier to read. The maximum *y* value is 1.0, which is the actual maximum for a cumulative distribution. The legend has been eliminated and each of the lines is more directly labeled. Also, the graphs are shown as lines only and not a combination of lines and markers, which makes reading the chart much easier.

Movement	Before		After	
	Queue length (ft)	Delay (sec/veh)	Queue length (ft)	Delay (sec/veh)
EBTH	72.7	33.6	30.1	15.1
EBRT	72.7	38.8	30.1	18.6
EBLT	52.2	69.0	12.1	24.5
NBRT	51.3	43.0	22.2	15.4
NBTH	51.3	33.2	22.2	19.6
NBLT	25.1	59.6	7.1	21.7
WBTH	74.1	35.9	27.4	14.9
WBRT	75.8	28.9	28.6	10.8
WBLT	29.5	64.4	6.1	23.8
SBLT	76.7	46.8	28.1	19.0
SBTH	76.7	43.8	28.1	19.0
SBRT	45.7	74.0	9.3	23.2
Average	58.7	47.6	21.0	18.8

Table 31. Performance measures (before and after)

Three presentations of the same data are shown on this page. Table 31 compares queue length and delay for before and after conditions. The data are shown for each movement at the intersection. The average for all movements is also shown at the bottom of the table. Delay is improved both for the intersection, as well as for each movement. In fact, the difference is significant, showing that the signal timing changes in the after condition have resulted in a measurable change for the user.

Movement	Queue Length (ft)		Delay (sec/veh)	
	Before	After	Before	After
EBTH	73	30	34	15
EBRT	73	30	39	19
EBLT	52	12	69	25
NBRT	51	22	43	15
NBTH	51	22	33	20
NBLT	25	7	60	22
WBTH	74	27	36	15
WBRT	76	29	29	11
WBLT	30	6	64	24
SBLT	77	28	47	19
SBTH	77	28	44	19
SBRT	48	9	74	23
Average	59	21	48	19

Table 32. Performance measures (before and after)

Table 32 makes the comparison easier by showing queue length data side by side (before and after); delay data are shown the same way. Also, the data are shown only to the nearest whole number, eliminating the tenth of a second for delay and tenth of a foot for queue length. The precision shown in Table 31 is not warranted.

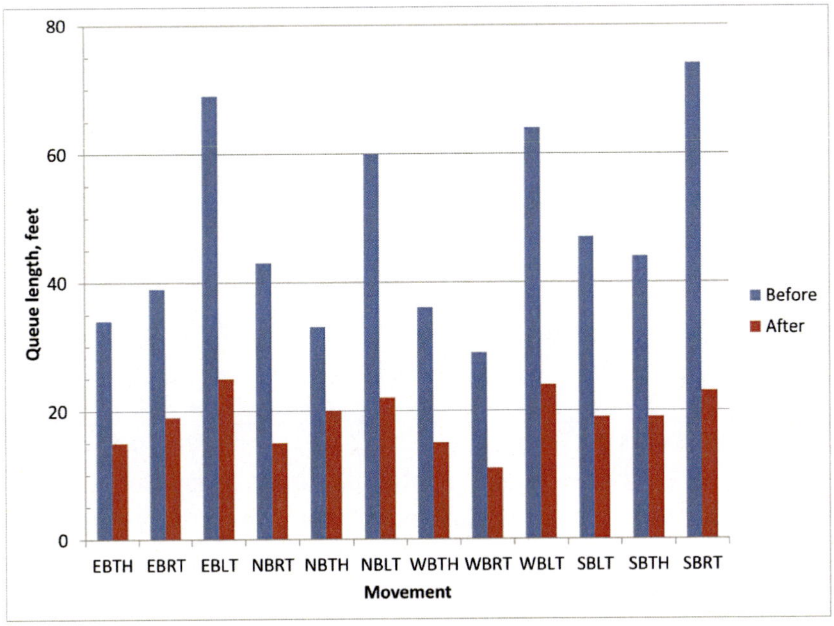

An even easier comparison between the before and after conditions for queue length can be made if a chart is used to compare the data. The reduction in queue length that results from the "after" case is clearly represented in Figure 187.

Figure 187. Queue length data (before and after)

Movement	Node	Permitted LT		Protected LT	
		aveQ	Delay	aveQ	Delay
W-E	1	46.9	7.6	15.5	5.4
W-N	1	496.4	178.4	76.9	39.3
E-W	1	24.9	6.9	101.4	19.1
E-N	1	0.1	1.4	0.8	4.0
N-E	1	49.0	34.2	47.4	35.7
N-W	1	49.0	36.1	47.4	35.0
All	1	100.8	24.3	47.2	17.7

Table 33. Performance data comparison

Table 33 compares the performance of the intersection for two cases, with permitted and protected left turns. However, it is difficult to identify the turning movements as they are shown in the default VISSIM notation (noted from one direction to another). Also, the "node" column is unnecessary as the data are all from node 1. Finally, the units are not given.

Movement	Permitted LT (sec)	Protected LT (sec)
EBTH	7.6	5.4
EBLT	178.4	39.3
WBTH	6.9	19.1
WBRT	1.4	4.0
SBLT	34.2	35.7
SBRT	36.1	35.0
All	24.3	17.7

Table 34. Delay data comparison

Table 34 provides for a more easily read comparison for delay between the permitted and protected left turn options. And, the movement label is the more traditional notation. But it is also worth looking at the differences. For example, the difference for the EBTH movement is negligible, only 2 seconds, a value too small to be perceived by the user. The differences are also small (and not operationally significant) for the WBRT, the SBLT, and the SBRT. However, differences are significant for the EBLT and the WBTH movements. The mean difference for all movements (24.3 vs. 17.7) is moderate and may not be perceivable by the users.

Experimental Results

We have tried to make the point regularly throughout this book that we want you to learn by observing and by learning to use a variety of data, synthesizing these observations and data into a decision about one element of your signal timing design. For example, in Chapter 6, you studied the effect of various passage time values on when a phase would terminate (given a specific detection zone length) and whether this termination would come just as the queue cleared, too early if the queue hadn't cleared, or too late if the phase continues on past when the queue has cleared. As we have said, this is a messy business. There is not one "right" answer that will apply to all conditions (short or long queues, low or high volumes). Our point is to get you to consider this messiness, as a regular part of the life of an engineer, and to learn to balance sometimes conflicting objectives.

But we also pointed you to the *Traffic Signal Timing Manual*, the standard guidebook for signal timing in the U.S., where you have been able to read about "practice", or guidelines or rules that can be used or referred to by practicing traffic engineers. Why not just turn to the chapter in the *Traffic Signal Timing Manual* on timing and see what the table says for the value of the passage time? A good question! And, if you ask many engineers in practice, they may say that they do this regularly because they don't have time to explore an issue in greater detail, or they may not understand the issues behind setting the passage time, or other timing parameters. We hope that by dealing with the results from your observations (both the visual observations as well as the numeric data that you collect in the field or generate from your simulation runs) that you will develop a better understanding of why timing parameters are set with certain values, or why a given range might be just as acceptable. Learn to use guidebooks (like the *Traffic Signal Timing Manual*) but take the time to use data that you have available to help you select signal timing design values for the particular case or set of conditions that you face.

How to Communicate Your Results

So how do you tell others what you have learned and what are the important parts of the design that you are recommending? And, of all of the data that you have generated and sifted through, which data do you include to justify and support your work? You will be asked to use two forms of communication to do this, each of which is briefly described here.

A written report is the most common and widely used method of communicating technical results. We will not attempt here to address all of the considerations that go into good technical writing. As you prepare your report (Activity #62 specifies requirements for your design report that you will need to follow), consider several issues, maybe rules of thumb, that are listed below to help you in this process:

1. As your writing improves and matures, your reports should change from what we call laboratory reports to professional reports. A laboratory report is filled with statements like: "I plugged the data into the software" or "We ran VISSIM." By contrast, a professional report includes statements like: "The results of the simulation analysis showed that..."
2. No one writes well the first time through preparing a document. While you have time constraints and many demands on your time as a student, you need to schedule time to write, then read and critique, then write again.
3. Read your writing out loud. There is no better way to literally hear how your writing sounds. Reading slowly points out poorly written sentences, points that are not well organized, and obvious mistakes in grammar and punctuation.
4. Take care with the first sentence in each paragraph: each should be strong and lead the ideas that follow in the paragraph.
5. You are telling a story, using your observations and data, a story that builds into a justification for a design that you are presenting.

You will also orally present the results of your work. It is common (almost standard) to use slides as a set of visual aids in your oral presentation. However, it has become all-too-common for the slides to become a barrier between the presenter and the audience. And too often, the slides are filled with text that the presenter (and the audience as well) simply reads. Pretty boring, and not very effective communication! We will not attempt to address all aspects of a good oral presentation. But what follows are some ideas to consider as you prepare your oral presentation. Tufte notes (in *Beautiful Evidence*) that "making a presentation is a moral act as well as an intellectual activity."

1. Learn to talk about your work in a manner that engages the audience and that tells the story that you want to get across. When you really talk to the audience, they can take on the role that you really want them to play: active listeners who want to learn about your design, how you have designed it, and how you justify each component of the design.
2. Talk about your design; don't just read from the text in your slides. In fact, use slides for what they do best: as visualizations of some form of information that is best seen and then talked about. Show an excerpt from an animation of VISSIM to illustrate a traffic problem that you solved or better managed. Show a chart that illustrates how increasing the maximum green time increases delay. Show a time space diagram to illustrate the issues in setting the yellow time.
3. Use text only when you want to list key points that you want readers to grasp. And don't just read the key points. Talk about what they mean. Tell a story about them.
4. Tufte (2001) describes the concept of "data-ink." He notes that "a large share of ink on a graphic should present data-information, the ink changing as the data change. Data-ink is the non-erasable core of a graphic, the non-redundant ink arranged in response to variations in the numbers presented." Look at your visuals carefully and only keep those parts of each that are necessary to make the point that you intend.

Summary

So as you begin to prepare your report and presentation, keep in mind the issues raised in this Reading and that are summarized below:

1. Which elements of the traffic signal control system did you affect in your analysis and design?
2. How can you integrate the variety of information that you have generated?
3. What measures of effectiveness best show the performance of your system?
4. How can you most effectively present your information?
5. How have you used your experimental results to analyze the various design options that you considered and to select your final design?
6. How can you make your written and oral reports as effective as possible?

ACTIVITY

59 Assembling Information For Your Timing Plan Design

Purpose

The purpose of this activity is to give you the opportunity to assemble information that you have prepared in previous design activities into a form that will help you to prepare your final report and presentation.

Learning Objective

- Prepare a timing plan for an isolated actuated signalized intersection based on an analysis of traffic flow quality and intersection performance for a range of timing parameter values and phasing alternatives

Required Resource

- Results from previous design activities

Deliverable

- Prepare an Excel spreadsheet that includes your design values, as well as the tables and charts that support the selection of your design values

  Tab 1: Title page with activity number and title, authors, and date completed

  Tab 2: Phase timing sheet that includes the timing parameters for each phase: minimum green time, vehicle extension time, maximum green time, yellow time, and red clearance time

  Tab 3: Ring barrier diagram showing your recommended phasing plan

  Tab 4: Intersection sketch showing geometry, vehicle movements, and phase numbering

  Tab 5: Performance data (delay and queue length data) that compares each step in your design process

Task 1

Assemble the design elements that you developed as part of the activities listed in Table 35.

Activity	Design elements
28	Base network conditions
36	Maximum allowable headway
37	Passage time
43	Maximum green time
50	Left turn treatment
56	Yellow and red clearance times

Table 35. Activities and design elements

Task 2

Prepare a side-by-side comparison of the performance of the base case and each of the iterations of your design. The comparison should include the performance measures (delay and queue length) that you used previously in these activities.

Student Notes:

ACTIVITY **60**

What Do You Know About the Signal Timing Design Process?

Purpose

The purpose of this activity is for you to review a report on signal timing that has been prepared by a practicing transportation engineer.

Learning Objectives

- Synthesize ideas from a professional engineering report

Required Resource

- Example report

Deliverable

- Prepare a document that includes your responses to the Critical Thinking Questions

Information

When you are doing something for the first time, it often helps to have examples of work that others have previously completed. The professional design report that you review will provide guidance and insights for you as you prepare your own final report.

Task 1

Read the design report that has been assigned to you.

Critical Thinking Questions

1. What were the primary conclusions of the report?

2. What were the strengths of the report?

3. What were the weaknesses of the report?

4. What aspects of the report will you attempt to model in your design report that you prepare in Activity #62?

Signal Timing Design In Practice

Purpose

The purpose of this activity is to give you the opportunity to compare your design results with recommended practice from the *Traffic Signal Timing Manual*.

Learning Objective

- Compare your design results with values recommended for practice

Required Resource

- *Traffic Signal Timing Manual*

Deliverable

- Prepare a document that includes your answers to the Critical Thinking Questions

Link to Practice

Use the *Traffic Signal Timing Manual* to complete the tasks below.

Information

You have completed a comprehensive analysis of five of the fundamental timing parameters used in an actuated traffic control system. As a result of this analysis, you have selected values for these timing parameters. Comparing your results with those recommended in practice will provide you with a perspective on the work that you have completed and give you a better idea of how professionals consider these same design issues.

Task 1

Review the discussions on minimum green time, maximum green time, passage time, yellow time, and red clearance time in the *Traffic Signal Timing Manual*.

Task 2

Compare your design values for the timing parameters listed in Task 1 with the recommended practice from the *Traffic Signal Timing Manual*.

CRITICAL THINKING QUESTIONS

When you have completed the reading, prepare answers to the following questions:

1. Compare the signal timing process described in the *Traffic Signal Timing Manual* with the design process that you have just completed in this course. How is it the same? How is it different?

2. How do your design values compare to the recommended settings in the *Traffic Signal Timing Manual*? Describe why you think your values are different than (or similar to) those from the manual.

IN MY PRACTICE...

by Tom Urbanik

Bill Kloos was the manager of the Signals and Street Lighting Division at the City of Portland, until his death in 2009. Bill was an innovative and inspirational leader in the field of traffic signal systems, and his opinions and wisdom were valued and used by practicing engineers throughout the world. Bill said:

"If you want to be outstanding in the field (of traffic signal control), you must be out standing in the field."

Our theory and models are only abstractions of reality. Each community and each intersection have their own peculiarities. These finer points can only be assessed by getting out to the intersection and observing.

ACTIVITY

62 Design Report

PURPOSE

The purpose of this activity is to give you the opportunity to prepare your final report.

LEARNING OBJECTIVES

- Integrate information into a professional style report and presentation
- Clearly communicate the timing plan design for an isolated actuated signalized intersection based on an analysis of traffic flow quality and intersection performance for a range of timing parameter values and phasing alternatives

REQUIRED RESOURCE

- Results from previous activities

DELIVERABLE

- Final written report using MS Word and an oral presentation using MS PowerPoint

TASKS

Your final report will have two components, a written report and an oral report.

TASK 1

Prepare your written report. Your final report should include the following information:

- Phasing plan shown in ring barrier diagram format
- Timing parameters (minimum green time, maximum green time, passage time, yellow time, and red clearance time), detector location and type, and other relevant controller settings. Justifications for each of your selected parameters including all relevant data should be given in the report.
- Evaluation of your plan using VISSIM with suitable measures of effectiveness and your visual observations of the simulation. Comparisons of the performance of existing or base conditions with each option considered.
- All options that you considered for various parts of your design, including those options that are a part of the final design and those that are not
- Comparison of your results with recommended practice from the *Traffic Signal Timing Manual*

Your report should include the following sections:

1. Title page, including title, authors, date
2. Table of contents
3. Executive summary
4. Introduction

5. Description of intersection (including geometry, demand, performance, base conditions)
6. Description and evaluation of the phasing and timing plans (including justifications for each element of the plan)
7. Appendices including all calculations for timing plan parameters and other supporting data

TASK 2

Prepare your oral report.

- Prepare a set of tables that include the data that you generated as part of Activity #59 and that describe the final signal timing plan
- Prepare a summary of the points that justify the selection of each element of your timing plan. Identify and construct the graphs or charts needed to support your key points
- Prepare a set of slides using PowerPoint that addresses the problem that you were assigned, the analysis that you have done supporting your design choices, a description of both the data analyzed and the observations that you have made, and the elements of your final design
- The presentation should include visualizations from VISSIM (both static and dynamic) that demonstrate the operation and performance of your intersection and how your results compared with recommendations from the *Traffic Signal Timing Manual*

63 Design Evaluations and Assessments

PURPOSE

The purpose of this activity is to give you the experience of assessing a report and presentation on traffic signal timing.

LEARNING OBJECTIVE

- Provide effective feedback to others

REQUIRED RESOURCE

- Results from previous activities

DELIVERABLE

- Complete evaluations as instructed

TASKS

Your final report will have two components, a written report and an oral report.

INFORMATION

Each team will be responsible for presenting their findings and participate in the review of the reports and presentations by other team members. You may be assigned one or more of the following tasks.

TASK 1

Responsibilities of presentation reviewers:

Carefully review the presentation as it is given. Make notes on the strengths and areas for improvement of the presentation while it is being given. Each member of the review team will be responsible for submitting their individual responses to the following questions:

1. What were the strengths of the presentation?
2. Were the elements of the design clearly presented?
3. Were the design elements supported or justified?
4. What suggestions would you make to the team to improve their presentation?

TASK 2

Responsibilities of questioners:

Carefully listen to the presentation as it is given. Make notes on the presentation as it is given. Consider the following questions as you consider questions that you will ask at the conclusion of the presentation:

1. What did you like about the presentation?
2. What points didn't you understand during the presentation?
3. What questions did you have regarding the design plan, its elements, and how it was justified?

Task 3

Responsibilities of the report reviewers:

Your task is to review the written report and other supporting materials. Refer to Activity #62 for the complete list of requirements as you complete your review. Please answer the following questions:

1. Did the report include the required information? If not, what was missing?
2. Did the report include the required sections? If not, what was missing?
3. Did the report clearly state the design elements? Provide a brief justification of your answer.
4. Did the report provide justification for each of the design elements? What is your assessment (strengths, areas for improvement) of the justification?
5. Was the report well written? Were you able to easily read through the report? What was the quality of the writing?
6. What suggestions would you make to improve the quality of the report?

Rubric

Students often ask: "How will my report be evaluated?" On the opposite page are the criteria (or rubric) that we (and you as peer reviewers) will use in evaluating your final reports. A *rubric* is an evaluation or scoring tool that lists the criteria for a piece of work or 'what counts.'

Table 36. Rubric for Evaluating Design Reports

Criteria	High quality performance	Acceptable performance	Unacceptable performance
Report contents	The report includes all of the required sections and displays them clearly and logically.	The report includes all required sections.	One or more required sections are not included in the report.
Timing plan	The report includes all of the required timing plan elements and the phasing plans for each intersection in both tables and supporting text.	The report includes the required timing plans and phasing plans.	The report does not include all of the required timing and phasing elements.
Optimization process	The report includes a description of the optimization process, and the supporting charts and calculations. The data are presented in clearly designed charts and tables, with text that elaborates and explains the charts and tables. The analysis is clearly described and supported by data.	The report includes a description of the optimization process and the supporting charts and calculations.	The optimization process is not described clearly, the supporting data are not included, and the results of the process are not shown.
Selection of timing parameters	The report includes the process by which all of the timing parameters were selected, as well as the supporting calculations justifying these parameters. The supporting calculations show all assumptions, steps, equations, and data used to justify the selection of the parameters.	The report includes the process by which all of the timing parameters were selected, as well as the supporting calculations justifying these parameters.	The process for selecting the timing parameters is not clearly described and the supporting data are not included.
Organization	The report is organized in a manner that allows the reader to follow the sequence of topics and decisions. The sequence of topics supports the arguments and conclusions presented.	The report is organized in a logical manner.	The report is not easy to follow because the organizational structure is not clear to the reader.
Readability	The writing style in the report is crisp and clear, and uses high standards of grammar and readability.	The writing in the report is of acceptable quality; that is, the writing is not so poor that it distracts the reader from understanding and agreeing with the points made in the report.	The writing is poor and does not clearly communicate the results.
Executive summary	The executive summary provides a complete overview of the key points that appear in the report in a way that provides the information that the reader needs to understand the design and how it was developed.	The executive summary provides a clear overview of the points that appear in the report.	The executive summary does not provide a summary of the important points made in the report.

Student Notes:

Made in the USA
San Bernardino, CA
23 August 2016